PREFACE

It has been the goal in this volume to summarize the current level of understanding in a number of areas relating to sweet potato production and utilization. To accomplish this, a group of authors, each with broad experience in the selected areas, was assembled. It is our hope that the volume will be useful as a beginning point — a foundation for further research efforts.

To the extent that this volume is successful in achieving its purpose, the authors are due the credit. The editor alone is responsible for its shortcomings and will be pleased to receive correspondence from readers advising as how to strengthen any future presentations of this nature.

It has been a pleasure to be involved in this undertaking. Many of the anticipated difficulties were avoided due to the excellent cooperation of the chapter authors and the editorial staff at CRC.

John C. Bouwkamp

THE EDITOR

Dr. John C. Bouwkamp is an Associate Professor at the University of Maryland at College Park. He obtained a B.S. majoring in Botany at Michigan State University in 1964, and M.S. and Ph.D. degrees in Horticulture in 1966 and 1969 at the same institution. He is actively engaged in several phases of research in sweet potatoes with special emphasis on breeding and production. His interests in sweet potato production and utilization in the tropics has been enhanced by opportunities to travel and work in production areas in China, Taiwan, the Philippines, and the Caribbean. His research efforts have been supported in part by grants from the U.S.D.A.

Sweet Potato Products: A Natural Resource for the Tropics

Editor

John C. Bouwkamp, Ph.D.
Associate Professor
Department of Horticulture
University of Maryland
College Park, Maryland

CRC Press, Inc.
Boca Raton, Florida

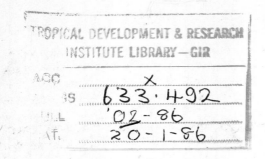

Library of Congress Cataloging in Publication Data

Bouwkamp, John C., 1942-
  Sweet potato products.

  Bibliography: p.
  Includes index.
  1. Sweet potato products--Tropics. 2. Sweet potatoes--
Tropics. I. Title.
TP444.S94B68   1985   641.3'522   85-3777
ISBN-0-8493-5428-5

This book represents information obtained from authentic and highly regarded sources. Reprinted material is quoted with permission, and sources are indicated. A wide variety of references are listed. Every reasonable effort has been made to give reliable data and information, but the author and the publisher cannot assume responsibility for the validity of all materials or for the consequences of their use.

All rights reserved. This book, or any parts thereof, may not be reproduced in any form without written consent from the publisher.

Direct all inquiries to CRC Press, Inc., 2000 Corporate Blvd., N.W., Boca Raton, Florida, 33431.

© 1985 by CRC Press, Inc.

International Standard Book Number 0-8493-5428-5
Library of Congress Card Number 85-3777
Printed in the United States

A dwarf standing on the shoulders of a giant may see further than a giant himself.

Lucan (39AD—65AD)
in De Bello Civili, 10, II

They lard their lean books with the fat of others' works.

Robert Burton (1577—1640)
in Anatomy of Melancholy

Dedication

To the giants and fat producers who provided the information necessary to inspire this effort and to those who provided the support and encouragement necessary to complete it.

CONTRIBUTORS

Der-Ming Chen
Manager
Personnel and General Service
Lu-Kuang, Inc.
Tainan City, Taiwan

S. Chiu
Asian Vegetable Research and
  Development Center
Shanhua, Tainan, Taiwan

Wanda W. Collins, Ph.D.
Associate Professor
Department of Horticultural Science
North Carolina State University
Raleigh, North Carolina

Alfred Jones, Ph.D.
Research Geneticist
U.S. Vegetable Laboratory
Agricultural Research Service
U.S. Department of Agriculture
Charleston, South Carolina

Stanley J. Kays, Ph.D.
Professor
Department of Horticulture
University of Georgia
Athens, Georgia

Steve S. M. Lin
Associate Plant Breeder
The Asian Vegetable Research and
  Development Center
Shanhua, Tainan, Taiwan

Hsiao-Feng Lo
Graduate Research Assistant
Department of Horticulture
Mississippi State University
Mississippi State, Mississippi

James W. Moyer, Ph.D.
Associate Professor
Department of Plant Pathology
North Carolina State University
Raleigh, North Carolina

Creighton Peet
Planning, Monitoring and Evaluation
  Advisor
USAID/HMG Nepal
Rapti Integrated Rural Development
  Project
Washington, D.C.

Satoshi Sakamoto
Chief, Sweet Potato Breeding
  Laboratory
National Agriculture Research Center
Yotsukaido, Chiba, Japan

James M. Schalk, Ph.D.
Research Leader
Vegetable Insects Unit
U.S. Vegetable Laboratory
Agricultural Research Service
U.S. Department of Agriculture
Charleston, South Carolina

Samson C. S. Tsou
Biochemist, Leader of Nutrition
Environment and Management
  Program
Asian Vegetable Research and
  Development Center
Shanhua, Tainan, Taiwan

Ruben L. Villareal
Director
Institute of Plant Breeding
College of Agriculture
University of the Philippines at Los
  Baños
College, Laguña, Philippines

William M. Walter, Jr., Ph.D.
Department of Food Science
North Carolina State University
Agricultural Research Service
U.S. Department of Agriculture
Raleigh, North Carolina

Tsa Pou Yeh, Ph.D.
Director
Animal Research Institute
Taiwan Sugar Corporation
Chu-nan, Miao-li, Taiwan

TABLE OF CONTENTS

PART I: PRODUCTION

Introduction — Part I 3
John C. Bouwkamp

Chapter 1
Production Requirements 9
John C. Bouwkamp

Chapter 2
Major Disease Pests 35
James W. Moyer

Chapter 3
Major Insect Pests 59
James M. Schalk and Alfred Jones

Chapter 4
The Physiology of Yield in the Sweet Potato 79
Stanley J. Kays

Conclusions — Part I 133
John C. Bouwkamp

PART II: UTILIZATION

Introduction — Part II 137
John C. Bouwkamp

Chapter 5
Sweet Potato Production and Utilization in Asia and the Pacific 139
Steve S. M. Lin, Creighton C. Peet, Der-ming Chen, and Hsiao-Feng Lo

Chapter 6
Relative Nutrient Costs and Nutritional Value 149
S. C. S. Tsou

Chapter 7
Fresh Roots for Human Consumption 153
Wanda W. Collins and William M. Walter, Jr.

Chapter 8
Sweet Potato Vine Tips as Vegetables 175
Ruben Villareal, S. C. S. Tsou, H. F. Lo, and S. C. Chiu

Chapter 9
Processing of Sweet Potatoes — Canning, Freezing, Dehydrating 185
John C. Bouwkamp

Chapter 10
Formulated Sweet Potato Products 205
Stanley J. Kays

Chapter 11
Industrial Products from Sweet Potatoes 219
Satoshi Sakamoto and John C. Bouwkamp

Chapter 12
Roots and Vines as Animal Feeds.. 235
Tsa Pou Yeh and John C. Bouwkamp

Conclusions — Part II... 255
John C. Bouwkamp

Index.. 263

Part I
Production

INTRODUCTION — PART I

Sweet potatoes (*Ipomoea batatas* (Lam.) L.) are presently an important crop in many areas of the world. Statistics derived from the 1977 FAO Production Yearbook indicate an average production (1975 to 1977) of 137 million metric tons, ranking sixth among food crops (Figure 1). The present importance and widespread distribution of sweet potatoes provide the basis for a greater role in that farmers, consumers, and scientists are familiar with the crop and its requirements. A fairly substantial body of information exists and is available to many researchers, although not in proportion to the importance of the crop.

Sweet potatoes are efficient producers of calories. We find that, among the ten leading food crops, sweet potatoes rank third in terms of calories produced per square meter (Table 1). Only minimal research efforts to increase yield through improved production techniques and breeding have been sustained (as compared to corn and potatoes). Thus, it is probable that calorie production could be substantially increased above these present high levels.

Sweet potatoes can (and do, in many parts of the world), make significant nutritional contributions to the diet. While starchy root crops are frequently considered to provide only calories to the diet, we find this not to be the case with sweet potatoes. If we calculate the nutrient per calorie provided as a percentage of nutrient per calorie required, we find sweet potato roots to be an excellent source of all nutrients, providing at least 90% of the requirement for all except protein and niacin (Table 2).

The center of origin for sweet potatoes is in the tropics. If we consider per capita income (since poor people are often hungry people), we find that 56% of the world population live in countries having a yearly income of less than U.S. $500. All but 6 of the 62 countries are in the tropics and all but 1 between 30°N and 30°S latitude. Sweet potatoes, already adapted to the tropics, can be expected to play an increasingly important role in tropical agriculture.

The adaptation to the tropics can be demonstrated by comparing yields of crops at various latitudes. Yield data for the 10 leading crops were obtained on a per country basis from the 1977 FAO Production Yearbook and 3-year average (1975 to 1977) yield was calculated for each of the 10 crops if the production was more than 5,000 metric tons. The latitude at the geographical center of the country was obtained and the average yield of the various crops was calculated of each latitude group. The results are presented in Figure 2. It may be noted that yields of most crops are higher in temperate than tropical regions but that sweet potatoes yield relatively better in the tropics than many crops of greater importance.

Yield in terms of calories per square meter in the tropical and temperate regions also demonstrate the importance and potential of sweet potatoes (Table 3).

Being a perennial, but grown as an annual, a sweet potato crop does not ripen or mature. While this may seem trivial to the casual observer, close examination of the consequences reveals its importance. The length of the growing season at any location may vary by 30 days or more. A sweet potato crop can frequently utilize the entire length of the growing season. A cereal crop may leave 30 days of growing season not utilized or may not ripen properly if the season is especially short. This nonripening feature also allows $1^1/_2$ or $1^3/_4$ crops to be produced instead of only one.

Sweet potato production may be adapted to high or low technology input agricultural systems. For example, we find good yields of sweet potatoes produced both in the U.S. and Japan with high levels of inputs and technology and in many lesser developed countries under a labor intensive, low technology system.

Regardless of the type of inputs, sweet potatoes do not usually require high levels of input. Since sweet potato vines grow rapidly and completely cover the ground within 4 to 6 weeks from planting, herbicides and cultivation may be used sparingly. Applications of insecticides and fungicides are frequently unnecessary since, with the exception of the sweet

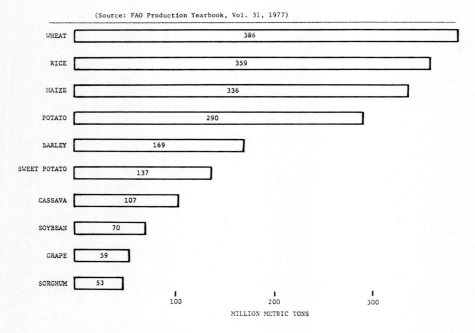

FIGURE 1. Average (1975 to 1977) annual production of the ten leading food crops.

Table 1
AVERAGE YIELD OF FRESH WEIGHT AND CALORIES FOR THE 10 LEADING CROPS

	Yield (kg/ha[a])	Calories (cal/kg[b])	(cal/M²)
Wheat	1,661	3,330	553
Rice	2,513	3,410	857
Maize	2,886	3,490	1007
Potato	13,658	710	970
Barley	1,877	3,270	614
Sweet potato	9,621	1,000	962
Cassava	8,729	990	864
Soybean	1,498	3,920	587
Grape	5,836	390	228
Sorghum	1,222	3,420	418

[a] Average yield 1975—1977. Source: FAO.[1]
[b] Source: Leung et al.[2]

potato weevil, most insect pest damage is cosmetic and not yield reducing and fungal diseases are usually not a problem in the growing crop. Sweet potatoes grow well in soil with a pH of 4.5 to 6.5. Thus, it may not be necessary to apply lime. Supplemental irrigation is often not required. Although sweet potatoes do show increased yields from supplemental irrigation in some years, Lambeth (1957) reports 12.5 mt/ha produced on only 10 cm of rainfall from planting to harvest.

Sweet potatoes are vegetatively propagated. This gives the crop an advantage in several ways. In areas where most or all of the crop can be produced by vine cuttings, the grower

Table 2
NUTRITIVE VALUE OF SWEET POTATOES EXPRESSED AS A PERCENTAGE OF THE NUTRIENT PER CALORIE SUPPLIED PER NUTRIENT PER CALORIE REQUIRED

Nutrient	Value[a]
Protein	73
Vitamin A	4160
Vitamin C	825
Niacin	81
Riboflavin	91
Thiamin	167
Phosphorus	139
Iron	161
Calcium	95

[a] Based on a male, age 23 to 50, at 70 kg, ht 172 cm. Source: Food and Nutrition Board, NAS-NCR[3] and Watt and Merrill.[4]

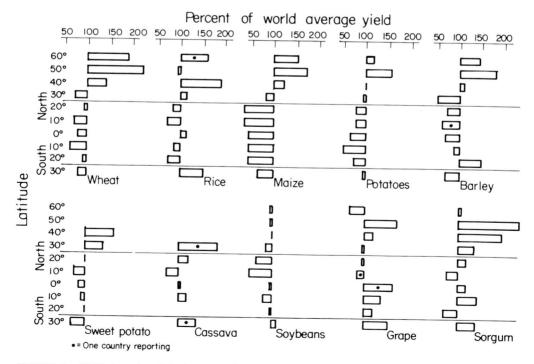

FIGURE 2. Yields in various latitude ranges of the ten leading world food crops as expressed as a percentage of world average yield (1975 to 1977).

Table 3
COMPARISON OF CALORIE YIELDS IN TEMPERATE AND TROPICAL REGIONS FOR THE 10 LEADING FOOD CROPS

	Temperate areas		Tropical areas	
	cal/m^2	Countries reporting	cal/m^2	Countries reporting
Wheat	763	57	432	25
Rice	1,303	38	670	69
Maize	1,177	44	686	65
Potato	1,125	58	636	44
Barley	688	55	461	12
Sweet potato	1,238	16	831	61
Cassava	1,386	2	817	68
Soybean	583	19	429	14
Grape	272	43	273	8
Sorghum	661	23	400	39

Table 4
SOIL AND WATER LOSSES FROM PLOTS PLANTED WITH VARIOUS CROPS AND LOSSES COMPARED TO SIMILAR PLOTS LEFT FALLOW

	Water loss (runoff)		Soil loss	
Crop(s)	cm/cm rainfall	% of uncropped plots	kg/ha/cm rainfall	% of uncropped plots
Sweet potatoes (1938)	0.022	13	26	2
Jack beans, corn, field beans, squash (1939)	0.140	45	355	22
Jack beans, sweet potatoes (1940)	0.026	8	104	6
Corn, lima beans, jack beans (1941)	0.41	21	234	14
Jack beans, velvet beans (1942)	0.47	9	100	7
Pigeon peas, Jack beans (1943)	0.088	24	416	47

is not faced with problems of seed storage, nor is he required to set aside part of his crop for the next season's seed supply. Since the farmer produces his own "seed" he is not required to buy seed each year (as is the case for hybrid corn) nor is progress limited by lack of seed technology skills and facilities.

Since the harvested portion of the crop is also vegetatively produced, flowering and pollination are not necessary. Thus the period, which for many crops is so important, and yield reductions due to adverse weather conditions so great, is avoided. As a result, adverse weather conditions rarely cause a complete crop loss. Farmers in some parts of the world intuitively realize this and frequently sweet potatoes are planted as an "insurance crop". If the preferred crop (frequently rice) fails, the farmer is assured of at least a partial crop of sweet potatoes to sustain himself and his family.

The production of annual crops often results in more soil erosion on sloping sites than the production of perennial crops. The results of Smith and Abruna[6] (Table 4) suggest that soils planted to sweet potatoes may be subject to less erosion than if other crops were planted. They planted a series of crops on Catalina and Cialitos clay soils on slopes of 40 to 45% and compared the soil and water losses to similar plots left fallow (bare soil).

We recognize that none of the points discussed is unique to sweet potatoes. However, the combination of characteristics provides an exciting potential for the increasing importance of this crop as a part of any plan for agricultural development.

REFERENCES

1. FAO, Production Yearbook, Vol. 31, Food and Agriculture Organization, Rome, Italy, 1977.
2. **Leung, W. W., Butrum, R. R., and Chang, F. H.,** *Food Composition Table for use in East Asia.* I. Proximate composition mineral and vitamin contents of East Asian foods, Food and Agriculture Organization, Rome, Italy, 1972.
3. Food and Nutrition Board, National Academy of Sciences, National Research Council, *Recommended Daily Allowances,* Washington, D.C., 1980.
4. **Watt, B. K. and Merrill, A. L.,** *Composition of Foods,* Handbook 8, U. S. Department of Agriculture, Washington, D. C., 1975.
5. **Lambeth, V. N.,** Sweet potato, a promising Missouri crop, *Mo. Agric. Exp. Stn. Bull.,* 686, 1957.
6. **Smith, R. M. and Abruna, F.,** Soil and water conservation research in Puerto Rico, 1938 to 1947, *P.R. Agric. Exp. Stn. Bull.,* 124, 1955.

Chapter 1

PRODUCTION REQUIREMENTS

John C. Bouwkamp

TABLE OF CONTENTS

I.	Edaphic Requirements		10
	A.	Soil Texture and Drainage	10
	B.	Soil Reaction (pH)	10
		1. Yield	10
		2. Quality	11
		3. Disease	11
	C.	Mineral Nutrition	11
		1. Deficiency Symptoms	11
		2. Elemental Concentration of Deficient and Normal Sweet Potato Tissue	13
		3. Nitrogen Studies	14
		4. Phosphorus Studies	16
		5. Potassium Studies	17
		6. Calcium Studies	18
		7. Magnesium Studies	18
		8. Sulfur Studies	18
		9. Minor Elements	18
		10. Organic and Complete Fertilizers	19
		11. Response to Adverse Soil Conditions	21
II.	Environmental Requirements		21
	A.	Temperature Requirements	21
	B.	Light Requirements	23
	C.	Moisture Requirements	23
III.	Production Practices		24
	A.	Propagating Materials	24
	B.	Transplanting	25
	C.	Weed Control	26
References			28

I. EDAPHIC REQUIREMENTS

A. Soil Texture and Drainage

Sweet potatoes may be successfully produced in a wide range of soil types although it is generally agreed that root shape and appearance are best when produced in light, sandy, or sandy loam soils. The soil should be friable enough to permit unimpeded root enlargement and have sufficient aeration porosity to provide oxygen to developing roots. Chua and Kays[1] have clearly demonstrated that oxygen is required for storage root induction and Watanabe[2] has reported that, on volcanic ash soil, harvest index and aeration porosity are correlated (r = 0.789**) in the range of 15 to 40% aeration porosity.

Several tests have been conducted to determine the effects of water table levels on sweet potato yields. Silva and Irizarry[3] found reduced yields of 'Gem' and 'Miguela' grown on a clay loam when the water table was maintained at 30 cm below the soil surface as compared to yields at a 45 cm water table. Greater yield reductions occurred if the water table was held at 15 cm below the surface. Ghuman and Lal[4] report less striking results in a sandy soil but conclude that "sweet potatoes with acceptable roots can be grown satisfactorily in soils with a water table at or below 50 cm". They also noted that vine growth in plots flooded twice weekly was greater than other treatments. Paterson et al.[5] conducted an experiment in which plants were flooded momentarily five times daily by subirrigation to the surface or to 10 cm below the surface of a very coarse silica sand. They found that the latter treatment resulted in a greater number and weight of storage roots and less vine weight.

Waterlogged soil during the growing season and shortly before harvest may result in loss of storage roots due to rotting in the soil[4] or during subsequent storage.[6,7]

We may conclude that regardless of the soil texture, it is important to maintain a relatively high level of aeration porosity. Since light soils are easy to manage in this regard, they are preferred if sufficient moisture for crop development is available. However, clayey soils can also be utilized if care is taken to promote aeration porosity by the provision of ridges or mounds and by light cultivation to prevent crusting and stimulate drying. Poorly drained soils should be avoided unless ridges or mounds can be constructed high enough to provide 30 to 50 cm (depending on the soil texture) above the free water table. Studies on the effects of ridges have generally shown that yields are higher on ridges. Woodard,[8] reporting on 7 years of experiments, found that yields on 20 cm high ridges were 3% greater than level culture and 30 cm ridges yielded 19% better than level culture. Edmond et al.[9] reporting on a cooperative study including 3 locations in 1 year and 4 locations for 2 years, found that, compared to low ridges (ca. 10 cm), medium ridges (ca. 20 cm) produced 10 and 13% greater yields and high ridges (ca. 35 cm) produced 13 and 21% greater yields. They also determined that harvest losses due to cut or damaged roots were reduced as the ridge height was increased.

B. Soil Reaction (pH)

1. Yield

Sweet potatoes are not especially exacting as to pH requirements. Geise[10] found lime to increase yield on a "strongly acid soil." He also found that lime addition to these soils increased the beneficial effect of other fertilizer application except when sodium nitrate was used. Abruna et al.[11] found that the effect of pH on yield depended on aluminum saturation and the Al/base ratio. With a pH range of 3.8 to 5.6 on 3 Ultisols they found the greatest yield at pH of 5.0 with generally less than 10% yield reduction at a pH of 4.5. Chew et al.[12] working with peat soils with a native pH of 3.5, found lime application to greatly increase both root yield and vine weight. The maximum root yield was obtained with applications of 18 metric tons per hectare, which resulted in a pH of about 5.7. Further additions decreased yield of roots but had little effect on vine weight.

Steinbauer and Beattie[13] reporting on experiments conducted in a Sassafras sandy loam with an initial pH of about 5.0, found annual lime applications of 2200 kg/ha to decrease yields by 0.9 mt/ha while increasing the pH to 7.1. They also noted that addition of calcium in the form of calcium chloride may increase yield in these soils. Watts and Cooper[14] found maximum yields at a pH range of 6.5 to 7.5 on Newtonia silt loam and 6.0 to 7.0 on Ruston fine sandy loam. They also noted that vine growth, vigor, and color were best at these pH ranges.

Navarro and Padda[15] reported that applications of sulfur to a Fredenborg clay loam at the rate of 2250 kg/ha increased yield from 9.8 to 16.6 mt/ha, and rates of 4500 kg/ha further increased yields to 21.2 mt/ha, although the soil reaction was only changed from 7.7 to 7.6. Although the pH was little affected, the authors speculate that "the addition of sulfur may have created some favorable pH condition in the microenvironment that is not discernible by the present method of pH measurements." They noted yellowing of new leaves as evidence of sulfur deficiency, but found that vine weight was not affected by sulfur treatments. Jones et al.[16] found that average yields of storage roots grown over an 8-year period (1966 to 1973) in soils amended with sulfur to maintain a pH of 4.5 to 5.0, or amended with lime to maintain a pH of 7.0 were nearly identical with the control (pH 5.5 to 6.0).

2. Quality

Constantin et al.[17] found no effect of pH in the range of 4.4 to 7.2 on carotene and protein concentration, on percent split roots when canned and only slight effects on fiber levels in processed roots. Percent dry matter of the roots and firmness of the canned roots generally deceased at higher pH levels.

3. Disease

Under conditions where soil rot or pox, caused by *Streptomyces ipomea* is a serious problem, reducing the pH to 5.2 — 5.4 has been effective in reducing the incidence of symptoms. Hartman and Gaylord[18] noted that 75% of the storage roots were affected by the disease at a pH of 7.5 but very little infection was noted at a pH of 5.4 or less.

Person[19] found that applications of sulfur to reduce soil pH to 5.2 on soils infested with the soil rot organism resulted in yield increases of 235% in dry years, 29% in normal years, and 1.2% in wet years. The large yield increases in the dry years may be attributed to the fact that the organism attacks the fibrous root system as well as the storage roots. Constantin et al.[13] noted that the incidence of soil rot reduced processing yield and grade of the processed product.

We may conclude that within the range of pH from 4.5 to 7.5, the soil reaction itself is not critical. Response of plants to treatments designed to change the pH may be related to the addition of essential elements (e.g., S or Ca). Since yield optima have been reported to be from 5.0 to 7.5 on various soils, it seems likely that deficiencies or imbalances of various minerals may have a greater influence on yield than pH.

C. Mineral Nutrition

1. Deficiency Symptoms

Deficiency symptoms have been reported by several authors.[20-23] These studies were conducted by growing plants in leached sand and watering with a series of nutrient solutions each lacking a specific element. A summary of the results is presented in Table 1. It may be noted that reduction in vine growth is not necessarily related to weight of storage roots. For example, nitrogen deficiency frequently results in a greater reduction of vine growth than root weight, while the case for potassium, magnesium, and calcium deficiency is the opposite. The results for the various elements are in general agreement with a few noteworthy exceptions. Cibes and Samuels[21] found that iron-deficient plants formed no storage roots

Table 1
VINE WEIGHT AND STORAGE ROOT WEIGHT OF SWEET POTATOES GROWN IN NUTRIENT SOLUTIONS COMPLETE EXCEPT FOR THE ELEMENTS NOTED RELATIVE TO THOSE WEIGHTS OF PLANTS GROWN WITH COMPLETE NUTRIENTS, EXPRESSED AS PERCENTAGE OF THE PLANTS GROWN ON COMPLETE SOLUTION

Treatment	Vine weight				Storage root weight			
-N	9[a]	10[b]	22[c]	8[d]	63[a]	1[b]	39[d]	= Control[e]
-P	35[a]	14[b]	30[c]	10[d]	98[a]	3[b]	33[d]	Moderate[e]
-K	24[a]	54[b]	96[c]	12[d]	27[a]	2[b]	5[d]	None[e]
-Ca	85[d]	121[b]	128[c]	68[d]	140[a]	67[b]	18[d]	None[e]
-Mg		94[b]	122[c]	69[d]	145[a]	63[b]	30[d]	None[e]
-S		77[b]	80[c]	31[d]		63[b]	46[d]	= Control[e]
-Mn		63[b]	114[c]	102[d]		10[b]	107[d]	
-Fe		37[b]				0[b]		= Control[e]
-B		154[b]				85[b]		

[a] Source: Edmond and Sefick.[20]
[b] Source: Cibes and Samuels.[21]
[c] Source: Bolle-Jones and Ismunadji[22] 67 days.
[d] Source: Bolle-Jones and Ismunadji[22] 150 days.
[e] Source: Spence and Ahmad.[23]

while Spence and Ahmad[23] reported that "root development in the -Fe treatments was similar to controls". Edmond and Sefick[20] found an increase storage root yield in the -Ca and -Mg treatments while others[21-23] have reported a moderate reduction to no roots formed for these treatments. It is possible that the water used by Edmond and Sefick[20] contained quantities of Ca and Mg sufficient for root production. Deficiency symptoms of the elements are fairly consistent and can be described as follows:

Nitrogen — Very restricted vine growth. The older leaves are the first to turn uniformly light green. Internode and petiole length are generally reduced. Increased reddish or purple pigmentation occurs on petioles, stems, and leaf margins. At later stages, increased abscission and dark green, smaller new leaves.

Phosphorus — Smaller, dark green leaves, purpling of petioles and veins on the underside of leaves. Occasionally light yellow patches on the leaf blade, turning purple then brown. Increased abscission at later stages.

Potassium — Interveinal chlorosis beginning on leaves starting at the tip of laminae and extending around the margins but not extending to petiole. Chlorosis extending inward between secondary veins giving a puckered appearance with scorched margins.

Calcium — Small chlorotic interveinal spots not at the margins, becoming necrotic and black. These spots later coalesce between veins. At later stages vine tips and developing axillary buds may die back. Flowering has been noted in cultivars that do not normally flower.

Magnesium — Interveinal chlorosis while leaves remain dark green, beginning at margins of older leaves and finally showing irregular necrotic areas. Less than normal purple pigmentation of petioles and stems. Flowering of some cultivars that normally do not flower.

Sulfur — Pale green leaves, yellow new leaves with greener vein arms. Some leaves showing purpling of margin and veins. Storage roots more round than normal.

Manganese — Mild interveinal chlorosis.

Table 2
CONCENTRATIONS OF VARIOUS ELEMENTS IN SWEET POTATO TISSUES AT WHICH VISIBLE DEFICIENCY SYMPTOMS BECOME APPARENT AND CONCENTRATION OF THE ELEMENTS IN LEAVES GROWN WITH A COMPLETE NUTRIENT SOLUTION

	Concentrations at which deficiency symptoms occur		Concentrations of elements when plants are grown in complete nutrient solution		
N%	1.5[a]	2.52[b]	3.32[a]	4.47[b]	2.20[c]
P%	0.10[a]	0.12[b]	0.24[a]	0.65[b]	0.43[c]
K%	0.5[a]	0.75[b]	2.28[a]	6.15[b]	2.04[c]
Ca%	0.2[a]	9.2[b]	1.12[a]	1.08[b]	1.00[c]
Mg%	0.05[a]	0.16[b]	0.15[a]	0.67[b]	1.67[c]
S%	0.08[a]	0.08[b]	0.15[a]	0.67[b]	0.41[c]
Mn ppm	2[a]		99[a]		40[c]
Fe ppm		35[b]		85[b]	42[c]
B ppm					6[c]

[a] Source: Bolle-Jones and Ismunadji[18] leaf laminae tissue.
[b] Source: Spence and Ahmad[19] all vine tissue including leaves, and stem above the 9th node.
[c] Source: Cibes and Samuels[17] leaf tissue.

Iron — Severe interveinal chlorosis. Younger leaves becoming pale yellow or nearly white if deficiency is severe.

Boron — Leaves with mottled appearance and burning of leaf margins. Vine tips shrunken and distorted.

2. Elemental Concentrations of Deficient and Normal Sweet Potato Tissue

The data presented in Table 2 should be considered as guidelines to provide workers with tissue concentration data which, together with visible symptoms and soil analysis data, may provide insight as to possible nutritional imbalances.

Elemental concentrations and uptake for a field grown crop are presented in Figures 1 and 2.[24] The data were obtained from a crop grown on a Norfolk sandy loam which, at the final harvest, had a dry matter yield of 10.7 mt/ha of storage roots and 3.5 mt/ha of vines. Several conclusions seem warranted. Concentration of the various elements in the roots is fairly constant, showing only slight increases or decreases during the growing season. Concentrations in the vines show a general pattern of decrease with the exception of Fe and Ca. Similar patterns are reported by Scott and Ogle[25] and Leonard et al.[26] With the exception of potassium and phosphorus, a greater proportion of the total elemental content of all elements was found in the vines than in the storage roots. If the vines are not removed from the field, these nutrients are returned to the soil. Thus, while sweet potatotes do have fairly high nutritional requirements for the production of a good crop, the soils are not rapidly depleted unless the vines are also removed.

Scott and Bouwkamp[24] noted significant differences among cultivars in the average concentration of N, P, K, Ca, Mg, Fe, and Mn in the vines and N, P, Ca, Fe, and Mn in the roots. If the concentration of these elements in well-fertilized plants represents a difference

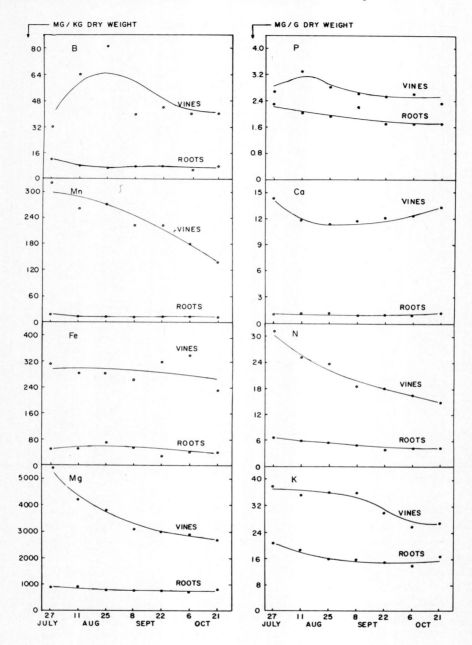

FIGURE 1. Concentration of mineral elements in sweet potato vines and roots at each sampling date.

in nutrient requirements, some of the discrepancies in Table 2 may be explained on the basis of cultivar differences.

3. Nitrogen Studies

The numerous field trials conducted to study the effects of nitrogen fertilization have produced a wide range of results. Johnson and Ware[27] found the addition of nitrogen to four soils at ranges of 22 to 90 kg/ha increased root yield at each increment on Chesterfield and

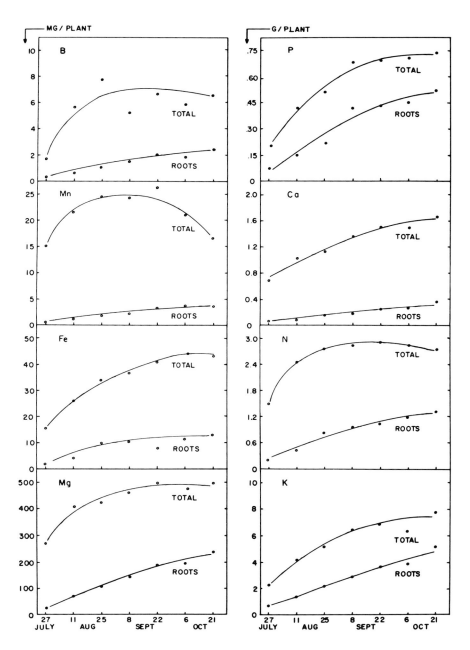

FIGURE 2. Accumulation of mineral elements in sweet potato roots and total uptake at each sampling date.

Norfolk soil and rates of 34 to 67 on Decatur and Hartsells soils. Rates of up to 235 kg/ha of applied nitrogen produced greater vine fresh weight for each 22 kg increment on each soil. Anderson[28] reported yield increases from nitrogen application of up to 72 kg/ha in Cahaba soil, 54 kg/ha on Ruston soil, and a yield reduction for rates higher than 18 kg/ha on Orangeburg soil. Bourke[29] noted that applications of nitrogen gave large yield increases in some trials, especially at grassland sites but reduced yields in a block continuously cropped with sweet potatoes. Nitrogen increased vine growth in all trials where this was measured.

Geise[30] reported that addition of nitrogen at rates of 38 to 54 kg/ha increased yields by an average of 35% over controls. Chew[31] found that nitrogen applied at rates of 90 to 180 kg/ha did not result in significantly higher yields of storage roots than an application of 45 kg/ha. However, vine growth increased with increasing nitrogen rate. Navarro and Padda[15] reported that addition of nitrogen to plots reduced yields and increased vine growth whether or not sulfur had been applied. Hammett[32] found no effects of late season applications of nitrogen on yield or quality but did find increased nitrogen concentration in the storage roots.

Constantin et al.[33] found nitrogen applications to significantly reduce percent fiber and carotene content of fresh and processed roots and to increase percent protein and firmness of the processed roots.

Knavel[34] found increased root yields associated with N levels of 140 and 280 ppm after 9 weeks, while Wilson[35] reported reduced root yields as N levels increased from 21 to 160 ppm and no storage roots formed at 210 ppm after 6 weeks. Both noted that storage roots were more elongated with increasing nitrogen concentrations.

The rather wide range of effects due to nitrogen application may be attributed to two major factors. First, the nitrogen supplying ability of the soil may be quite different among soils used in the various studies. The data from Scott and Bouwkamp[24] indicate that about 100 kg/ha of nitrogen is taken up by a good crop. Smaller crops would require somewhat less and some soils may be able to supply the requirements of such a reduced crop without additional applications. Secondly, there is a wide variation in the response of cultivars to nitrogen applications. Tsunoda[36] suggested that cultivars of sweet potatoes could be classified as adapted to low, medium or high optimum levels of fertilization. Cultivars adapted to low optimum levels tend to have a higher leaf area per plant than those adapted to high optimum levels. Haynes et al.[37] demonstrated this phenomenon with two West Indian cultivars. The low leaf area cultivar, "C9", produced low yields with low nitrogen and high yields with nitrogen applications. The high leaf area cultivar, "049", produced high yields without additional nitrogen and excessive vine growth and low yields with nitrogen application. Watanabe[2] in summarizing number of experiments conducted between 1957 and 1970 concluded that cultivars could be classified into two groups: (1) those which do not produce luxuriant growth easily, which are adaptable to fertile soil, and in which good management practices resulted in a proportional increase of both vines and storage root yield; and (2) those which easily produce luxuriant vine growth under good management practices, frequently resulting in reduced storage root yield.

We may conclude that specific recommendations depend on several factors such as the nitrogen supplying capacity of the soil, the type of cultivar to be grown and the economics of producing a high response cultivar as opposed to a low response cultivar. Each of these variables must be addressed and resolved before recommendations to producers can be formulated.

4. Phosphorus Studies

Research has frequently indicated that sweet potato yields are not increased by the application of phosphorus.[15,23,31,38,42] Hammett et al.[42] also found no effect of phosphorus applications at rates of 0 to 68 kg/ha in 17 kg/ha increments on percent dry matter, percent carotenoids, percent protein, percent fiber or on firmness and percent splits of canned roots. Hotchkiss[43] and Speights and Paterson[44] reported increased yields of 54 and 18% resulting from the application of about 24 kg/ha of phosphorus in east Texas. Geise[30] reported a 38% increase in yield with application of 47 kg/ha of phosphorus. Stino and Lashin[45] found significant increases in vine and storage root yield for each increment of phosphorus applied up to 12 kg/ha in one year and 18 kg/ha in another.

Several factors may explain these results. First, sweet potatoes do not require large amounts of phosphorus to achieve good yields. Scott and Bouwkamp[24] estimate a total uptake of about 26 kg/ha. Secondly, Nishimoto et al.[46] found that a phosphorus concentration of 0.003 ppm in the soil solution (the lowest levels they tested) resulted in a yield of 70% of optimum. Thus, some responses may go undetected since responses of less than 1.5 mt/ha are frequently not statistically significant. Thirdly, agricultural soils are frequently fairly well supplied with phosphorus. Jones et al.[47] quite clearly demonstrate the effect of phosphorus levels in the soils on yield responses to the applications of additional phosphorus. They found no significant yield response to applications of 15 kg/ha in cases where soil tests indicated more than 52 ppm of available phosphorus in the soil, and only 1 case of a significant response to the second 15 kg when soil tests indicated more than 20 ppm of available phosphorus. Lastly, there is some evidence that vesicular arbuscular micorrhizae may be present on the fibrous root system of sweet potato. These fungi are generally believed to aid in phosphorus uptake, particularly in low phosphorus soils.

5. Potassium Studies

The response of sweet potatoes to applications of potassium is fairly consistent among the various reports. Although there are several studies that report a small but insignificant response,[38,44,48,50] a great majority have found addition of potassium to result in significant yield increases. An important feature of potassium addition is that the response is frequently found to be linear. Zimmerley,[41] reporting on a 9-year study found an average yield increase of 1.3 t/ha for each increment of 28 kg/ha of applied K up to 140 kg/ha. Duncan et al.[51] in a 3-year study, obtained yield increases of 1.6, 2.7, 4.4, and 7.3 t/ha (as compared to the check) from potassium applied at the rates of 56, 112, 224, and 448 kg/ha, respectively. Jones et al.[47] noted that the response to applied potassium was nearly linear in those cases where a significant response was obtained, averaging 1.2 t/ha per 28 kg/ha of applied potassium. In cases in which applied potassium did not result in significant increases, they obtained a yield increase averaging 0.15 t/ha per 28 kg/ha of applied potassium. Stino and Lashin[45] obtained successive yield increases of 1.8, 2.4, and 2.2 t/ha corresponding to potassium applications in 17 kg/ha increments. An additional 17 kg/ha did not result in increased yields. Speights and Paterson[44] reported in a yield increase of 4.2 t/ha resulting from an application of 93 kg/ha of potassium and an additional 2.2 t/ha with an additional 93 kg/ha increment.

Geise[30] reported an average yield increase of 3.6 t/ha resulting from the application of 93 kg/ha in various studies conducted over a 5-year period. Knavel[34] reported that the yield of potatoes fertilized with a solution containing 195 mg/ℓ of K was more than double the 0 controls but that 390 mg/ℓ did not further increase yields. Root numbers for the 0, 195, and 390 mg/ℓ treatments were 4.1, 6.1, and 7.9, respectively. Bourke,[20] in a continuous cropping study, found potassium application to increase yield and storage root numbers, but obtained no significant yield increases at former forest sites.

Godfrey-Sam-Aggrey[52] found potassium sulfate to be effective in increasing yields at rates up to 448 kg/ha in intensively cropped soils but noted that 112 kg/ha was suffficient in newly reclaimed sites. A number of factors probably contribute to the response of sweet potatoes to potassium application. The requirement is rather high (the data of Scott and Bouwkamp[24] indicate a requirement of 280 kg/ha for a good crop) and probably in excess of the release capabilities of many soils.

The importance of potassium on photosynthesis, translocation,[53,54] and starch synthesis[55,56] has been documented. Tsuno and Fujise[57] have shown that additional potassium supplies can counteract the effect of high nitrogen on excessive vine growth and decreased yield. Finally, the data of Duncan et al.[50] and Fujise and Tsuno[58] show that higher potassium may result in lower dry matter content of the roots, thus dry weight yields may not be increased to the extent of fresh weight yields. This last point should be considered since, in the final analysis, the main objective is to produce dry matter and thus calories.

6. Calcium Studies

Few studies have been conducted on crop response to application of calcium other than its effect as a constituent of liming materials. Steinbauer and Beattie[23] applied roughly equal amounts of calcium from two sources, calcium chloride and hydrated lime. They found no significant difference between the lime treatments and the control treatments and a slight (ca. 1.2 t/ha) but insignificant increase in the calcium chloride treatment as compared to the control. Hammett[32] applied calcium as a foliar spray and found a small but insignificant yield increase in the treated plots. He found no effect on percent cracking, intercellular space, or in the concentration of N, Ca, Na, or Mn in the storage roots. Landrau and Samuels[59] obtained yield increases of 16% with the application of calcium at the rate of 160 kg/ha in the form of calcium sulfate. Scott and Bouwkamp[24] found that a good crop requires about 60 kg/ha. It may be that many soils are able to supply this amount and the crop does not respond to additional supplies.

7. Magnesium Studies

Although sweet potatoes do not require large amounts of magnesium (18 to 20 kg/ha),[24] deficiency symptoms may be observed when the crop is grown in soils fairly low in available magnesium under conditions of heavy potassium applications. Hester et al.[60] obtained yield increases averaging 0.8 t/ha with the application of magnesium at the rate of 21.5 kg/ha derived 50% from soluble sources and 50% from dolomitic limestone. Greater applications did not result in greater yields than the 21.5 kg/ha rate. Jackson and Thomas[61] found that magnesium applied as dolomitic limestone at the rate of 1.1 t/ha of limestone resulted in yield increases of 2.6 t/ha. An application rate of 2.2 t/ha did not result in a further increase in yield. They noted that magnesium concentrations in the leaves were reduced by the application of potassium and resulted in deficiency symptoms at the higher potassium rates but that "this did not, however, result in as severe reduction of root yield as of tops". Duncan and Stark[62] did not obtain yield increases from magnesium at rates of 7.5 and 15 kg/ha applied as magnesium sulfate and Landrau and Samuels[50] obtained nonsignificant yield increases with the application of 200 kg/ha of Mg in the form of MgO.

8. Sulfur Studies

Adams[68] reported yield increases resulting from the application of sulfur at rates of 220 to 340 kg/ha even under conditions where pox was not a problem. Navarro and Padda[15] found yield of 9.8, 16.6 and 21.2 t/ha with the application of 0, 2246 and 4492 kg/ha of sulfur although the pH was 7.72, 7.67, and 7.62, respectively. They also noted sulfur deficiency symptoms in plots not receiving sulfur or ammonium sulfate.

9. Minor Elements

Anderson et al.[63] conducted a total of 13 trials over 5 years on 7 different soil types to determine the effects of manganese application. Their results indicate a consistent yield increase at rates of 4.1 to 24.5 kg of Mn per ha. Average yield increases as compared to the control, were 1.9, 2.4, 2.5, and 2.0 t/ha for rates of 4.1, 8.2, 12.2, and 24.5 kg/ha of applied manganese, respectively. Although, increases associated with manganese application were frequently not statistically significant it is noteworthy that of 34 treatments, 29 resulted in increased yields (average increase 2.6 t/ha) and 5 resulted in decreased yields (average decrease 0.7 t/ha) as compared to the controls. Mishra and Kelley[64] in a study on early growth response to manganese concentration in solution culture found that tops and fibrous roots both responded optimally to a concentration of 0.031 ppm of manganese in the solution.

Boron applications of 0.6 kg/ha by Duncan and Stark[62] and 2.3 and 6.8 kg/ha by Lutz et al.[65] did not result in yield increases. Landrau and Samuels[58] found that boron application at a rate of 3.8 kg/ha on a Sebana Seca sandy clay (Oxic Plinthaqualts) resulted in significant

Table 3
SWEET POTATO ORGANIC FERTILIZER TRAILS IN PAPUA NEW GUINEA

Organic fertilizer	Level of app (t/ha)	Crop yield (t/ha)	Level of significance	Ref.
Pig manure	0	13.4	*	71
	22	17.0		
	44	15.9		
Chicken manure	0	40.7	ns	72
	15	42.7		
Chicken manure[a]	0	18.2	**	73
	5	16.6		
	10	24.9		
	15	19.2		
	20	18.8		
Coffee pulp[b]	0	9.3	**	74
	15	12.1		
	30	13.0		
	45	12.9		
	60	15.8		
	75	15.6		
Compost placed in mounds	0	8.9	$p = 0.1$	75
	10	10.3		
	20	12.7		
	30	12.0		
Compost placed in mounds	0	6.6	**	164
	18	10.1		

[a] Manure applied to a preceding crop of winged bean.
[b] Average of three trials.

yield increases but that applications of 6.5 kg/ha on Lares clay and on Catano loamy sand resulted in nonsignificant yield increases. Miller and Nielsen[66] found that boron applications at the rate of 1.2 kg/ha would eliminate a disorder called blister, described as "protruding necrotic hypertrophies under the periderm of affected roots that appeared only after at 30 days of storage." Badillo-Feliciano and Lugo-Lopez[67] found no significant effect zinc application on yield at rates up to 9 kg/ha.

Russell and Manns[69] obtained yield increases of 6% and 7.9% of Sassafras and Portsmouth sandy loams, respectively, with the application of 22 kg/ha of copper in the form of copper sulfate. Landrau and Samuels[59] did not obtain yield increases with the application of copper or manganese.

10. Organic and Complete Fertilizers

In contrast to application of chemical fertilizers, the addition of organic materials nearly always results in yield increases. Bourke[70] has summarized the results of several such trials conducted in Papua New Guinea (Table 3). Jana[76] commented that in East Africa "farmyard manure always gives good results but is seldom applied." Leng[77] reports good results from using sweet potato vines as composting material and describes a technique for adding this material to mounds. This technique should be attempted only with great caution if a severe weevil infestation exists.

Organic material, in addition to supplying plant nutrients, is likely to have other beneficial effects on soil structure, water holding capacity, cation exchange capacity, and nutrient release rate. Some of these effects can be inferred from the series of experiments conducted

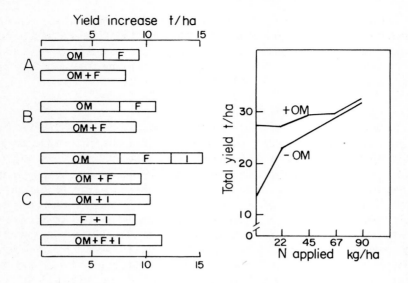

FIGURE 3. Yield increases obtained by the application of organic matter (OM), irrigation (I) or additional fertilizer (F) singly and in combination. A, organic matter, 13.5 mt/ha of green manure, fertilizer 0.45 mt/ha of 4-4.4-5.8 in addition to a standard application of 0.9 mt/ha. B organic matter, 13.5 mt/ha of animal manure, fertilizer 0.56 mt/ha of 6-3.5-3.3 in addition to a standard application of 0.56 mt/ha, irrigation, 2.5 cm/week if not supplied by rainfall.

by Ware and Johnson[78] on the combined effects of organic materials and other inputs such as chemical fertilizers and irrigation (Figure 3). They obtained yield responses to organic materials, chemical fertilizers, and irrigation. The response to a combination of treatments is to a considerable extent additive (the sum of responses to individual treatments) suggesting an independence of action. The consistent shortfall to additivity is evidence of similarity of action. Thus the benefits of compost cannot be explained simply in terms of nutrients or water holding capacity but as a combination of several factors. It also seems clear that the addition of organic matter can partially substitute for other inputs such as chemical fertilizer or irrigation if these are not available.

Similar results were obtained by Houghland[79] who reported that green manure in conjunction with organic fertilizers resulted in a yield increase of 1.7 t/ha, while in conjunction with inorganic fertilizers, a yield increase of 2.6 t/ha was measured.

As animal manures became less frequently used in the U.S., due, in part, to their unavailability and to the relative ease with which chemical fertilizers could be applied, it became apparent that sweet potatoes would respond to fairly high levels of complete chemical fertilizers. Optimum timing and placement of fertilizers have also been studied. Although responses to specific treatments were quite variable, several factors were identified which affect response to the timing and placement of fertilizers. Application of high rates of fertilizer at or before transplanting may result in high levels of soluble salts, considerable plant mortality and yield reductions.[80,81] Leaching losses during the season,[61,82] especially in the sandy soils preferred for production, may result in deficiencies later in the growing season. Excessive fertilizer (especially nitrogen) may result in excessive vine growth and disruption of optimum distribution of assimilates. Early vigor and quick establishment of a full canopy could be expected to result in good yields and reduce weed control problems.

Several techniques have been tested to resolve one or more of the above factors. Delayed application of the fertilizer until 2 to 4 weeks after transplanting resulted in less mortality[80] but not necessarily greater yield.[81,82] Placement of the fertilizer in bands along the rows generally failed to provide consistent yield increases[83,84] although it was observed that placement too close (<6 cm) to the rows may decrease yields.[84,85] For a specific set of circumstances, the optimum schedule and rate of application will depend on the fertility of the soil, the potential and/or likelihood for significant leaching losses, and customs and procedures normally practiced by the growers in the area.

11. Response to Adverse Soil Conditions

Greig and Smith[86-88] in a series of experiments, conducted after it was observed that sweet potatoes grew poorly in previously flooded soils, found that soil solution osmotic pressures of 1.42 to 1.89 atmospheres (depending on the cation) reduced vine growth. High concentrations of K and Na killed plants and storage root development was completely inhibited at more moderate concentrations. Root growth seemed more sensitive to Na ions than Ca or K although vine growth was not reduced. Bernstein[89] reported that sweet potatoes were moderately tolerant to salinity, showing a small reduction in yield when the electrical conductivity was 3 to 5 mmohs/cm^2. Johnson et al.[90] found yield increases with the application of up to 1120 kg/ha of NaCl or Na_2SO_4 and Worley and Harmon[91] reported that substitution of Na for K did not significantly reduce yields over a 4 year test. Yields in plots where 25 to 75% of the K had been replaced by Na were slightly higher than plots receiving Na or K only.

Sweet potatoes are quite tolerant of low pH soil. Abruna et al.[11] found yields to be reduced 20% or less by ca. 50% Al saturation of the soil. They also noted that sweet potatoes were more tolerant to low pH than sugar cane, tobacco, corn, greenbeans, or soybeans tested on the same soils. Jones et al.[16] found that sulfur applications increased extractable Al in the soil from about 60 to 130 ppm but yields were not affected. Munn and McCollum[92] found differences among cultivars in tolerance to Al as measured by root growth on cuttings. Jones et al.[16] observed high Mn concentrations in the leaves from sulfur treated plots (5,000 vs. 600 ppm). Early vine growth was inhibited but there were no significant yield reductions. Abruna et al.[11] noted high yields in a Cota sandy clay with high levels of soluble Mn.

Studies on the adaptability of sweet potatoes to dry compacted soil following rice culture[93,94] have indicated considerable differences among genotypes and that it may be possible to select cultivars suitable for these adverse conditions.

One may conclude that sweet potatoes may be successfully produced under several types of adverse soil conditions. Yields will not be optimum but sweet potatoes may be expected to produce a worthwhile crop even under these adverse conditions.

II. ENVIRONMENTAL REQUIREMENTS

Although the origin of the sweet potato is presumed to be in the American tropics, it is fairly well adapted to many ecological zones. In reference to climatic adaptation, Wilson[95] noted that the "sweet potato is the most widely distributed of the tropical (root) crops". Macroclimatic effects on sweet potato yields may be noted from Figure 4. The increased yields as one proceeds northward may be a result of longer day lengths, cooler night temperatures, lower soil temperatures, or a combination of two or more of these factors. Badillo-Feliciano and Lugo-Loupez[96] found that in Puerto Rico, yields of most cultivars are highest when planted in November but that successful crops may be planted from July through December. They also note that "Gem", an orange-fleshed cultivar developed in North Carolina, produced heavy yields the year around.

A. Temperature Requirements

Being a tropical crop, it is not surprising that sweet potatoes are sensitive to low temperatures. Harter and Whitney[97] reported that sweet potatoes would not survive at temper-

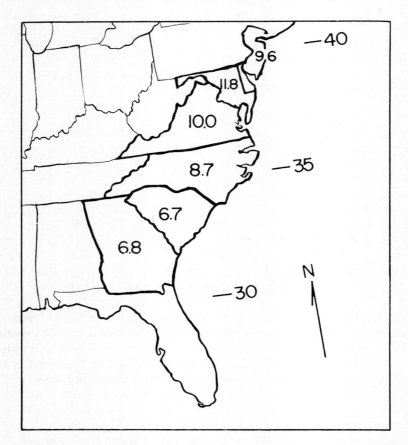

FIGURE 4. Yields (average 1909 to 1982) of sweet potatoes in southeastern U.S.

atures less than 12°C. At 15° the plants were able to survive but did not grow. Above 15°C growth increased with increasing temperature up to 35°C. Plants at 38°C did not grow quite as well at 35°C. Spense and Humphries[98] report that one cultivar produced no storage roots at 10 or 15°C and another when grown at 15° resulted in only about 10% of the production at optimum temperature. Sakr[99] reported that sweet potatoes grown in a 10 to 15°C greenhouse had much reduced vine growth as compared to those grown at 21 to 25°C. Vine growth of the two treatments became equal after transfer to the field, but the yield of storage roots of the cold-treated plants was only 75% of the controls.

Tsuno and Fujise[53] found that temperature in the range of 23 to 33°C had no affect on photosynthetic rate. Sekoika[110] reported that translocation rate was optimum at air temperatures of 20 to 30°C and that distribution of assimilates to the storage roots was greater at 15°C and to vines at 25°C although soil temperatures in the range of 15 to 25°C had relatively little effect on the distribution of assimilates. Maximum storage root growth was obtained at temperatures of 25°C day, 20°C night. Hasegawa and Yahiro[101] determined that soil temperatures of 32°C inhibited the growth of storage roots although vine growth was stimulated and that night temperatures were more important than day temperatures. They also found that high night temperatures for the first 20 days were much less detrimental than high night temperatures later in the season. Kim[102] also found that a constant 29°C temperature resulted in inhibition of storage root growth but not vine growth and that a temperature regime of 29°/20°C (day/night) or 25°/20°C resulted in good storage root growth, slightly superior to a regime of 26°/14°C. The major effect of high temperature seems to be in its

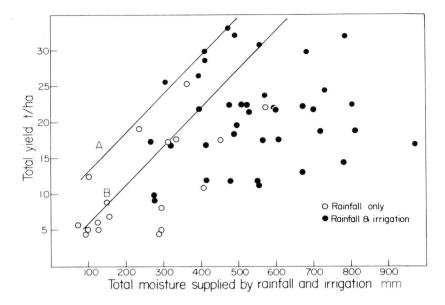

FIGURE 5. Yield response of sweet potatoes as influenced by moisture supply. (A) Optimum observed responses. (B) calculated response 54 kg/ha/mm of irrigation.

effect on the translocation and distribution of assimilates, thus in part accounting for the excessive vine growth often noted in the tropics. The observation that one cultivar, "Gem", is less affected[96] suggests the possibility of identifying similar cultivars.

B. Light Requirements

Individual detached leaves were found by Tsuno and Fujise[53] to be saturated at approximately 30 klux, but field canopies were not saturated at intensities up to 1 g·cal/cm^2/sec. They also reported[103] that shading increased the apparent rate of photosynthesis, even though long term light reduction greatly decreased storage root growth with little effect on the aerial parts. Sekoika[104] found that progressively more photosynthate was translocated to the roots and less to the leaves and upper stems as the light intensity was decreased. Zara et al.[105] reported that sweet potatoes grown under a canopy of coconut trees produced less yield than in an open field and that the difference was greater in wet seasons than dry seasons and vine growth was less affected in wet seasons than dry. They also identified several cultivars with potential for production under coconut in the dry season. Moreno[106] noted that sweet potato as an intercrop is more successful when not competing for light.

C. Moisture Requirements

The results of many experiments[42,93,94,107–112] conducted in the U.S. and elsewhere are presented in Figure 5. Line A represents observations thought to be near optimum with regard to both time of distribution of moisture and utilization by the plant of available moisture (probably including soil reserves). Line B is the average response of sweet potato yield to applications of water. This response is the average of the seven experiments reported by Hernandez et al.[107] and the eight years of experiments reported by Hammett et al.[42]

Yields are affected by the timing and distribution of moisture supply as well as the amount. Miller[113] covered sweet potato plants to protect them from rain for the first 80 days, the first 40 days, and 55 days beginning 40 days after transplanting. He obtained yields of 11.2, 19.9, 30.3, and 31.9 t/ha for the respective treatments and the control. This suggests that the first 40 days are more critical to high yield than later periods, although Ogle and Scott[114] found that protecting the plants for a period of 30 to 40 days beginning 40 days after transplanting resulted in yields of 16.1 to 16.4 t/ha as compared to 24 t/ha for the control. Hernandez et al.[107] state, "drought approximately 40 days after transplanting . . . caused the greatest reduction in yield." It seems likely that if sufficient moisture is supplied for the plant to become well established it will be better able to use reserve moisture under dry conditions later in the season.

The amount of water required has been reported by several authors. Hernandez et al.[107] estimates 18 mm/week in the early season gradually increasing to as much as 44 mm/week in midseason. Lambeth[112] estimates 27 to 36 mm/week at midseason. Jones[115] in a carefully conducted study found that evapotranspiration (average of 2 years) was 2.6 mm/day for the first 45 days, 3.9 for the next 45 days, and 2.5 mm/day for the last 30 days.

The sweet potato is considered by many to be drought tolerant. There are several reasons for this reputation. First, it is quite deep rooted. Weaver[116] found sweet potato roots to penetrate to 175 cm with a working depth of 130 cm. This permits the plant to survive and the vines to appear more normal than many shallow-rooted crops. Secondly, fairly good yields have been obtained under low moisture conditions. Lambeth[112] reported a yield of 12.5 t/ha in a season with 101 mm of rainfall. Carpena[93] obtained a yield of 15 t/ha in a season of 69 mm of rainfall and 8 to 70 cultivars tested produced more than 10 t/ha in 90 days in a season of 98 mm rainfall. Villareal et al.[94] reported a yield of 13 t/ha in a season of 129 mm of rainfall. In spite of these high yields under dry conditions, irrigation experiments clearly demonstrate that yields can be significantly increased by irrigation in areas where rainfall distribution is erratic or insufficient.

The effects of excessive rainfall resulting in waterlogged soil have been previously mentioned. It also seems likely that rainfall affects the partitioning of assimilates. Ehara and Sekoika[117] found that in high humidity more assimilate was translocated to the aerial parts and less to the roots than in low humidity. They also noted more assimilate transferred to roots held at low soil moisture (40%) than high soil moisture (100%) but this may have been an oxygen effect. The data of Gollifer[118] indicated that rainfall has considerable effect on the distribution of assimilates. Periods of high rainfall result in a greater vine-to-root ratio.

III. PRODUCTION PRACTICES

A. Propagating Materials

Various propagating materials are used to establish a sweet potato crop. In areas where sweet potato production cannot be carried on continuously, storage roots from the previous crop are bedded and the sprouts that develop on the roots are used in propagating materials. This procedure is generally used in temperate and subtropical regions and is adequately described by Steinbauer and Kushman[119] and Edmond and Ammerman.[120]

Vine cuttings are used in both temperate and tropical regions. Vine tip cuttings are generally preferred, although mid-vine cuttings may be used. Cuttings from near the base of the vines are generally not recommended (Boswell[121] and Godfrey-Sam-Aggrey[122]). Bouwkamp and McArdle (Table 4) found cuttings from the base of the vine to root as well as tip cuttings when leaves were intact but less when when the leaves were removed. deKraker and Bolhuis[123] found that cuttings from near the base grew about as well as tip cuttings. They also noted that longer cuttings generally grew faster than shorter cuttings. Godfrey-Sam-Aggrey[122] reported that cuttings 45 or 61 cm long produced a higher yield than 23 cm cuttings. Chen and Allison[124] found that 40 to 50 cm sprouts produced a larger yield than 20 to 25 cm sprouts. Beattie et al.[125] testing 3 cultivars for 3 years, found a small (1.4 t/ha) statistically insignificant increase in yield from sprouts 23 to 25 cm long as compared to sprouts 15 to

Table 4
EFFECT OF VINE POSITION ON ROOTING OF 20 CM VINE CUTTINGS OF CV "5-113"

Vine position (tip)	With leaves		Leaves removed	
	Roots/cutting	Root dry wt (g)	Roots/cutting	Root dry wt (g)
1st 20 cm	33	69	34	29
2nd 20 cm	35	72	29	32
3rd 20 cm	34	75	27	28
4th 20 cm	36	59	32	22
5th 20 cm	32	58	26	18
6th 20 cm (base)	35	65	22	14

18 cm long. El-Kattan and Stark[126] and Cordner and Galeotti[127] obtained smaller yields (0.5 and 4.9 t/ha) from sprouts with the below ground portion removed as compared to intact sprouts, suggesting the importance of early establishment of rooting.

It should be noted that longer cuttings are more difficult to handle, transport, and plant. The profitability of larger propagation materials will depend on local circumstances and economics.

Obtaining propagating material is frequently a labor intensive procedure, often occurring concurrent with other labor intensive tasks. In order to reduce labor requirements, several groups,[128,129] have studied the potential of using cut root pieces in a system similar to that used in the production of white potatoes (*Solanum tuberosum* L.). The system is still in the experimental stage but many of the procedures have been worked out. This system has not been tested in the tropics using freshly harvested roots.

Sweet potatoes may also be propagated from rooted leaves or leaf laminae as described by Isbell.[130] Rooted leaves have been suggested by several authors[98,131] as a useful research model. Meristem culture may be employed to establish virus free stocks[132] and plants have been regenerated from several plant parts in vitro.[133,134]

B. Transplanting

Sprouts are generally planted upright, with the base 5 to 10 cm below the soil. Vine cuttings are generally planted at an angle or horizontal to the surface with 3 to 4 nodes covered. Chen et al.[135] have reported on a modification of a mechanical transplanter that permits larger propagules to be planted horizontally resulting in a greater yield.

Studies on the effects of plant density on yield have been reported by many authors.[82,85,125,136-150] These tests were performed at many locations over many years at various spacings ranging from 15 to 110 cm using various cultivars. Since it is difficult to summarize such diverse data, the data was transformed to facilitate analysis. Each report was considered to be a single test, irrespective of the number of years, locations, and cultivars reported. The density index of the least dense spacing was considered to be 1 and all other density indexes a multiple of (e.g., if the spacing were 10, 20, 40, 60 cm the density indices would be 6, 3, 1.5, and 1). Likewise, the yield index was calculated by dividing the yields obtained at each spacing treatment by the yield of the treatment with the lowest plant density. The results are presented in Figure 6.

One may conclude on the basis of the data presented that the total yield is likely to be greatest at high plant densities (40,000 to 50,000 plants/ha). This phenomenon is fairly general and may be shown for all cultivars thus far studied. As plant densities are increased, root set per plant is decreased but not equivalently, resulting in more storage roots being

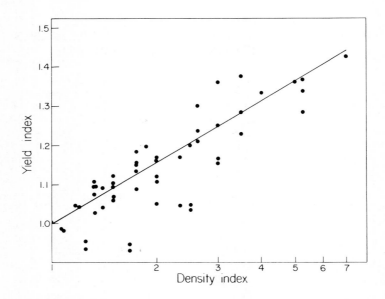

FIGURE 6. Effects of increasing plant density on relative total yield of sweet potato roots.

produced per unit area.[151] Thus a greater sink potential per unit area is available to accept assimilates. At higher plant densities, the surface area is covered more quickly with canopy. The optimum canopy is available for photosynthesis for a greater portion of the growing season. It is likely that both mechanisms contribute to greater yield at higher plant densities.

Therefore producers would be advised to plant at close spacings and control size distribution (if necessary) by adjusting planting and harvest dates. If manipulation of growing days is impractical (for reasons of restriction of the length of the growing season, processing plant schedules or other time constraints), yield distributions of the various size grades can be controlled by changes in plant density. Increased plant density usually results in a greater percentage of smaller and a smaller percentage of very large root yield although the magnitude of response may be different among cultivars.[151]

Plant density has been shown to have an effect of the percentage of yields in the various size grades. The exact nature of these effects are different among cultivars. Thus, in cases where control of yield distribution of the various size grades is important, recommendations should be preceded by experimentation to determine plant density effects on locally adapted cultivars. The optimum density will, of course, depend on local conditions and practices. It should be noted that the spacing experiments summarized were conducted in temperate areas using sprouts as propagating material.

C. Weed Control

Sweet potato vines grow quickly and may reach full canopy in 6 weeks, after which time new weeds generally cannot compete effectively. Weeds that develop before the canopy is developed may persist although Harris[152] reported that a crop of sweet potatoes will practically eliminate an infestation of nutsedge *Cyprus rotundus*.

A summary of the results of several tests of weed control methods is presented in table 5. It may be noted that some form of weed control usually results in higher yields. Cultivation or chemical herbicides result in similar effects on yield. Cultivation in combination with herbicides resulted in higher yields than either.[153] Complete control of weeds generally resulted in the highest yields.

Table 5
TOTAL YIELDS (t/ha) OF STORAGE ROOTS OBTAINED BY VARIOUS METHODS OF WEED CONTROL

No weed control	Cultivation	Herbicides	Cultivation and herbicides	Hand weeding weeding or hoeing	Ref.
11.3	16.3	14.8	17.5		153
11.8		16.7		18.7	154
	21.6	21.3			155
2.6		12.4		14.4	156
13.2		20.7		22.8	157
	15.3	15.3		16.1	158
2.8		24.4			159
7.6		18.8		24.4	113

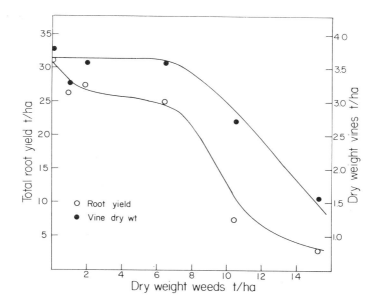

FIGURE 7. Yields and vine weight as influenced by weed infestation.

Although the highest yields are generally obtained with the best weed control it is not always economically feasible to try to obtain complete control since potatoes will produce fairly good yields with some weeds present. This is demonstrated by the data of Greig and Al-Trikriti[159] (Figure 7). Neither yields or vine weights are greatly reduced until weed weights became greater than 6.5 t/ha, nearly two times the vine weights. This is not an argument that weed control is unimportant but that, since sweet potatoes are good competitors, they can tolerate some weed competition and still produce worthwhile yields.

Several herbicides are commonly used in the U.S. They are DCPA at the rate of 7.8 to 10 kg/ha, Diphenamid at the rate of 3.4 to 6.7 kg/ha, and Chloramben at the rate of 3.4 to 4.5 kg/ha. All rates are on the basis of active ingredient. Herbicides are mixed with water and applied after transplanting. Herbicides have been found to not effect storage quality[160,161] or processing quality,[154,161] although slight effects on flavor[155,162] have been noted. Welker[155] noted that roots from plots treated with CIPC did not sprout normally and Robbins and Peterson[163] found that Dacthal inhibited rooting at vine nodes.

REFERENCES

1. **Chua, L. K. and Kays, S. J.**, Effect of soil oxygen concentration on sweet potato storage root induction and/or development, *HortScience*, 16, 71, 1981.
2. **Watanabe, K.**, Agronomic studies on the mechanism of excessive vegetation growth in sweet potato *(Ipomoea batatas)*, *J. Central Agric. Exp. Stn.*, 29, 1, 1979.
3. **Silva, S. and Irizarry, H.**, Effect of depth of water table on yields of two cultivars of sweet potatoes, *J. Agric. Univ. P. R.*, 65, 114, 1981.
4. **Ghuman, B. S. and Lal, R.**, Growth and plant water relations of sweet potato *(Ipomoea batatas)* as effected by soil moisture regimes, *Plant and Soil*, 70, 95, 1983.
5. **Paterson, D. R., Earhard, D. R., and Fuqua, M. C.**, Effects of flooding level on storage root formation, ethylene production, and growth of sweet potato, *HortScience*, 14, 739, 1979.
6. **Ahn, J. K., Collins, W. W., and Pharr, D. M.**, Influence of preharvest temperature and flooding on sweet potato roots in storage, *HortScience*, 15, 261, 1980.
7. **Corey, K. A. and Collins, W. W.**, Effects of vine removal prior to flooding on ethanol concentration and storage quality of sweet potato roots, *HortScience*, 17, 631, 1982.
8. **Woodward, O.**, Sweet potato production on the coastal plain of Georgia, *Ga. Coastal Plain Exp. Stn. Bul.*, 17, 1932.
9. **Edmond, J. B., Garrison, O. B., Wright, R. E., Woodard, O., Steinbauer, C. E., and Deonier, M. T.**, Cooperative studies on the effects of height of ridge, nitrogen supply, and time of harvest on yield and flesh color of the 'Porto Rico' sweet potato, *U.S. Dep. Agric. Circ.*, 832, 1950.
10. **Geise, F. W.**, Fertilizer studies and the production of sweet potatoes, *Md. Agric. Exp. Stn. Bull.*, 311, 1929.
11. **Abruna, F., Vincente-Chandler, J., Rodriguez, J., Badillo, J., and Silva, S.**, Crop response to soil acidity factors in Ultisols and Oxisols in Puerto Rico. V. Sweet potato, *J. Agric. Univ. P. R.*, 63, 250, 1979.
12. **Chew, W. Y., Joseph, K. T., Ramli, K., and Majid, A. B. A.**, Liming needs of sweet potato in Malaysian peat, *Exp. Agric.*, 18, 65, 1981.
13. **Steinbauer, C. E. and Beattie, J. B.**, Influence of lime and calcium chloride applications on the growth and yield of sweet potatoes, *Proc. Am. Soc. Hortic. Sci.*, 36, 526, 1939.
14. **Watts, V. M. and Cooper, J. R.**, Influence of varied soil reactions on growth and yield of vegetable crops on Newtonia silt loam and Ruston fine sandy loam soils, *Ark. Agric. Exp. Stn. Bull.*, 433, 1943.
15. **Navarro, A. A. and Padda, D. S.**, Effects of sulfur, phosphorous and nitrogen application on the growth and yield of sweet potatoes grown on Fredensborg clay loam, *J. Agric. Univ. P. R.*, 67, 108, 1983.
16. **Jones, L. G., Constantin, R. J., Cannon, J. M., Martin, W. J., and Hernandez, T. P.**, Effects of soil ammendment and fertilizer applications on sweet potato growth, production and quality, *La. Agric. Exp. Stn. Bull.*, 704, 1977.
17. **Constantin, R. J., Jones, L. G., and Hernandez, T. P.**, Sweet potato quality as affected by soil reaction (pH) and fertilizer, *J. Am. Soc. Hortic. Sci.*, 100, 604, 1975.
18. **Hartman, J. O. and Gaylord, F. C.**, Soil acidity for muskmelons and sweet potatoes on sand soil, *Proc. Am. Soc. Hortic. Sci.*, 37, 847, 1939.
19. **Person, L. H.**, The soil rot of sweet potato and its control with sulfur, *Phytopathology*, 36, 869, 1946.
20. **Edmond, J. B. and Sefick, J. J.**, A description of certain nutrient deficiency symptoms of the sweet potato, *Proc. Am. Soc. Hortic. Sci.*, 36, 544, 1938.
21. **Cibes, H. and Samuels, G.**, Mineral deficiency symptoms displayed by sweet potato plants grown under controlled conditions, *P. R. Agric. Exp. Stn. Tech. Pap.*, 20, 1957.
22. **Bolle-Jones, E. W. and Ismunadji, M.**, Mineral-deficiency symptoms of the sweet potato, *Emp. J. Exp. Agric.*, 31, 60, 1963.
23. **Spence, J. A. and Ahmad, N.**, Plant nutrient deficiencies and related tissue composition of the sweet potato, *Agron. J.*, 59, 59, 1967.
24. **Scott, L. E. and Bouwkamp, J. C.**, Seasonal mineral accumulation by the sweet potato, *HortScience*, 9, 233, 1974.
25. **Scott, L. E. and Ogle, W. L.**, The mineral uptake by the sweet potato, *Better Crops with Plant Food*, 36 (8), 12, 1952.
26. **Leonard, D. A., Anderson, W. S., and Geiger, M.**, Field studies on the mineral nutrition of the sweet potato, *Proc. Am. Soc. Hortic. Sci.*, 53, 387, 1949.
27. **Johnson, W. A. and Ware, L. M.**, Effects of rates of nitrogen on the relative yields of sweet potato vines and roots, *Proc. Am. Soc. Hortic. Sci.*, 52, 313, 1948.
28. **Anderson, W. S.**, The influence of nitrogen on grade and shape of 'Triumph' sweet potatoes in Mississippi, *Proc. Am. Soc. Hortic. Sci.*, 36, 605, 1938.

29. **Bourke, R. M.,** Sweet potato *(Ipomoea batatas)* fertilizer trials on the Gazelle Peninsula of New Britain: 1954-1976, *Papua New Guinea Agric. J.,* 28, 73, 1977.
30. **Geise, F. W.,** The influence of nitrogen, phosphorous and potash separately and in combination on sweet potato production, *Proc. Am. Soc. Hortic. Sci.,* 22, 363, 1925.
31. **Chew, W. Y.,** Effects of length of growing season and NPK fertilizers on the yield of 5 varieties of sweet potatoes *(Ipomoea batatas* Lam) on peat, *Malaysian Agric. J.,* 47, 453, 1970.
32. **Hammett, L. K.,** Effects of late season nitrogen and foliar calcium applications on sweet potatoes, *HortScience,* 16, 336, 1981.
33. **Constantin, R. J., Hernandez, T. P., and Jones, L. G.,** Effects of irrigation and nitrogen fertilization on quality of sweet potatoes, *J. Am. Soc. Hortic. Sci.,* 99, 308, 1974.
34. **Knavel, D. E.,** The influence of nitrogen and potassium nutrition on vine and root development of the 'Allgold' sweet potato at early stage of storage root development, *J. Am. Soc. Hortic. Sci.,* 96, 718, 1971.
35. **Wilson, L. A.,** Effect of different levels of nitrate-nitrogen supply on early tuber growth of two sweet potato cultivars, *Trop. Agric. (Trinidad),* 50, 53, 1973.
36. **Tsunoda, S.,** A development analysis of yielding ability in varieties of field crops. I. Leaf area per plant and leaf area ratio, *Jpn. J. Breeding,* 9, 161, 1959.
37. **Haynes, P. H., Spence, J. A., and Walter, C. J.,** The use of physiological studies in the agronomy of root crops, *Proc. Int. Symp. Trop. Root Crops,* Vol. 1, Sect. III, University of the West Indies, St. Augustine, Trinidad, 1969, 1.
38. **Anderson, W. S.,** The influence of fertilizers upon the yield and starch content of the 'Triumph' sweet potato, *Proc. Am. Soc. Hortic. Sci.,* 34, 449, 1936.
39. **Kimber, A. J.,** Fertilizing sweet potato on mineral soils in the highlands, *Harvest,* 8, 71, 1982.
40. **Kyle, J. K., Speights, D. E., Paterson, D. R., and Walker, H. J.,** Sweet potato fertilizer and variety trials on the high plains, 1969, *Texas Agric. Exp. Stn. Prog. Rep.,* 2177, 1961.
41. **Zimmerley, H. H.,** Sweet potato fertilizers, *Va. Truck Exp. Stn. Bull.,* 66, 1929.
42. **Hammett, H. L., Constantin, R. J., Jones, L. G., and Hernandez, T. P.,** The effect of phosphorus and soil moisture levels on yield and processing quality of 'Centennial' sweet potatoes, *J. Am. Soc. Hortic. Sci.,* 107, 119, 1982.
43. **Hotchkiss, W. S.,** Sweet potato fertilizer experiments at substation No. 2, Troup, *Texas Agric. Exp. Stn. Bull.,* 277, 1921.
44. **Speights, D. W. and Paterson, D. R.,** Sweet potato fertilizer and variety trials in northeast Texas, 1969, *Texas Agric. Exp. Stn. Prog. Rep.,* 219, 1961.
45. **Stino, K. R. and Lashin, M. E.,** Effect of fertilizers on the yield and vegetative growth of sweet potatoes, *Proc. Am. Soc. Hortic. Sci.,* 61, 367, 372.
46. **Nishimoto, R. K., Fox, R. L., and Parvin, P. E.,** Response of vegetable crops to phosphorus concentration in soil solution, *J. Am. Soc. Hortic. Sci.,* 102, 705, 1977.
47. **Jones, L. G., Constantin, R. J., and Hernandez, T. P.,** The response of sweet potatoes to fertilizer phosphorus and potassium as related to levels of these elements available in the soil, *La. Agric. Exp. Stn. Bull.,* 727, 1979.
48. **Boswell, V. R. and Beattie, J. H.,** Grade and shape of sweet potatoes in response to potash in South Carolina, *Proc. Am. Soc. Hortic. Sci.,* 34, 451, 1936.
49. **Boswell, V. R., Beattie, J. H., and McCown, J. D.,** Effect of potash on grade, shape and yield of certain varieties of sweet potatoes grown in South Carolina, *U.S. Dep. Agric. Circ.,* 498, 1939.
50. **Anderson, W. S.,** The influence of potash on grade and shape of 'Triumph' sweet potatoes in Mississippi, *Proc. Am. Soc. Hortic. Sci.,* 35, 709, 1937.
51. **Duncan, A. A., Scott, L. E., and Stark, F. C.,** Effect of potassium chloride and potassium sulfate on yield and quality of sweet potatoes, *Proc. Am. Soc. Hortic. Sci.,* 71, 391, 1958.
52. **Godfrey-Sam-Aggrey, W.,** Effects of potash fertilizers on sweet potatoes in Sierra Leone, *Exp. Agric.,* 12, 87, 1976.
53. **Tsuno, Y. and Fujise, K.,** Studies on the dry matter production of sweet potato, *Bull. Natl. Inst. Agric. Sci. (Jpn.),* Series D, No. 13, 1965.
54. **Tsuno, Y.,** Dry matter production of sweet potatoes and yield increasing technics, *Fertilite,* 38, 3, 1970.
55. **Tsuno, H. and Fujise, K.,** Studies on the dry matter production of sweet potato. IX. The effect of potassium on the dry matter production of sweet potato, *Proc. Crop Sci. Soc. Jpn.,* 33, 236, 1965.
56. **Murata, T. and Akagawa, T.,** Enzymatic mechanism of starch synthesis in sweet potato roots. I. Requirement of potassium ions for starch synthetase, *Arch. Biochem. Biophys.,* 126, 873, 1968.
57. **Tsuno, Y. and Fujise, K.,** Studies on the dry matter production in sweet potato. IV. The relation between the contribution of mineral nutrients in plant distribution ratio over dry matter produced, *Proc. Crop Sci. Soc. Jpn.,* 32, 301, 1964.
58. **Fujise, K. and Tsuno, Y.,** Effect of potassium on the dry matter of sweet potato. *Proc. Inter. Symp. Trop. Root Crops,* Vol. 1, Sect. II, University of the West Indies, St. Augustine, Trinidad, 1969, 20.

59. **Landrau, P., Jr. and Samuels, G.,** The effect of fertilizers on the yield and quality of sweet potatoes, *J. Agric. Univ. P.R.*, 35, 71, 1951.
60. **Hester, J. B., Isaacs, R. L., and Shelton, F. A.,** Magnesium and potassium nutrition for sweet potatoes in the coastal plain, *Better Crops with Plant Food*, 36 (7), 21, 1952.
61. **Jackson, W. A. and Thomas, G. W.,** Effects of KCl and dolomitic limestone on growth and ion uptake of the sweet potato, *Soil Sci.*, 89, 437, 1960.
62. **Duncan, A. A. and Stark, F. C.,** The influence of potassium, magnesium and boron on the yield of the 'Maryland Golden' sweet potato, *Trans. Peninsula Hortic. Soc.*, 1, 1951.
63. **Anderson, O. E., Dempsey, A. H., and Wentworth, J.,** Sweet potato yields and quality as affected by field applications of manganese sulfate, *Ga. Agric. Exp. Stn. Mimeo Ser.*, N.S. 132, 1962.
64. **Mishra, U. N. and Kelley, J. D.,** Manganese nutrition of sweet potatoes in relation to manganese content, deficiency symptoms, and growth, *Agron. J.*, 59, 578, 1967.
65. **Lutz, J. M., Deonier, M. T., and Walters, B.,** Cracking and keeping quality of 'Puerto Rico' sweet potatoes as influenced by rate of fertilizer, nitrogen ratio, lime and borax, *Proc. Am. Soc. Hortic. Sci.*, 54, 407, 1949.
66. **Miller, C. H. and Nielsen, L. W.,** Sweet potato blister, a disease associated with boron nutrition, *J. Am. Soc. Hortic. Sci.*, 95, 685, 1970.
67. **Badillo-Feliciano, J. and Lugo-Lopez, M. A.,** Differential response of corn and sweet potatoes to Zn applications in an Oxisol in northwestern Puerto Rico, *J. Agric. Univ. P.R.*, 64, 482, 1980.
68. **Adams, J. F.,** The use of sulphur as a fungicide and fertilizer for sweet potatoes, *Phytopathology*, 14, 411, 1924.
69. **Russell, R. and Manns, T. F.,** The value of copper sulphate as a plant nutrient, *Trans. Peninsula Hortic. Soc.*, 51, 1933.
70. **Bourke, R. M.,** Sweet potato in Papua New Guinea, in *Sweet Potato Proc. of the First Int. Symp.*, Villareal, R. L. and Griggs, T. D., Eds., Asian Vegetable Research and Development Center, Shanhua, Taiwan, 1982, 45.
71. **Kimber, A. J.,** The effect of pig manure on sweet potato yields, *Harvest*, 8, 81, 1982.
72. **Velayutham, K. S., Pondrilei, K. S., and Natera, E. A.,** Effect of soil native nutrients and applied nutrients on sweet potato, in *Proc. Second Papua New Guinea Food Crops Conf.*, Bourke, R. M. and Kesvan, V., Eds., Department of Primary Industries, Port Moresby, 1981, 382.
73. **Thaigalingam, K. and Bourke, R. M.,** Utilization of organic wastes in crop production, in *Proc. Second Papua New Guinea Food Conf.*, Bourke, R. M. and Kesvan, V., Eds., Department of Primary Industries, Port Moresby, 1981, 218.
74. **Siki, B. F.,** Coffee pulp as manure on sweet potato, *Harvest*, 6, 1, 1980.
75. **Bourke, R. M. and D'Souza, E.,** Intensification of subsistence agriculture on the Nembi Plateau: preliminary results, in *Proc. Second Papua New Guinea Food Conf.*, Bourke, R. M. and Kesvan, V., Eds., Department of Primary Industries, Port Moresby, 1981, 202.
76. **Jana, R. K.,** Status of sweet potato cultivation in East Africa, in *Sweet Potato Proc. of the First Int. Symp.*, Villareal, R. L. and Griggs, T. D., Eds., Asian Vegetable Research and Development Center, Shanhua, Taiwan, 1982, 73.
77. **Leng, A. S.,** Maintaining fertility by putting compost into sweet potato mounds, *Harvest*, 8, 83, 1982.
78. **Ware, L. M. and Johnson, W. A.,** Effects of organic materials, fertilizers, and irrigation on the yield of sweet potatoes, *Proc. Am. Soc. Hortic. Sci.*, 52, 317, 1947.
79. **Houghland, G. V. C.,** Fertilizer studies with sweet potatoes, *Soil Sci.*, 26, 291, 1928.
80. **Skinner, J. J., Williams, C. B., and Mann, H. B.,** Fertilizers for sweet potatoes based on investigations in North Carolina, *U.S. Dep. Agric. Tech. Bull.*, 335, 1932.
81. **Tiedjens, V. A. and Schermerhorn, L. G.,** Recent studies on fertilizer applications for 'Gold Skin' sweet potatoes, *Proc. Am. Soc. Hortic. Sci.*, 37, 857, 1930.
82. **Hollar, V. E. and Haber, E. S.,** Cultural and fertilizer studies with sweet potatoes, muskmelons and watermelons on Buckner coarse sand, *Iowa Agric. Exp. Stn. Bull.*, P56, 801, 1943.
83. **Phillips, C. E.,** Sweet potato fertilizer placement, *Proc. Ann. Meeting Natl. Joint Comm. Fert. Appl.*, 17, 99, 141.
84. **Mahoney, C. H., Cummings, G. A., and Redit, W. A.,** Results of fertilizer placement studies on sweet potatoes in Maryland in 1939, *Trans. Peninsula Hortic. Soc.*, 1939.
85. **Anderson, W. S. and Randolph, J. W.,** Sweet potato production: fertilization and hill spacing studies, *Miss. Agric. Exp. Stn. Bull.*, 402, 1944.
86. **Greig, J. K. and Smith, F. W.,** Some effects of various levels of calcium, potassium, magnesium and sodium on sweet potato plants grown in nutrient solutions, *Proc. Am. Soc. Hortic. Sci.*, 75, 561, 1969.
87. **Greig, J. K. and Smith, F. W.,** Sweet potato growth, cation accumulation and carotene content as affected by cation level in the growth medium, *Proc. Am. Soc. Hortic. Sci.*, 77, 463, 1961.
88. **Greig, J. K. and Smith, F. W.,** Salinity effects on sweet potato growth, *Agron. J.*, 54, 309, 1962.

89. **Bernstein, L.,** Salt tolerance of vegetable crops in the West, *U.S. Dep. Agric. Inform. Bull.*, No. 205, 1959.
90. **Johnson, T. C., Zimmerley, H. H., and Geise, F. W.,** Effect of certain sodium and potassium salts on sweet potato production in eastern Virginia, *Proc. Am. Soc. Hortic. Sci.*, 20, 155, 1923.
91. **Worley, R. E. and Harmon, S. A.,** Effect of substituting Na for K on yield, quality and leaf analysis of sweet potatoes grown on Tifton loamy sand, *HortScience*, 9, 580, 1974.
92. **Munn, D. A. and McCollum, R. E.,** Solution culture evaluation of sweet potato cultivar tolerance to aluminum, *Agron. J.*, 68, 989, 1976.
93. **Carpena, A. L., Rebancos, E. T., Jr., and Estolano, M. P.,** Screening sweet potato varieties for adaptability to paddy field cultivation, *Philipp. J. Crop Sci.*, 2, 209, 1977.
94. **Villareal, R. L., Lin, S. K., and Lai, S. H.,** Variations in the yielding ability of sweet potato under drought stress and minimum input conditions, *HortScience*, 14, 31, 1979.
95. **Wilson, L. A.,** Root crops, in *Ecophysiology of Tropical Crops*, Alvim, P. de T. and Kouzlowski, T. T., Eds., Academic Press, New York, 1977.
96. **Badillo-Feliciano, J. and Lugo-Lopez, M. A.,** Sweet potato production in Oxisols under a high level of technology, *P.R. Agric. Exp. Stn. Bull.*, 256, 1977.
97. **Harter, L. L. and Whitney, W. A.,** Influence of soil temperature and soil moisture on the infection of sweet potato by the black rot fungus, *J. Agric. Res.*, 32, 1153, 1926.
98. **Spence, J. A. and Humphries, E. C.,** Effect of moisture supply, root temperature and growth regulators on photosynthesis of isolated rooted leaves of sweet potato *(Ipomoea batatas)*, *Ann. Bot.*, 36, 115, 1972.
99. **Sakr, E. S. M.,** Effect of temperature on yield of the sweet potato, *Proc. Am. Soc. Hortic. Sci.*, 42, 517, 1943.
100. **Sekoika, H.,** Effect of temperature on the translocation of sucrose-^{14}C in the sweet potato plant, *Proc. Crop. Sci. Soc. Jpn.*, 30, 27, 1961.
101. **Hasegawa, H. and Yahiro, T.,** Effects of high soil temperatures on the growth of the sweet potato plant, *Proc. Crop. Sci. Soc. Jpn.*, 26, 37, 1957.
102. **Kim, Y. C.,** Effects of thermoperiodism on tuber formation in *Ipomoea batatas* under controlled conditions, *Plant Physiol.*, 36, 680, 1961.
103. **Tsuno, T. and Fujise, K.,** Studies on the production of the sweet potato. VIII. The factors influencing the photosynthetic activity of the sweet potato leaf, *Proc. Crop. Sci. Soc. Jpn.*, 33, 230, 1965.
104. **Sekoika, H.,** The influence of light intensity on the translocation of sucrose-^{14}C in the sweet potato plant, *Proc. Crop. Sci. Soc. Jpn.*, 31, 159, 1962.
105. **Zara, D. L., Cuevas, S. E., and Carlos, J. T., Jr.,** Performance of sweet potato cultivars grown under coconuts, in *Sweet Potato Proc. of the First Int. Symp.*, Villareal, R. L. and Griggs, T. D., Eds., Asian Vegetable Research and Development Center, Shanhua, Taiwan, 1982, 233.
106. **Moreno, R. A.,** Intercroping with sweet potato *(Ipomoea batatas)* in Central America, in *Sweet Potato Proc. of the First Int. Symp.*, Villareal, R. L. and Griggs, T. D., Eds., Asian Vegetable Research and Development Center, Shanhua, Taiwan, 1982, 243.
107. **Hernandez, T. P., Hernandez, T. P., Miller, J. C., and Jones, L. G.,** The value of irrigation in sweet potato production in Louisiana, *La. Agric. Exp. Stn. Bull.*, 607, 1965.
108. **Peterson, L. E.,** The varietal response of sweet potatoes to changing levels of irrigation, fertilizer and plant spacing, *Proc. Am. Soc. Hortic. Sci.*, 77, 452, 1961.
109. **Lana, E. P. and Peterson, L. E.,** The effect of fertilizer-irrigation combinations on sweet potatoes in Buckner coarse sand, *Proc. Am. Soc. Hortic. Sci.*, 68, 400, 1956.
110. **Bowers, J. L., Benedict, R. H., and McFerran, J.,** Irrigation of sweet potatoes, snap beans, and cucumbers in Arkansas, *Ark. Agric. Exp. Stn. Bull.*, 578, 1956.
111. **Hernandez, T. P. and Hernandez, T. P.,** Irrigation to increase sweet potato production, *Proc. Int. Symp. Trop. Root Crops*, Vol. 1, Sect. III, University of the West Indies, St. Augustine, Trinidad, 1969, 31.
112. **Lambeth, V. N.,** Sweet potato, a promising Missouri crop, *Mo. Agric. Exp. Stn. Bull.*, 686, 1957.
113. **Miller, J. C.,** Sweet potato breeding and yield studies, *La. Agric. Exp. Stn. Circ.*, 41, 1958.
114. **Ogle, W. L. and Scott, L. E.,** The effect of fumigation, nitrogen level, and soil moisture conditions upon cracking of sweet potatoes, *Trans. Peninsula Hortic. Sci.*, 1951.
115. **Jones, S. T.,** Effect of irrigation at different levels of soil moisture on yield and evapotranspiration rate of sweet potatoes, *Proc. Am. Soc. Hortic. Sci.*, 77, 458, 1961.
116. **Weaver, J. E. and Bruner, W. E.,** *Root Development of Vegetable Crops*, McGraw-Hill, New York, 1927, 229.
117. **Ehara, K. and Sekoika, H.,** Effect of atmospheric humidity and soil moisture on the translocation of sucrose-^{14}C in the sweet potato plant, *Proc. Crop. Sci. Jpn.*, 31, 41, 1962.
118. **Gollifer, D. E.,** A time of planting trial with sweet potatoes, *Trop. Agric. (Trinidad)*, 57, 363, 1980.
119. **Steinbauer, C. E. and Kushman, L. J.,** Sweet Potato Culture and Diseases, Handbook 338, U.S. Department of Agriculture, Washington, D.C., 1971.

120. **Edmond, J. B. and Ammerman, G. R.,** *Sweet Potatoes: Production, Processing, Marketing,* AVI Publishing, Westport, Conn., 1971, 81.
121. **Boswell, V. R.,** Commercial growing and harvesting of sweet potatoes, *U.S. Dept. Agric. Farmers Bull.,* 2020, 1950.
122. **Godfrey-Sam-Aggrey, W.,** Effects of cutting lengths on sweet potato yields in Sierra Leone, *Exp. Agric.,* 10, 33, 1974.
123. **deKraker, J. P. and Bolhuis, G. G.,** Propagation of sweet potato with different kinds of cuttings, *Proc. Int. Symp. Trop. Root Crops* 1969, 1, Sect III, 131.
124. **Chen, L. H. and Allison, M.,** Horizontal transplanting increases sweet potato yield, *Miss Agric. Forestry Exp. Stn. Res. Highlights,* 45(4), 1, 1982.
125. **Beattie, J. H., Boswell, V. R., and McCown, J. D.,** Sweet potato propagation and transplanting studies, *U.S. Dept. Agric. Circ.,* 502, 1938.
126. **El-Kattan, A. A. and Stark, F. C., Jr.,** A comparison of sprouts and vine cuttings for sweet potato production, *Trans. Peninsula Hortic. Soc.,* 1949.
127. **Cordner, H. B. and Galeotti, C.,** Effects of length of growing season and method of propagating on yields of sweet potatoes, *Okla. Agric. Exp. Stn. Processed Series,* P 392, 1961.
128. **Kobayashi, M.,** Studies on breeding and vegetative propagation of sweet potato varieties adapted to direct planting, *Bull. Chugoku Agric. Exp. Stn., A. Crop. Div.,* 16, 245, 1968.
129. **Bouwkamp, J. C.,** Research on production of sweet potatoes from cut root pieces, in *Sweet Potato Proc. of the First Int. Symp.,* Villareal, R. L. and Griggs, T. D., Eds., Asian Vegetable Research and Development Center, Shanhua, Tainan, Taiwan, 1982, 191.
130. **Isbell, C. L.,** Regeneation in leaf cuttings of *Ipomoea batatas, Bot. Gaz.,* 91, 441, 1931.
131. **Martin, F. W.,** Technique for rooting sweet potato leaves, *HortScience,* 17, 395, 1982.
132. **Alconero, R., Santiago, A. G., Morales, R., and Rodriguez, F.,** Meristem tip culture and virus indexing of sweet potatoes, *Phytopathology,* 65, 769, 1975.
133. **Sihachakr, D.,** Premiers resultats concernant la multiplication vegetative *in vitro* de la patate douce (*Ipomoea batatas* L. Lam., convolvulance), *Agronomie Tropicale,* 37, 142, 1982.
134. **Hwang, L. S., Skirvin, R. M., Casyao, J., and Bouwkamp, J.,** Adventitious shoot formation from sections of sweet potato grown in vitro, *Scientia Hortic.,* 20, 119, 1983.
135. **Chen, L. H., Younis, T. S., and Allison, M.,** Horizontal Planting of Sweet Potatoes, Paper 82-1023 presented to summer meeting of the American Society of Agricultural Engineers, 1982.
136. **Anderson, W. S., Cochran, H. L., Edmond, J. B., Garrison, O. B., Wright, R. E., and Boswell, V. R.,** Regional studies of time planting and hill spacing of sweet potatoes, *U.S. Dep. of Agric. Circ.,* 72, 1945.
137. **Anderson, W. S., Currey, E. A., Ferris, E. B., and Robert, J. C.,** Sweet potato plant spacing, *Miss. Agric. Exp. Stn. Bull.,* 358, 1941.
138. **Anderson, W. S. and Randolph, J. W.,** Sweet potato production: time of planting and hill spacing studies, *Miss. Agric. Exp. Stn. Bull.,* 378, 1943.
139. **Anonymous,** *N. Mex. Agric. Exp. Stn. 46th Ann. Rep.,* 1935.
140. **Beattie, J. H., Boswell, V. R., and Hall, E. E.,** Influence of spacing and time of planting on the yield and size of 'Porto Rico' sweet potato, *U.S. Dep. of Agric. Circ.,* 327, 1934.
141. **Cochran, H. L.,** Influence of time of planting and spacing on the yield of 'Porto Rico' and 'Triumph' sweet potatoes, *Ga. Agric. Exp. Stn. Bull.,* 230, 1943.
142. **Currey, E. A.,** Growing sweet potatoes in the Yazoo-Mississippi delta, *Miss. Farm Res.,* 6(12), 6, 1943.
143. **Kattan, A. A. and Bryan, B. B.,** Irrigation and twin spacing improve yield and grade of sweet potatoes, *Ark. Farm. Res.,* 9(6), 8, 1960.
144. **Kimbrough, W. D.,** Studies of the production of sweet potatoes for starch or food purposes, *La. Agric. Exp. Stn. Bull.,* 348, 1942.
145. **Miller, J. C. and Kimbrough, W. C.,** Sweet potato production in Louisiana, *La. Agric. Exp. Stn. Bull.,* 281, 1936.
146. **Poole, C. F.,** Seedling improvement in sweet potatoes, *Hawaii Agric. Exp. Stn. Tech. Bull.,* 17, 1952.
147. **Price, J. C. C.,** The production and storage of sweet potatoes, *Mississippi Agric. Exp. Stn. Bull.,* 279, 1930.
148. **Stark, F. C., Jr. and Scott, L. E.,** Planting sweet potatoes and treating sprouts, *Md Agric. Exp. Serv. Fact Sheet* 24, 1955.
149. **Woodard, O.,** Sweet potato culture in the Coastal Plain of Georgia, *Ga. Coastal Plain Exp. Stn. Bull.,* 17, 1932.
150. **Zimmerley, H. H.,** The effect of plant spacing on the development of sweet potato storage roots, *Proc. Am. Soc. Hortic. Sci.,* 32, 494, 1934.
151. **Bouwkamp, J. C. and Scott, L. E.,** Effect of plant density on yield and yield components of sweet potato, *Ann. Trop. Res.,* 2, 1, 1980.

152. **Harris, V. C.**, Nutgrass controlled by competition, *Miss. Farm Res.*, 22(8), 1, 1958.
153. **Glaze, N. C., Harmon, S. A., and Phatak, S. C.**, Enhancement of herbicidal weed control in sweet potato (*Ipomoea batatas*) with cultivation, *Weed Sci.*, 29, 275, 1981.
154. **Hernandez, T. P., Constantin, R. J., Barry, J. R., and Wascom, B. W.**, Herbicides for sweet potatoes, *La. Agric.*, 12(3), 10, 1969.
155. **Welker, W. V., Jr.**, Effect of herbicides on quality and yield of sweet potatoes, *Weeds*, 15, 112, 1967.
156. **Talbert, R. E.**, Weed control studies in sweet potatoes and southern peas, *Proc. South. Weed Conf.*, 19, 176, 1966.
157. **Velez-Ramos, A. and Morales, A.**, Chemical weed control in sweet potatoes, *J. Agric. Univ. P.R.*, 61, 197, 1977.
158. **Peterson, L. E., Robbins, M. L., and Weigle, J. L.**, Herbicidal control of weeds in sweet potato, *Ipomoea batatas* Poir, *HortScience*, 7, 65, 1972.
159. **Greig, J. K. and Al-Tikriti, A. S.**, Effects of herbicides on some chemical components of sweet potato foliage and roots, *Proc. Am. Soc. Hortic. Sci.*, 88, 466, 1966.
160. **Hammett, L. K. and Monaco, T. J.**, Effect of oryzalin and other herbicide treatments on selected quality factors of sweet potatoes, *J. Am. Soc. Hortic. Sci.*, 107, 432, 1982.
161. **Constantin, R. J., Hernandez, T. P., and Wascom, B. W.**, Effects of herbicides on quality of sweet potatoes, *Proc. South. Weed Conf.*, 28, 194, 1975.
162. **Boggess, T. S., Jr., Marion, J. E., and Dempsey, A. H.**, The effects of herbicides on organoleptic qualities of baked sweet potatoes, *Ga. Agric. Exp. Stn. Res. Rep.*, 56, 1969.
163. **Robbins, M. L. and Peterson, L. E.**, Reduction of adventitious root formation on vines of sweet potato (*Ipomoea batatas* Poir) in Dacthal treated soil, *Hortic. Res.*, 10, 151, 1970.

Chapter 2

MAJOR DISEASE PESTS

J. W. Moyer

TABLE OF CONTENTS

I.	Introduction		36
II.	Control — General Considerations		36
III.	Sweet Potato Pathogens		38
	A.	Bacterial and Mycoplasma Pathogens	38
		1. Witches' Broom	38
		2. *Erwinia chrysanthemi*	38
		3. *Streptomyces ipomoea*	39
	B.	Fungal Pathogens	39
		1. *Ceratocystis fimbriata*	40
		2. *Diplodia gossypina*	40
		3. *Monilochaetes infuscans*	41
		4. *Macrophomina phaseolina*	41
		5. *Sclerotium rolfsii*	42
		6. *Fusarium oxysporum* f. sp. *batatas*	44
		7. *Fusarium oxysporum* and *Fusarium solani*	44
		8. *Rhizopus nigricans* (syn. *stolonifer*)	44
		9. *Sphaceloma batatas*	45
		10. Other Leaf Spot Pathogens	45
	C.	Nematode Pathogens	45
		1. *Meloidogyne* spp.	46
		2. *Rotylenchulus reniformis*	48
	D.	Viral Pathogens	48
		1. Aphid-Transmitted Viruses	49
		a. Sweet Potato Feathery Mottle Virus	49
		b. Sweet Potato Vein Mosaic Virus	51
		c. Others	51
		2. White-Fly-Transmitted Viruses	51
		a. Sweet Potato Mild Mottle Virus	51
		3. Others	53
		a. Sweet Potato Virus N	53
Acknowledgments			53
References			53

I. INTRODUCTION

The pathogens that attack sweet potatoes can influence production in a wide variety of ways. Yield, i.e., mass per unit area of production, is a primary consideration for evaluating disease loss and/or control recommendations for many crops. Root and foliar pathogens may suppress yield without directly attacking the edible portion for which the crop is being grown. The disease may reduce the size and number of fruits or seeds; however, those that are produced may still be used as they are not directly affected. Although the disease may have reduced some of the quality factors, the major portion of the edible portions are salvageable. Quality is an important component of marketable sweet potato yield because the pathogens frequently attack the organ that is to be utilized. Root pathogens may attack the fibrous roots suppressing the total mass produced and then attack the fleshy roots rendering them unsuitable. A single necrotic lesion or crack on a sweet potato root will decrease its value and will significantly reduce its potential storage life regardless of the intended use. Likewise, foliar pathogens that may not cause a significant loss in root production may significantly reduce the quality of the foliage for consumption. Thus, the evaluation of potential control strategies is not only dependent upon the particular disease to be controlled, but also the intended use for which the sweet potatoes are being grown.

These strategies, of course, must be integrated into local sweet potato production practices. The importance of this aspect of disease control would make the presentation of specific control recommendations in this treatise inappropriate due to the great differences among methods used to produce sweet potatoes in the tropics. Rather, the information presented in this chapter should be used as a guide in considering control "options". A discussion of generally accepted control procedures and their application to the control of specific sweet potato diseases is presented in the second section.

The major pathogen groups and the diseases they cause will be covered in the third section. Emphasis is given to diseases which have been described in the tropics; however, other diseases which might become of greater importance with anticipated changes in usage patterns are included. Among these are the diseases of roots that occur during transit and storage. In that section specific information will be presented on the causal agent, disease symptoms, losses, diagnostic procedures, and geographic distribution. The current status of available controls for each disease will also be presented; however, emphasis is given to disease control in Section II.

II. CONTROL — GENERAL CONSIDERATIONS

There are several approaches which have been used for the control of plant diseases. These have generally included sanitation measures, alteration of cultural practices, chemicals, resistant cultivars, and quarantines. The primary objective of these procedures has been to prevent the introduction of inoculum into the cropping system and/or slow the rate of disease spread.

The primary emphasis in controlling sweet potato diseases, as with diseases of many other crops that directly affect the product, is to prevent infection. There are no practical therapeutic methods that can restore infected plant tissues to a healthy state. When tissues become infected the infection continues either until the organ is destroyed or, if the organ is to be used as propagative material, it serves as an inoculum source for subsequent crops.

The importance of disease prevention as a guiding philosophy for sweet potato disease control may be accentuated by the relatively low disease thresholds for this crop, i.e., the maximum acceptable level of disease. This threshold is a function of the organs of the plants that are used relative to the organs that are diseased and the importance of pathogen transmission in propagative material. Frequently the plant organ, usually the root, which becomes infected is the organ which is to be consumed. Thus a seemingly low level of infection may result in the entire root being discarded. Furthermore, the disease will frequently continue to develop during storage or transit. Thus, a single lesion can result in the destruction of the entire root if it must be stored prior to consumption or prior to use as propagative material. The loss may be a direct result of the primary pathogen or by secondary infections.

The parameters used to quantify the amount of disease should be carefully considered. The importance of choosing appropriate parameters cannot be over stated because they are the criteria used to evaluate control measures including host resistance. Disease incidence (number of lesions per root or number of infected roots) would provide a useful measure for those diseases, such as bacterial soft rot or rootknot nematodes, where one or two lesions per root, irrespective of size, would be unacceptable. Alternatively, disease severity (incidence × lesion area) would be an appropriate measure for foliar diseases such as scab or root diseases such as scurf or russet crack. Marketable yield is, of course, the final measure of successful disease control. However, yield alone can be misleading in that it may only represent tolerance or escape and not resistance or actual control.

Vegetative propagation of sweet potatoes provides an important means of spread for several pathogens.[1,2] Thus it provides one of the few opportunities in the production cycle that diseased material can be purged from the system. The utilization of infected plant material for propagative stock not only provides a source of inoculum for that crop but it frequently reduces the efficiency with which transplants; i.e., sprouts or vine cuttings, are produced. Thus, greater inputs of "seed stock" are required making fewer vines or roots available for sale or consumption. One also risks introducing the pathogens into previously noninfested locations.

The desirability of maintaining strict disease tolerances at or near zero in "seed stock" is not only based upon the practical considerations cited above but also on epidemiological theory. Diseases tend to increase over time at a geometric rate. Thus, the probability that economic losses will occur becomes high when the initial incidence of disease is above some low threshold level. For example, with diseases that spread rapidly, and on those caused by certain foliar pathogens or insect vectored viruses, there may be little benefit from reducing the initial incidence of disease from 10 to less than 1%. Similarly, low tolerances are required for storage diseases that spread rapidly where one infected root in a wash tank can provide enough inoculum to infect hundreds of bushels of sweet potatoes.

Chemicals have been used to control some sweet potato diseases in the temperate growing regions.[2-7] However, they may be of limited practical value in many tropical growing situations due to their expense and the requirement for specialized equipment for application. Sprouts and seed roots dipped in appropriate fungicides aid in the control of some pathogens which infect or are spread by propagative material. Diseases such as scurf, black rot, and some Fusarium diseases can be controlled in this manner. Soil fumigants have been used as part of field preparation to control nematodes and *Streptomyces ipomoea*.[7,8] There are no major foliar diseases in temperate growing areas that have warranted the application of chemicals for disease control. However, the widespread occurrence of scab in some tropical areas may require chemical control if sweet potatoes are to be economically produced in these regions. Frequently, fungicides[3] and/or bactericides[9] are applied to roots that are being packaged and then shipped for fresh market consumption. These compounds protect the roots from organisms such as *Rhizopus* spp. and *Erwinia chrysanthemi* which have the potential to decay an entire root in 48 hr under the appropriate environmental conditions.

The use of resistant cultivars is the most desirable form of disease control.[10-15] However, the hexaploid genetics of sweet potatoes often makes the task of incorporating disease resistance along with all the desirable horticultural traits a difficult one. The development of efficient screening procedures for evaluating disease resistance on a large scale is a prerequisite for any program aimed at developing disease-resistant sweet potato cultivars.[10]

Frequently, quantitative resistance[14,15] or tolerance[13] has been the only available form of resistance for many of the sweet potato diseases. The effectiveness of these forms of resistance are subject to variation in the growing environment and changes in strains of the pathogen that may occur between growing regions.[11] In spite of these limitations, new cultivars often contain combined resistance to many of the major sweet potato diseases such as fusarium wilt, internal cork, root-knot nematodes, and the sweet potato virus disease complex. However, exhaustive searches of the sweet potato germplasm have failed to identify significant levels of resistance to other diseases such as scurf.[16] We continue to rely on cultural and chemical alternatives for control of these and other diseases.

Strict quarantine regulations govern the international movement of the vegetative sweet potato organs.[17] The objective of these programs is to reduce the probability of spreading pests into areas where they do not already occur. Seeds of *Ipomoea batatas* can be transported between countries relatively easily. However, sweet potato is a vegetatively propagated crop and it is often desirable to move selected clones. This necessitates transporting the vegetative organs, either roots or sprouts. Most sweet potato pathogens are easily detected in the vegetative tissues by visual inspection. However, virus-infected roots and vines are not always symptomatic. Thus, the introduction of viruses as well as mycoplasma-like organisms are the primary targets of these quarantine programs. To insure that plants are not harboring these pathogens, the plants are assayed either by grafting onto virus-sensitive indicator plants or by serological methods. The relatively low titer of viruses in sweet potato has limited the acceptance of the serological assays.[18,19]

"Healthy" plants (i.e., free of known viruses) are obtained by meristem-tip culture.[20-22] Plants regenerated from meristematic tissues are repeatedly assayed and those which test negative for viruses are then distributed. It should be noted that there is no single assay that can detect all known viruses. Procedures, assay hosts, etc. should be specified when certifying that a plant is "healthy". Because sweet potato viruses are so poorly understood, plants are generally held one to two years following meristem-tip culture and are assayed a minimum of two to three times to insure that viruses are not inadvertently distributed.

III. SWEET POTATO PATHOGENS

A. Bacterial and Mycoplasma Pathogens

Relatively few bacterial pathogens have been shown to cause economically important diseases on sweet potato. The diseases caused by mycoplasma-like organisms are also included here.

1. Witches' Broom Agent

This disease is reported from southeast Asia and Oceania[23-28] and it has been referred to by a variety of common names: witches' broom,[24-27] little leaf,[25,26] and Ishuku-byo.[28] There is one report of a proliferation disease in the U.S.[29] Excessive proliferation of young shoots from the leaf axils, an abnormally erect growth habit, stunting of leaf development and an overall dwarfing of the plant are the predominant symptoms of the disease. Selection of healthy planting material is aided by these distinct symptoms. Phloem cells of plants expressing the witches' broom symptoms have been shown to contain mycoplasma-like bodies.[30,31] The agent is vectored by leafhoppers, *Orosius* spp.[26,32] Disease incidence has been correlated to the natural range of the vector.[25] Thus, roguing of diseased plants is an effective control for areas where the vector does not occur.

2. Erwinia chrysanthemi

This organism causes diseases on many crops throughout the tropics, although it has not been recognized as a major sweet potato pathogen in this region. It has caused significant losses in the southeastern U.S.[9,33] It is included here because of the potential hazard it presents to sweet potato production in the tropics. *E. chrysanthemi* causes a severe root and stem rot on sweet potato. Infected roots develop a rapid, moist decay somewhat similar to that caused by *Rhizopus*. The stem symptoms are similar to Fusarium wilt. High temperatures and free moisture are required for infection.[9,33] The organism can survive in plant debris.[9] This provides an excellent source of inoculum for infection of roots during the harvesting process. Controls have included rotation with crops which are not susceptible to *E. chrysanthemi* and use of healthy planting material. Cultivar differences in disease reaction have been observed; however there are no known resistant cultivars.[9]

3. Streptomyces ipomoea

The disease caused by this organism, commonly referred to as pox or soil rot, was first described in 1890.[34] However, the etiology of the disease was not determined until much later.[35,36] Infection by *S. ipomoea* results in extensive necrosis of fibrous roots which significantly reduces yield. In addition, dark brown to black lesions develop on fleshy roots around the root hairs. Early infections inhibit storage root development resulting in a girdled appearance. This disease may be a potential problem only for limited areas of the tropics. Disease development is favored by warm temperatures: however, acid soils (below pH 5.2)[37] and periods of high soil moisture[38] are inhibitory to disease development. Controls have traditionally relied on avoiding severely infested fields or expensive soil fumigation.[39] Recently, cultivars have been introduced with sufficiently high levels of resistance[40,41] to reduce the need for fumigation.[8] Improved methods for evaluating clones for resistance have also been developed.[10]

B. Fungal Pathogens

The sweet potato disease literature is dominated by investigations of diseases caused by fungi. It is interesting to note that most of these diseases were described before 1900 with only a few of the major diseases caused by fungi having been reported since 1930.[34,42] These diseases have been reported throughout the tropics as well as the temperate sweet potato growing regions.[43-45] The vast majority of this literature covers disease descriptions, distribution, and control. In addition, research on *Ceratocystis fimbriata* and other fungal pathogens have contributed to our understanding of sweet potato diseases as well as to general plant disease physiology. The sweet potato-*Ceratocystis fimbriata* interaction has been used as a model system to study phytoalexins,[46-52] altered metabolism in diseased plants[53,54] and to examine host-pathogen specificity via common antigens[55] and agglutination.[56] There remains considerable opportunity for basic research that will provide the basis for more efficient methods of evaluating disease resistance in sweet potato germplasm and into the epidemiology of soilborne sweet potato pathogens.

Adequate controls are available for many but not all of these diseases. For those diseases that have been controlled in certain growing regions, specific tactics may require modification for use in other areas necessary to fit local cultural practices. The causal organisms are usually endemic[17] having a wide host range and/or the capability of surviving on *Ipomoea* weed species.[28] Thus, once the organism is introduced into a given area, and many of these pathogens already occur wherever sweet potatoes are grown, controls must be considered each time the crop is to be planted.

The majority of these organisms attack both the roots and the underground portion of the stem of sweet potato plants.[34,42,58-64] This compounds the effects of these pathogens on sweet potato production. Yield is suppressed from a loss of plant vigor due to poor stands and an inadequate fibrous root system. Lesions on the fleshy roots diminish their desirability for fresh market consumption and significantly increases losses due to rotting during storage. In addition, vegetative propagation material may serve as the initial source of inoculum for each new crop. There are numerous examples of fungal pathogens that are spread on sprouts produced from infected roots which are to be used as transplants for subsequent crops.[58-63]

Primary considerations for the recommendation of existing disease control strategies or the development of new strategies are the effectiveness of the control measure and the likelihood of its acceptance. As previously stated with regard to sweet potato diseases in general, control of many sweet potato fungal diseases should emphasize prevention of infection rather than slowing disease progress. This is essential for those fungal diseases which occur on the roots. Thus, controls should be targeted at or before those events in the production cycle where infection is most likely to occur. This is crucial for control of storage diseases.[64] Careful handling at harvest and during storage are essential for maintenance of roots after digging. Curing roots immediately after harvest in high humidity at 30°C for 5 to 7 days also greatly increases storage longevity.

Acceptance is dependent upon the expected benefit, i.e., the value of increased yield as compared to the increased inputs required for control, and whether the recommended controls can be reasonably integrated into existing cultural practices. In many instances, sound cultural practices such as selecting resistant cultivars or "seed" free of diseases may be all that is necessary for control of some diseases such as scurf. There are numerous detailed reviews[34,42,44,64] of the fungal diseases of sweet potato. Only a brief description of the etiology, diagnostic symptoms, and general control considerations are presented here for the major fungal diseases of sweet potato.

1. Ceratocystis fimbriata

For a period in the earlier literature this organism was referred to as *Ceratostomella fimbriata*.[65] This organism causes a disease of the roots and underground stems of sweet potato plants.[66] The disease is generally called black rot or occasionally black shank when referring to the disease on the underground stem. It has been reported to occur wherever sweet potatoes are extensively grown. Isolates of *C. fimbriata* also cause diseases of coffee, cacao, rubber, mango, pimento, and *Prunus* spp. However, these isolates do not necessarily infect sweet potato.[67]

Infection occurs most readily on injured roots. Symptoms include dark circular lesions on fleshy roots that may penetrate to the vascular ring. Secondary infection can result in rotting of the entire root. The disease may occur on roots prior to harvest[68,69] or on roots in storage[70] with spores being transmitted by mites and in wash water. Infected roots used for the production of sprouts for the next crop become a major source of primary inoculum.[6,71] The basal portion of sprouts produced from diseased roots become infected; in severe infections these sprouts may be killed. The organism may also survive in plant debris for up to two years.[68]

This disease has been virtually eliminated from many growing areas by selecting "seed" roots free of disease, by rotating fields with crops that are not susceptible to *C. fimbriata* and sanitation in storage facilities.[6,71] The excellent control of this disease using sound cultural practices has resulted in relatively little effort being expended to develop new cultivars resistant to black rot even through methods are available for evaluating resistance and differences in cultivar reaction have been reported.[11,73,74]

2. Diplodia gossypina

Java black rot is widespread throughout the tropics.[42] Early research suggested that the organism may have a wide host range.[75] Isolates collected from dasheen and mango caused a disease on sweet potato identical to that caused by isolates from sweet potato. It is primarily a storage disease, resulting in a slow rot of the entire root. The root becomes dry and brittle with small black protuberances on the surface which are pycnidia. The interior of the root is black and hard (Figure 1). The base of sprouts may also become infected when infected

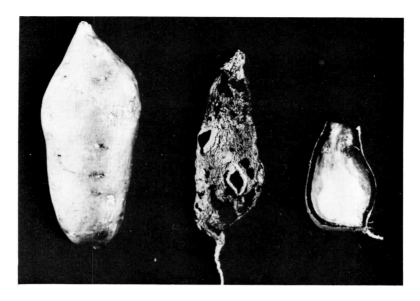

FIGURE 1. Sweet potato storage roots which are healthy (left) or infected (middle, right) with *Diplodia gossypina*. Notice black pycnidia on the external surface of the middle root. Root on right shows early stages of infection.

roots are used for propagation. Foliar symptoms and those at the base of sprouts are very similar to those of the black rot disease, but can be differentiated on the basis of fruiting structures at the base of the stem.[76]

The recommended controls for Java black rot are the same as for black rot.[61]

3. Monilochaetes infuscans

The disease called scurf is also widespread.[43] The organism is only known to infect sweet potatoes and related species.[57,59] Symptoms of scurf begin as small, brown to black spots on the roots and underground stems of sweet potato (Figure 2A). These spots will eventually coalesce (Figure 2B).[34,42,77] The disease may be confused with the raised, brown, blister-like symptoms of boron deficiency.[79] Scurf differs from the two previous diseases in that infection is limited to the periderm.[78] The quality of the flesh is not significantly affected except in severe cases when increased moisture loss occurs. However, the lesions on the root surface are unsightly to the consumer.

Controls for scurf are the same as for black rot.[59,71] The organism can survive in the soil and on other *Ipomoea* spp. When diseased seed roots are used, it is possible to reduce the spread into the current crop by cutting sprouts above the soil line and being careful not to contaminate the sprouts or cutting instruments with infested soil and by fungicidal dips.[1,62] Thermotherapy is possible but may not be practical under commercial growing conditions.[80]

4. Macrophomina phaseolina

M. phaseolina, the pycnidial stage of *Rhizoctonia bataticola*, which was originally described on sweet potato, is the causal agent of charcoal rot.[81,82] Charcoal rot is another disease of the roots and underground stems of sweet potatoes. The flesh of infected roots first becomes brown then black or charcoal-like then sclerotia are formed. Lesions may also occur at the soil line on plant stems in the field. Sclerotia are produced in the xylem. These lesions may extend down to the storage roots by harvest time. The roots then become infected either directly or are inoculated during the harvesting process.[42] Plants under stress favor the development of this disease under field conditions.

FIGURE 2. Symptoms of underground stems and roots infected with *Monilochaetes infuscans*. (A) Roots developing from an infected transplant. (B) Symptoms developing on roots from the proximal (stem) end towards the distal end.

There is little information available for the control of charcoal rot on sweet potato or for sources of resistant germplasm. Presumably care in handling, sanitation in storage facilities, and selection of propagation material free of the disease will minimize losses due to charcoal rot.

5. Sclerotium rolfsii

This organism infects the roots and stems of sweet potato plants[42] as well as many other tropical crops such as cocoyam and peanuts.[82] The disease on sweet potato is known by several common names such as circular spot (Figure 3), sclerotial blight and southern stem rot. The disease occurs in plant beds at about the time sprouts are large enough for transplanting. The fungus attacks the stem at the soil line resulting in wilting and eventual death of the plant. Circular, barren areas with the white mycelial mats on the soil surface are typical characteristics of this disease in plant beds. Circular, light brown lesions with distinct margins occur on roots at harvest time (Figure 3).[83,84] Older lesions can sometimes be confused with lesions caused by *Streptomyces ipomoea*.

Once *S. rolfsii* becomes established in a field it is very difficult to control and problem

FIGURE 2B.

FIGURE 3. Circular lesions on sweet potato storage roots caused by *Sclerotium rolfsii*.

fields should be avoided. Long term rotation with resistant crops may be of some value. Methods have been developed for identifying resistant germplasm. Jewel is a relatively resistant cultivar.[85]

6. *Fusarium oxysporum* f. sp. *batatas*

The disease caused by this organism has several common names including Fusarium wilt, stem rot, blue stem, and yellow blight.[34,42] The organism initially causes a yellowing of the foliage followed by wilting and death in susceptible cultivars. These symptoms are accompanied by extensive vascular necrosis. Internal necrosis may also be observed in storage roots following harvest. The vascular necrosis and wilting can be used as a diagnostic aid. However, caution is warranted as similar symptoms are also caused by *Erwinia chrysanthemi*.[33]

F. oxysporum f. sp. *batatas* is relatively long-lived in agricultural soils even in the absence of sweet potato production. Excellent control of this disease has been achieved by introducing high levels of resistance into new cultivars. A major source for this resistance has been P.I. 153655 (Tinian).[12] The resistance has been shown to be quantitatively inherited.[86] Although resistant clones become infected, colonization and multiplication of *F. oxysporum* was significantly less in resistant that in susceptible clones. In one study, six to nine days after inoculation the vascular system of susceptible clones was colonized throughout while the organism was isolated from less than 50% of the stem sections from resistant plants.[87]

This disease has become much less significant with the availability of resistant cultivars. However, it remains a constant threat in soils conducive to *Fusarium oxysporum*. Susceptible cultivars should not be planted in these areas.

7. *Fusarium oxysporum* and *Fusarium solani*

These two *Fusarium* spp. are widespread in the tropics and are treated together here due to similarities in the diseases they cause, their ecology and their control measures.[64,88-92] They are both relatively "weak" wound-invading pathogens requiring a wound to initiate infection.[89,91,93] The wounds may be caused either by mechanical damage, such as those incident to harvesting, or by other organisms, such as nematodes.[89,90]

Fusarium oxysporum causes a disease called surface rot on storage roots.[92] It is a nonspecialized form not to be confused with *Fusarium oxysporum* f. sp. *batatas*. The disease usually occurs during storage as a result of rough handling during harvest.[89] The lesions continue to enlarge during storage but the disease usually does not spread to other roots in storage except when additional wounding occurs. Infection may also occur prior to harvest when injuries occur due to growth cracks, nematodes, insect injury, and rodents. The circular lesions are dark brown in color and seldom extend below the periderm.[92]

The disease syndrome caused by *F. solani* differs somewhat from that caused by *F. oxysporum*.[91,93] Symptoms on storage roots are superficially similar except dark brown rings are occasionally present on the surface of the lesion.[91] In addition, some isolates have the capacity to decay the entire root, while infection by other isolates will extend only to the vascular ring.[91] The freshly decayed tissue is brown to orange and is sharply delimited from the healthy tissue. White mycelium is often visible within elliptical cavities present in the lesion.[91,93] Lesions may also be present on the basal portion of sprouts produced from infected roots.[58] The decay does not extend more than a few centimeters from the base, however, storage roots produced from these plants may also become infected.[58]

Differences have been reported for varietal responses to both of these pathogens.[91,94] Control of these diseases, however, has relied on careful handling during harvest and storage, proper curing immediately following harvest and culling infected roots from "seed" stocks.[89-91] Growing conditions that favor preharvest injuries or where curing facilities are not practical may stimulate renewed interest in resistance to these nonspecialized Fusaria. Details of storage disease management have been reviewed.[64,89-91]

8. *Rhizopus nigricans* (syn. *stolonifer*)

Root tissue infected with *R. nigricans* becomes soft, moist and is stringy,[34] and the root may be destroyed within a few days after infection. The disease is referred to as soft rot,

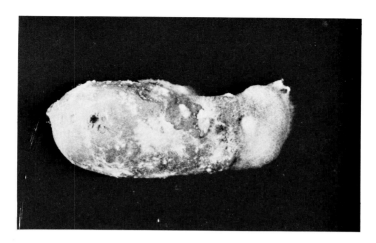

FIGURE 4. Sweet potato root infected with *Rhizopus nigricans*. Extensive mycelial growth (whiskers) on the exterior of the root is a diagnostic sign of "Rhizopus" soft rot.

ring rot or collar rot.[96] However, it is distinct from soft rots caused by other organisms such as *Erwinia chrysanthemi*.[9] The presence of dense aerial mycelia with dark brown spores "whiskers" on infected roots is a diagnostic sign for Rhizopus soft rot (Figure 4). At this stage of infection the organism sporulates profusely providing inoculum that can be disseminated in storage by fruit flies, rodents, and air currents. An abundance of fruit flies in storage facilities is another sign indicating the presence of Rhizopus soft rot. Although, infection is favored at relatively low temperatures (20°C)[95] the disease reportedly occurs in the tropics.

Control of this disease consists of the same management practices[64] (curing, sanitation, etc.) as used for other storage root diseases. Treatment with a fungicide, such as 2,6 dichloro-4-nitroaniline (DCNA or dichloran), will provide added protection when stored roots are to be washed prior to marketing.[3]

9. Sphaceloma batatas

This organism causes a severe disease, commonly called scab, of sweet potato stems and foliage (Figure 5).[97] Recent studies have confirmed earlier observations that the perfect stage of this organism is an *Elsinoe* sp.[98] Symptoms on the foliage begin as small brown lesions on the veins which eventually assume a corky texture. The veins shrink, the leaf curls and then stops growing. Lesions with slightly raised purple-brown centers and light borders occur on the stems. These lesions coalesce to form a scab-like structure on the stem. It has been reported to occur from Taiwan to Hawaii in the Pacific region and in Brazil.[44,97,99] It has resulted in certain areas being abandoned from sweet potato production. The disease is most severe in areas where there is frequent rain, fog, or dew.[97] Although published information describing controls for this disease on sweet potato is not readily available, measures which have been used to control diseases caused by other *Sphaceloma* and *Elsinoe* spp. on other crops may be appropriate.

10. Other Leaf Spot Pathogens

Members of several genera of fungi cause necrotic lesions on sweet potato foliage. However, none of them have been considered to cause major losses except in isolated instances. Some of the genera include: Alternaria, Cercospora, Phyllosticta, and Septoria.[42,44]

C. Nematode Pathogens

Members of the Meloidogyne (root-knot)[100] and Rotylenchulus[101,102] genera are the known

FIGURE 5. Symptoms on sweet potato leaves, stems and petioles caused by *Sphaceloma batatas*, the causal agent of scab.

major nematode pests of sweet potatoes in the tropics, however, many other plant parasitic nematodes have been isolated from tropical soils used for sweet potato production. Both genera of nematodes are widely distributed throughout the tropics. They, like other root pathogens of sweet potato, damage roots in several ways. They attack the fibrous roots and may cause a reduction in yield. Injury to the fleshy roots reduces quality and provides wounds through which other pathogens may become established causing further damage.[103,104] Control of nematodes in the temperate growing regions have emphasized the use of chemicals to reduce initial population levels. Concomitantly, many breeders evaluate promising clones for resistance. Rotation with resistant crops to maintain nematode populations at relatively low levels[103] is an effective adjunct to chemical control and genetic resistance.

1. Meloidogyne spp.

There are four species of *Meloidogyne* which are the most prevalent in cultivated crops: *M. arenaria*, *M. hapla*, *M. incognita*, and *M. javanica*. *M. incognita* and *M. javanica* are the only agriculturally important *Meloidogyne* spp. widely distributed throughout the tropics.[103] However, there are many other *Meloidogyne* spp. that have been collected from warm climates.[100] These nematode species are favored in areas of relatively low rainfall and high soil pH.[100]

Symptoms of root-knot nematode infection on sweet potatoes include galls on fibrous roots (Figure 6A) and cracking of fleshy roots as well as yield suppression. An additional sign of root-knot infection is the presence of eggs or larvae in the fleshy roots (Figure 6B).[104-106]

It is important that populations be collected from problem areas and accurately identified because of the variation in host specificity that may exist between *Meloidogyne* spp. and between populations of the same species. This is a crucial requisite for the development of root-knot-resistant cultivars and identification of resistant crops for rotations.[107] *Meloidogyne* spp. can be identified on the basis of morphological characters such as larval length and perineal pattern. An effective approach for the identification of root-knot nematodes has been suggested for those less acquainted with nematode taxonomy. ''(1) Collect widely from

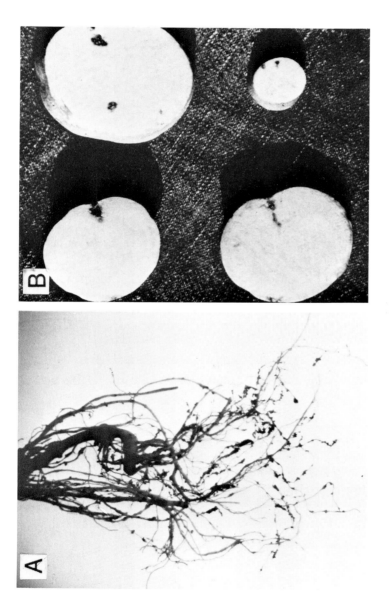

FIGURE 6. Symptoms on fibrous and fleshy sweet potato roots caused by *Meloidogyne incognita*, the root-knot nematode. (A) Dark bodies on fibrous roots are galls typical of infection by root-knot nematodes. (B) Cross sections of fleshy sweet potato roots containing eggs or larvae of *M. incognita*.

the crop plants of the region, taking care to get good representative samples from each field. (2) Identify the root-knot nematodes by the North Carolina Differential Host Test . . . ."[103] This test will identify *M. arenaria*, *M. hapla*, *M. incognita*, and *M. javanica* as well as detect races and mixed infestations.[103]

Nematicides have been used extensively to reduce nematode populations in agricultural soils. However, these are expensive materials and often require specialized equipment for effective application. Crop rotation is an important component of an integrated control strategy. A thorough knowledge of the species and races present is important to derive the greatest benefit from this control measure.[103,107] Peanuts are a good rotation crop if *M. hapla* or race 1 of *M. arenaria* are not present. Cotton and fescue are other crops that will reduce populations of root-knot nematodes. Sound weed control practices should accompany any crop rotation scheme. It has recently been shown that several *Ipomoea* spp. will support multiplication of root-knot nematodes.[57]

The development of resistant cultivars is the most promising alternative for control of root-knot nematodes in sweet potato plantings.[107] Germplasm, designated W-13, W-178, and W-51, has been developed with a high level of resistance to *M. incognita*. W-51 also has a high level of resistance to *M. javanica* and *M. hapla*.[108] In addition, it has been shown that this resistance can be incorporated with other horticultural characteristics using mass selection breeding techniques.[14,109]

2. Rotylenchulus reniformis

This nematode, like the root-knot nematodes, is widely distributed throughout the tropics.[101,102] They are known to occur in the same fields and their interaction has been the subject of investigation. Field studies indicated that either *M. incognita* or *R. reniformis* have the potential to predominate over the other depending on the conditions.[105] Additional investigations under greenhouse conditions suggested that *R. reniformis* is the more aggressive of the two species.[106]

R. reniformis feeding on sweet potato results in necrotic lesions on fibrous roots. High populations result in severe root pruning and suppression of yield. It has been demonstrated that the most sensitive sweet potato cultivars, i.e., those that suffered the greatest yield suppression, do not necessarily support the highest populations of *R. reniformis*. The investigators hypothesized that sensitive hosts have fewer feeding sites as a result of the severe root pruning. Thus, late season population levels may be negatively correlated with resistance levels in sweet potato to *R. reniformis*.[110]

Another type of injury caused by *R. reniformis* is cracking in fleshy roots. These cracks are deeper than cracks due to other stimuli such as *M. incognita*. Larvae are not usually associated with these cracks as with *M. incognita*. Most sweet potato cultivars will respond by cracking to lower population levels of *R. reniformis* than are necessary to cause a suppression of yield. However, one breeding line has been identified as resistant to *R. reniformis* induced cracking.[110]

Control of *R. reniformis* is similar to that of root-knot nematodes. Rotation with a resistant crop, such as corn or leaving the ground fallow are possible alternatives that will reduce the population of *R. reniformis*.[111] Application of soil fumigants has also shown promise.[112,113] However, less attention has been given to the development of resistant cultivars.

D. Viral Pathogens

The viruses which cause diseases of sweet potatoes are the most poorly understood group of sweet potato pathogens. Surprisingly, they are also the most ubiquitous in commercial sweet potato production. Virtually all commercial sweet potatoes that are grown without the benefit of a virus indexing program are infected by one or more viruses. This condition is not unique to sweet potatoes, but rather is a characteristic of vegetatively propagated crops.

The diseases are primarily chronic; however, acute virus diseases have been reported in the tropics. The severity of losses is highly dependent upon the tolerance to the specific virus(es) in the cultivar being grown. The primary symptoms being one or more types of foliar chlorotic patterns, such as mosaic, vein-clearing, mottling, leaf spots or ringspots, and stunting. Root symptoms have also been associated with some virus diseases.

The identity of the viruses that cause these diseases is not well established.[117,118] The properties and diseases caused by the known viruses are described below. Much of the confusion concerning the identity of sweet potato viruses has resulted from the practice of naming a new disease based on symptoms induced in one or a few hosts that are prevalent in a particular region. Then as time passed the word "virus" was added as a matter of common usage. In addition, it is difficult to make direct comparisons between viruses collected in different countries because of strict quarantines.[17] Recently, greater attention has been given to etiological investigations of these diseases and scientists have been reluctant to add new names to an already confused literature.[117,120-124] As more individual viruses are characterized using accepted virological protocol, host differentials and virus-specific antiserum will become available that will allow easier comparison of these viruses.

Control of sweet potato viruses has been limited to the use of virus-indexed propagation material[2,20-22] and to the development of resistant[15] or tolerant cultivars.[13] The use of virus-indexed propagative material requires a knowledge of the important viruses, an efficient and reliable method for detecting these viruses, a source of healthy plants for the cultivars being grown and an acceptable means of distributing the healthy plants to the growers. Once all of these factors have been achieved, the certification program must then be uncompromisingly administered if this approach is to be successful. However, the logistics of such a program may be prohibitive in the tropics where the clientele may consist of many growers with relatively small plots. In addition, a high density of alternate hosts and insect vectors during the growing season may result in an unacceptably high reinfection rate.

The development of virus resistant[15] or tolerant cultivars[13] has been the most effective means of reducing sweet potato losses due to virus infection. Germplasm is often evaluated under field conditions to take advantage of natural inoculum sources and insect vectors to inoculate test plots.[15] Viruses such as feathery mottle which are easily mechanically transmitted to other experimental hosts, are not efficiently mechanically transmitted to sweet potato. The only reliable method of inoculation is by grafting. This method consists of implanting tissue from infected plants into roots or stems of the sweet potato clones that are to be tested.[117,125]

1. Aphid-Transmitted Viruses

This presentation of sweet potato viruses is divided into sections based on their natural vectors. This is done only for convenience due to the lack of confirmed etiological information for many of the virus diseases and does not imply any virological relationships other than vectors.

a. Sweet Potato Feathery Mottle Virus (FMV)

The name feathery mottle was first used to describe a disease typified by an irregular chlorotic pattern (Figure 7A) associated with the veins of sweet potato leaves.[126] This causal virus is now known to be a long flexous rod shaped virus (Figure 7B) approximately 850 nm × 12 nm[120,122,123] and contains a single, plus sense strand of RNA measuring approximately 3.7×10^6 daltons which is encapsidated in a protein coat with subunits of approximately 3.5×10^4.[127] It is aphid-transmitted in a nonpersistent manner.[128] Its known host range is limited to the Convolvulaceae, however, some strains have been experimentally transmitted to *Chenopodium* spp.[119,120] Electron micrographs have revealed the presence of pinwheel inclusions in infected tissues.[129] It is serologically related to SPVA reported in Taiwan and to aphid-transmitted viruses isolated in Nigeria[141] and Zaire.[130]

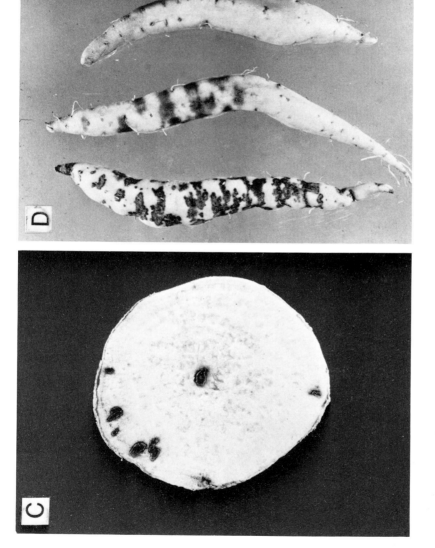

FIGURE 7. Symptoms and virions of sweet potato feathery mottle virus (FMV). (A) One of the chlorotic patterns which may occur on sweet potato leaves infected with FMV. (B) An electron micrograph of FMV virions. The rod shaped particles are 850 nm long. (C) A cross-section of a fleshy root exhibiting symptoms of internal cork. (D) Necrotic lesions on 'Jersey' sweet potato roots symptomatic of the russet crack disease. Root on right is from a noninfected plant.

Strains of this virus have been shown to be the causal agents of several of the virus diseases of sweet potato.[117,119,120] Many of the foliar disease symptoms reported in the U.S. that have been attributed to viruses are probably caused by this virus. Some of these symptoms may be caused by other viruses although this is the only virus that has been described in the U.S. In addition, the russet crack disease (Figure 7C)[117,120] and internal cork disease (Figure 7D)[119] have also been shown to be caused by strains of this virus. However, there is an unconfirmed report of internal cork being caused by a spherical virus.[131]

The effect of this virus on sweet potato production is highly dependent upon the cultivar. Significant marketable losses have been reported for cultivars which express russet crack or internal cork symptoms. Many new cultivars have tolerance to these two diseases even though the virus can be consistently isolated from them. Most infected cultivars express various combinations of chlorotic patterns in the foliage. Stunting is not detectable in most cultivars and reliable yield loss estimates are not available.

b. Sweet Potato Vein Mosaic Virus

This virus was isolated in Argentina. It too, has a host range limited to the Convolvulaceae.[132] The isolates tested were not transmitted to *Chenopodium amaranticolor*. It has a normal length of 761 nm[123] and is nonpersistently transmitted by aphids. It is distinct from FMV based on direct comparison of particle morphology.[123] Infected sweet potato plants exhibit a general chlorosis and stunting due to shortened internodes. Leaves are distorted with a distinct vein clearing and interveinal areas have a diffuse mosaic.[132]

c. Others

Many other virus-like diseases of sweet potato have been reported throughout the tropics as being caused by an aphid-transmitted agent.[133-137] The aphid-transmitted component of SPVD has been reported from Nigeria[116] and a similar agent was reported from East Africa.[136] Aphid-transmitted viruses have also been reported from Taiwan,[135] Israel,[133] and from Brazil.[134] However, for reasons previously stated here and reviewed by others, direct comparisons have not been possible.

2. White-Fly-Transmitted Viruses

White-fly-transmitted agents have been implicated in several sweet potato diseases: sweet potato virus disease complex (Nigeria),[116] virus B (East Africa),[140] sweet potato mild mottle virus (East Africa),[121] unnamed disease characterized by leaf rolling and crinkling (Taiwan),[135] mosaic and yellow dwarf (USA),[138,139] and vein clearing (Israel).[133] *Bemisia tabaci* has been reported as the vector of these diseases in the tropics. However, *Trialeurodes abutilonea* was cited as a vector in the U.S.[138] These diseases can be severe, resulting in a suppression of yield of up to 80%. However, in at least one disease (SPVD), both the white-fly-transmitted component and an aphid-transmitted component must be present for the disease to occur.[116] The white-fly component may be latent unless the aphid-transmitted virus is also present. These complex diseases have remained poorly understood.

a. Sweet Potato Mild Mottle Virus

White-fly transmission has only be associated with one described virus, sweet potato mild mottle virus (SPMMV).[121] SPMMV is a filamentous rod ca. 950 nm in length. Preliminary evidence suggests that SPMMV is an RNA virus and its coat protein is comprised of 37.7 kD subunits. The reported host range of SPMMV includes 45 species in 14 plant families. Symptoms in sweet potato include mottling, stunting, and a loss of yield. It has been suggested that SPMMV may be the same agent as Sheffield's Virus B.[121,136]

3. Other
a. Sweet Potato Virus N

This virus is a long flexous rod 750 to 800 nm long isolated in Taiwan.[135] Other characteristics have not yet been determined. The virus is latent in many cultivars, however, infection may result in mild mottling and vein yellowing in some cultivars. It has been mechanically transmitted to *Chenopodium* spp. and *Nicotiana* spp. as well as *Ipomoea* spp. The virus is not transmitted by aphids (*M. persicae*) and it is not known if white-fly transmission was attempted.[135]

ACKNOWLEDGMENTS

Journal series paper No. 9204 of the North Carolina Agricultural Research Service.

REFERENCES

1. **Cook, H. T.,** Control of sweet potato scurf by vine cuttings, *Phytopathology,* 38, 568, 1948.
2. **Nielsen, L. W.,** Elimination of the internal cork virus by culturing apical meristem of infected sweet potatoes, *Phytopathology,* 50, 840, 1960.
3. **Martin, W. J.,** Effectiveness of fungicides in reducing soft rot in washed, cured sweet potatoes, *Plant Dis. Rep.,* 48, 606, 1964.
4. **Martin, W. J.,** Evaluation of fungicides for effectiveness against the sweet potato black rot fungus, *Ceratocystis fimbriata, Plant Dis. Dep.,* 55, 523, 1971.
5. **Martin, W. J.,** Further evaluation of thiabendazole as a sweet potato "seed" treatment fungicide, *Plant Dis. Rep.,* 56, 219, 1972.
6. **Daines, R. H.,** Sprout dip treatment for control of black rot of sweet potatoes, *Plant Dis. Rep.,* 46, 253, 1962.
7. **Brathwaite, C. W. D.,** Effect of DD soil fumigant on nematode population and sweet potato yields in Trinidad, *Plant Dis. Rep.,* 58, 1048, 1974.
8. **Clark, C. A. and Watson, B.,** Control of sweet potato soil rot with varietal resistance and chloropicrin, *Fungic. & Nematic. Test 1980,* 36, 77, 1981.
9. **Schaad, N. W. and Brenner, G.,** A bacterial wilt and root rot of sweet potato caused by *Erwinia chrysanthemi, Phytopathology,* 67, 302, 1977.
10. **Moyer, J. W., Campbell, C. L., Echandi, E., and Collins, W. W.,** Improved methodology for evaluating resistance in sweet potato to *Streptomyces ipomoea, Phytopathology,* 74, 494, 1984.
11. **Martin, W. J.,** Varietal reaction to *Ceratostomella fimbriata* in sweet potato, *Phytopathology,* 44, 383, 1954.
12. **Steinbauer, C. E.,** A sweet potato from Tinian Island highly resistant to Fusarium wilt, *Proc. Am. Soc. Hortic. Sci.,* 52, 304, 1948.
13. **Nielsen, L. W. and Pope, D. T.,** Resistance in sweet potato to the internal cork virus, *Plant Dis. Rep.,* 44, 342, 1960.
14. **Dukes, P. D., Jones, A., Cuthbert, F. P., Jr., and Hamilton, M. G.,** W-51 root-knot resistant sweet potato germplasm, *HortScience,* 13, 201, 1978.
15. **Hahn, S. K., Terry, E. R., and Leuschner, K.,** Resistance of sweet potato to virus complex, *HortScience,* 16, 535, 1981.
16. **Nielsen, L. W. and Yen, D. E.,** Resistance in sweet potato to the scurf and black rot pathogens, *N. Z. J. Agric.,* 9, 1032, 1966.
17. **Nielsen, L. W. and Terry, E. R.,** Sweet potato, in *Plant Health and Quarantine in International Transfer of Genetic Resources,* Hewitt, W. B. and Chiarappa, L., Eds., CRC Press, Cleveland, 1977, 346 p.
18. **Cadena-Hinojosa, M. A. and Campbell, R. N.,** Serological detection of feathery mottle virus strains in sweet potatoes and *Ipomoea incarnata, Plant Dis.,* 65, 412, 1980.
19. **Esbenshade, P. R. and Moyer, J. W.,** An indexing system for sweet potato feathery mottle virus in sweet potato using the enzyme linked immunosorbent assay, *Plant Dis.,* 66, 911, 1982.

20. **Alconero, R., Santiago, A. G., Morales, F., and Rodriguez, F.,** Meristem tip culture and virus indexing of sweet potatoes, *Phytopathology,* 65, 769, 1975.
21. **Frison, E. A. and Ng, S. Y.,** Elimination of sweet potato virus disease agents by meristem tip culture, *Trop. Pest Manage.,* 27, 452, 1981.
22. **Over de Linden, A. J. and Elliott, R. F.,** Virus infection in *Ipomoea batatas* and a method for its elimination, *N. Z. J. Agric. Res.,* 14, 720, 1971.
23. **Anonymous,** Experiments on the bud-atrophy disease of sweet potatoes, *Taiwan Res. Inst. Annu. Rep.,* 1, 42, 1955.
24. **Dabek, A. J. and Sagar, C.,** Witches' broom chlorotic little-leaf of sweet potato in Guadalcanal, Solomon Islands, possibly caused by mycoplasma-like organisms, *Phytopathol. Z.,* 92, 1, 1978.
25. **Jackson, G. V. H. and Zettler, F. W.,** Sweet potato witches' broom and legume little-leaf diseases in the Solomon Islands, *Plant Dis.,* 67, 1141, 1983.
26. **Kahn, R. P., Lawson, R. H., Monroe, R. L., and Hearson, S.,** Sweet potato little-leaf (witches' broom) associated with a mycoplasmalike organism, *Phytopathology,* 62, 903, 1972.
27. **So, I. Y.,** Studies on the mycoplasmic witches' broom of sweet potato in Korea. I. Symptoms and pathogen, *Korean J. Microbiol.,* 11, 19, 1973.
28. **Summers, E. M.,** "Ishuku-byo" (dwarf) of sweet potato in Ryukyu Islands, *Plant Dis. Rep.,* 35, 266, 1951.
29. **Nyland, G., Campbell, R. N., and Raju, B. C.,** A proliferation disease of sweet potato, *Phytopathology,* 69, 918, 1979.
30. **Tsai, P., Shinkai, A., Mukoo, H., and Nakamura, S.,** Distribution of mycoplasma-like organisms and ultrastructural changes in host cells in witches' broom of mycoplasm-like organisms and ultrastructural changes in host cells in witches' broom diseased sweet potato and morning glory, *Ann. Phytopathol. Soc. Jpn.,* 38, 81, 1972.
31. **Lawson, R. H., Kahn, R. P., Hearon, S., and Smith, F. F.,** The association of mycoplasmalike bodies with sweet potato little-leaf (witches' broom) disease, *Phytopathology,* 60, 1016, 1970.
32. **Yang, I. L. and Chou, L. Y.,** Transmission of sweet potato witches' broom by *Orosius orientalis* in Taiwan, *J. Agric. Res. China,* 31, 169, 1982.
33. **Martin, W. J. and Dukes, P. D.,** Bacterial stem and root rot of sweet potato, *Plant Dis. Rep.,* 61, 158, 1977.
34. **Halsted, B. D.,** Some fungous diseases of the sweet potato, *N. J. Agric. Exp. Stn. Bull.,* No. 76, 1890.
35. **Person, L. H. and Martin, W. J.,** Soil rot of sweet potatoes in Louisiana, *Phytopathology,* 30, 913, 1940.
36. **Adams, J. F.,** An actinomycete, the cause of soil rot or pox in sweet potatoes, *Phytopathology,* 19, 179, 1929.
37. **Person, L. H.,** The soil rot of sweet potatoes and its control with sulfur, *Phytopathology,* 36, 869, 1946.
38. **Poole, R. F.,** The relations of soil moisture to the pox or ground-rot disease of sweet potatoes, *Phytopathology,* 15, 287, 1925.
39. **Lorbeer, J. W.,** The pathogenesis and control of *Streptomyces ipomoea* in California, *Phytopathology,* 52, 18, 1962.
40. **Hernandez, T., Constantin, R. J., Hammett, H., Martin, W. J., Clark, C., and Rolston, L.,** 'Travis' sweet potato, *HortScience,* 16, 574, 1981.
41. **Martin, W. J., Hernandez, T. P., and Hernandez, T. P.,** Development and disease reaction of Jasper, a new soil rot-resistant sweet potato variety from Louisiana, *Plant Dis. Rep.,* 59, 388, 1975.
42. **Harter, L. L. and Weimer, J. L.,** A monographic study of sweet potato diseases and their control, *U.S.D.A. Tech. Bull.,* No. 99, 1929, 117 p.
43. **Arene, O. B. and Nwankiti, A. O.,** Sweet potato diseases in Nigeria, *PANS,* 24, 294, 1978.
44. **Cook, A. A.,** *Diseases of Tropical and Subtropical Vegetables and Other Plants,* Macmillan, New York, 1978, 381.
45. **Baker, R. E. D.,** Distribution of fungous diseases of crops plants in Caribbean Region, *Trop. Agric.,* 17, 90, 1940.
46. **Akazawa, T., Uritani, I., and Akazawa, Y.,** Biosynthesis of ipomeamarone. I. The incorporation of acetate-2-C^{14} and mevalonate-2-C^{14} into ipomeamarone, *Arch. Biochem. Biophys.,* 99, 52, 1962.
47. **Hyodo, H., Uritani, I., and Akai, S.,** Production of furanoterpenoids and other compounds in sweet potato root tissue in response to infection by various isolates of *Ceratocystis fimbriata, Phytopathol. Z.,* 65, 332, 1969.
48. **Kubota, N. and Matsuura, T.,** Chemical studies on black rot disease of sweet potato, structure of ipomeamarone, *J. Chem. Soc. Jpn.,* 74, 248, 1953.
49. **Akazawa, T.,** Chromatographic isolation of pure ipomeamarone and reinvestigation of its chemical properties, *Arch. Biochem. Biophys.,* 90, 82, 1960.

50. **Kato, N., Imaseki, M., Nakashima, N., and Uritani, I.,** Structure of a new sesquiterpene, ipomeamaronal, in diseased sweet potato root tissue, *Tetrahedron Lett.*, 13, 843, 1971.
51. **Kojima, M. and Uritani, I.,** Possible involvment of furanoterpenoid phytoalexins in establishing host-parasite specificity between sweet potato and various strains of *Ceratocystis fimbriata, Physiol. Plant Pathol.*, 8, 97, 1976.
52. **Oba, K., Oga, K., and Uritani, I.,** Metabolism of ipomoeamarone in sweet potato, *Ipomoea batatas*, root slices before and after treatment with mercuric chloride or infection with *Ceratocystis fimbriata, Phytochemistry*, 21, 1921, 1982.
53. **Akazawa, T. and Uritani, I.,** Pattern of the carbohydrate breakdown in sweet potato roots infected by *Ceratostomella fimbriata, Agric. Biol. Chem.*, 25, 873, 1961.
54. **Imaseki, H., Takei, S., and Uritani, I.,** Ipomoeamarone accumulation and lipid metabolism in sweet potato infected by the black rot fungus *(Ceratocystis fimbriata), Plant Cell Physiol.*, 5, 119, 1964.
55. **Kawashima, N., Hyodo, H., and Uritani, I.,** Investigations on antigenic components produced by sweet potato roots in response to black rot infection, *Phytopathology*, 54, 1086, 1964.
56. **Kojima, M. and Uritani, I.,** The possible involvement of a spore agglutinating factors in various plants in establishing host specificity by various strains of black rot fungus, *Ceratocystis fimbriata, Plant Cell Physiol.*, 15, 733, 1974.
57. **Clark, C. A. and Watson, B.,** Susceptibility of weed species of Convolvulaceae to root-infecting pathogens of sweet potato, *Plant Dis.*, 67, 907, 1983.
58. **Moyer, J. W., Campbell, C. L., and Averre, C. W.,** Stem canker of sweet potato induced by *Fusarium solani, Plant Dis.*, 66, 65, 1982.
59. **Kantzes, J. G. and Cox, C. E.,** Nutrition, pathogenicity, and control of *Monilochaetes infuscans* Ell. & Halst. ex Harter, the incitant of scurf of sweet potatoes, *Univ. Md. Agric. Exp. Stn. Bull.*, No. 95, 1958.
60. **Daines, R. H.,** The influence of bedding stock source, plant bed temperature, and fungicide treatment on the development of black rot of sweet potato sprouts and on the resulting crop during storage, *Phytopathology*, 49, 249, 1959.
61. **Daines, R. H.,** The effect of plant bed temperature and fungicide treatments on the occurence of Java black rot disease of sweet potato sprouts, *Phytopathology*, 49, 252, 1959.
62. **Daines, R. H.,** Effect of plant bed, prebedding air, and fungicide dip temperature in controlling scurf on sweet potatoes, *Phytopathology*, 60, 1474, 1970.
63. **Daines, R. H., Brennan, E., and Leone, I. A.,** Effect of plant bed temperature and seed potato dip treatments on incidence of sweet potato sprout decay caused by *Diaporthe batatatis, Phytopathology*, 50, 186, 1960.
64. **Moyer, J. W.,** Postharvest disease management, in *Sweet Potato Proc. First Int. Symp.* Villareal, R. L. and Griggs, T. D., Eds., Asian Vegetable Research and Development Center, Shanhua, Tainan, Taiwan, 1981, 481.
65. **Elliot, J. A.,** A cytological study of *Ceratostomella fimbriata* (E. & H.) Elliott, *Phytopathology*, 15, 417, 1925.
66. **Chester, F. D.,** The black-rot of the sweet potato, *Ceratocystis fimbriata* Ell. and Hals, *Del. Agric. Exp. Stn. Ann. Rep.*, No. 3, 90, 1890.
67. **Martin, W. J.,** Coffee and sweet potato strains of *Ceratostomella fimbriata, Assoc. South. Agric. Wkrs. Proc.*, 46, 127, 1949.
68. **Goto, K., Suzuki, N., Kondo, S., and Miyajima, M.,** On the soil infection of blackrot of sweet potato and its transmission by field mice, *Tekai-Cinki Agric. Exp. Stn. B*, 1, 138, 1954.
69. **Harter, L. L. and Whitney, W. A.,** Influence of soil temperature and soil moisture on the infection of sweet potatoes by the black-rot fungus, *J. Agric. Res.*, 32, 1153, 1926.
70. **Lauritzen, J. I.,** Infection and temperature relations of black rot of sweet potatoes in storage, *J. Agric. Res.*, 33, 663, 1926.
71. **Nielsen, L. W.,** You can control scurf and black rot, *N. C. Agric. Exp. Stn. Bull.*, No. 406, 1958.
72. **Martin, W. J., Lutz, J. M., and Ramsey, G. B.,** Control of black rot in washed, uncured sweet potatoes, *Phytopathology*, 39, 580, 1949.
73. **Harter, L. L., Weimer, J. L., and Lauritzen, J. I.,** The comparative susceptibility of sweet-potato varieties to black rot, *J. Agric. Res.*, 32, 1135, 1926.
74. **Cheo, Pen Ching,** Varietal differences in susceptibility of sweet potato to black rot fungus, *Phytopathology*, 43, 78, 1953.
75. **Harter, L. L.,** Storage rots of economic aroids, *J. Agric. Res.*, 6, 549, 1916.
76. **Clendenin, I.,** Lasiodiplodia E. & E., N., *Gen. Bot. Gaz,.* 21, 92, 1896.
77. **Taubenhaus, J. J.,** Soil-stain, or scurf, of the sweet potato, *J. Agric. Res.*, 5, 995, 1916.
78. **Lawrence, G. W., Moyer, J. W., and Van Dyke, C. G.,** Histopathology of sweet potato roots infected with *Monilochaetes infuscans, Phytopathology*, 71, 312, 1981.
79. **Nielsen, L. W.,** Blister, a new disease of sweet potato, *Plant Dis. Rep.*, 49, 97, 1965.

80. **Nielsen, L. W.,** Thermotherapy to control sweet potato sprout-borne root knot, black rot, and scurf, *Plant Dis. Rep.,* 61, 882, 1977.
81. **Halsted, B. D. and Fairchild, D. G.,** Sweet potato black-rot, *J. Mycol.,* 7, 1, 1891.
82. **Harter, L. L., Weimer, J. L., and Adams, J. M. R.,** Sweet potato storage rots, *J. Agric. Res.,* 15, 337, 1918.
83. **Higgins, B. B.,** Physiology and parasitism of *Sclerotium rolfsii* Sacc., *Phytopathology,* 17, 417, 1927.
84. **Harter, L. L.,** Rhizoctonia and *Sclerotium rolfsii* on sweet potatoes, *Phytopathology,* 6, 305, 1916.
85. **Dukes, P. D., Jones, A., and Schalk, J. M.,** Evaluating sweet potato for reaction to sclerotial blight (*Sclerotium rolfsii* Sacc) in field plant beds, *Phytopathology,* 72, 998, 1982.
86. **Jones, A.,** Quantitative inheritance of Fusarium wilt resistance in sweet potatoes, *J. Am. Soc. Hortic. Sci.,* 94, 207, 1969.
87. **Collins, W. W. and Nielsen, L. W.,** Nature of Fusarium wilt resistance in sweet potatoes, *Phytopathology,* 66, 489, 1976.
88. **Nelson, P. W. E., Toussoun, T. A., and Cook, R. J., Eds.,** *Fusarium: Diseases, Biology, and Taxonomy,* The Pennsylvania State University Press, University Park, 1981, 457.
89. **Lauritzen, J. I.,** Factors affecting infection and decay of sweet potatoes by certain storage rot fungi, *J. Agric. Res.,* 50, 285, 1935.
90. **Nielsen, L. W.,** Harvest practices that increase sweet potato surface rot in storage, *Phytopathology,* 55, 640, 1965.
91. **Nielsen, L. W. and Moyer, J. W.,** A Fusarium root rot of sweet potatoes, *Plant Dis. Rep.,* 63, 400, 1979.
92. **Harter, L. L. and Weimer, J. L.,** The surface-rot of sweet potato, *Phytopathology,* 9, 465, 1919.
93. **Clark, C. A.,** End rot, surface rot, and stem lesions caused on sweet potato by *Fusarium solani*, *Phytopathology,* 70, 109, 1980.
94. **Scott, L. E., Kantzes, J., and Bouwkamp, J. C.,** Clonal differences in the incidence of surface rot (Fusarium spp.) on sweet potato, *Plant Dis. Rep.,* 56, 783, 1972.
95. **Lauritzen, J. I. and Harter, L. L.,** The influence of temperature on the infection and decay of sweet potatoes by different species of Rhizopus, *J. Agric. Res.,* 30, 793, 1925.
96. **Harter, L. L., Weimer, J. L., and Lauritzen, J. I.,** The decay of sweet potatoes *(Ipomoea batatas)* produced by different species of Rhizopus, *Phytopathology,* 11, 279, 1921.
97. **Jenkins, A. E. and Viegas, A. P.,** Stem and foliage scab of sweet potato *(Ipomoea batatas)*, *J. Wash. Acad. Sci.,* 33, 244, 1943.
98. **Lao, F. O.,** Morphology of the sweet potato scab fungus *(Sphaceloma batatas* Saw.), *Ann. Trop. Res.,* 2, 40, 1980.
99. **Goto, K.,** Outbreak of shoot scab of sweet potato in Amami Islands, *Phytopathol. Soc. Jpn.,* 7, 143, 1937.
100. **Taylor, A. L., Sasser, J. N., and Nelson, L. A.,** *Relationship of Climate and Soil Characteristics to Geographical Distribution of Meloidogyne Species in Agricultural Soils,* North Carolina State University Graphics, Raleigh, 1982, 65.
101. **Gapasin, R. M.,** Survey and identification of plant parasitic nematodes associated with sweet potato and cassava, *Ann. Trop. Res.,* 1, 121, 1979.
102. **Brathwaite, C. W. D.,** Preliminary studies on plant-parasitic nematodes associated with selected root crops at the University of West Indies, *Plant Dis. Rep.,* 56, 1077, 1972.
103. **Taylor, A. L. and Sasser, J. N.,** *Biology, Identification and Control of Root-knot Nematodes (Meloidogyne Species),* North Carolina State University Graphics, Raleigh, 1978.
104. **Giamalva, M. J., Martin, W. J., and Hernandez, T. P.,** Sweet potato varietal reaction to species and races of root-knot nematodes, *Phytopathology,* 53, 1187, 1963.
105. **Thomas, R. J. and Clark, C. A.,** Population dynamics of *Meloidogyne incognita* and *Rotylenchulus reniformis.* Alone and in combination, and their effects on sweet potato, *J. Nematol.,* 15, 204, 1983.
106. **Thomas, R. J. and Clark, C. A.,** Effects of concomitant development on reproduction of *Meloidogyne incognita* and *Rotylenchulus reniformis* on sweet potato, *J. Nematol.,* 15, 215, 1983.
107. **Sasser, J. N. and Kirby, M. F.,** *Crop cultivars resistant to root-knot nematodes, Meloidogyne species,* North Carolina State University Graphics, Raleigh, 1979.
108. **Jones, A., Dukes, P. D., and Cuthbert, F. P., Jr.,** W-13 and W-178 sweet potato germplasm, *HortScience,* 10, 533, 1975.
109. **Jones, A., Dukes, P. D., and Cuthbert, F. P., Jr.,** Mass selection in sweet potato: Breeding for resistance to insects and diseases and for horticultural characteristics, *J. Am. Soc. Hortic. Sci.,* 101, 701, 1976.
110. **Clark, C. A. and Wright, V. L.,** Effect and reproduction of *Rotylenchulus reniformis* on sweet potato selections, *J. Nematology,* 15, 197, 1983.
111. **Brathwaite, C. W. D.,** Effect of crop sequence and fallow on populations of *Rotylenchulus reniformis* in fumigated and untreated soil, *Plant Dis. Rep.,* 58, 259, 1974.
112. **Birchfield, W. and Martin, W. J.,** Evaluation of nematocides for controlling reniform nematodes on sweet potatoes, *Plant Dis. Rep.,* 52, 127, 1968.

113. **Gapasin, R. M.,** Control for *Meloidogyne incognita* and *Rotylenchulus reniformis* and its effect on the yield of sweet potato and cassava, *Ann. Trop. Res.*, 3, 92, 1981.
114. **Hahn, S. K.,** Effects of viruses (SPVD) on growth and yield of sweet potato, *Exp. Agric.*, 15, 1, 1979.
115. **Mukiibi, J.,** Effect of mosaic on the yield of sweet potatoes in Uganda, *Proc. 4th Symp. Int. Soc. Trop. Root Crops*, CIAT, Cali, Colombia, 1977, 169.
116. **Schaefers, G. A. and Terry, E. R.,** Insect transmission of sweet potato disease agents in Nigeria, *Phytopathology*, 642, 1976.
117. **Campbell, R. N., Hall, D. H., and Mielinis, N. M.,** Etiology of sweet potato russet crack disease, *Phytopathology*, 64, 210, 1974.
118. **Mukiibi, J.,** Synonymy in sweet potato virus diseases, *Proc. 4th Symp. Int. Soc. Trop. Root Crops*, CIAT, Cali, Colombia, 1977, 163.
119. **Cadena-Hinojosa, M. A. and Campbell, R. N.,** Characterization of isolates of four aphid-transmitted sweet potato viruses, *Phytopathology*, 71, 1086, 1981.
120. **Cali, B. B. and Moyer, J. W.,** Purification, serology, and particle morphology of two russet crack strains of sweet potato feathery mottle viruses, *Phytopathology*, 71, 302, 1981.
121. **Hollings, M., Stone, O. M., and Bock, K. R.,** Purification and properties of sweet potato mild mottle, a white-fly borne virus from sweet potato *(Ipomea batatas)* in East Africa, *Ann. Appl. Biol.*, 82, 511, 1976.
122. **Moyer, J. W. and Kennedy, G. G.,** Purification and properties of sweet potato feathery mottle virus, *Phytopathology*, 68, 998, 1978.
123. **Nome, S. F., Shalla, T. A., and Petersen, L. J.,** Comparison of virus particles and intracellular inclusions associated with vein mosaic, feathery mottle, and russet crack diseases of sweet potato, *Phytopathology*, 79, 169, 1974.
124. **Liao, C. H., Chien, I. C., Chung, M. L., Han, Y, H., and Chiu, R. J.,** Two sap-transmissible sweet potato viruses occuring in Taiwan. a. Identification and some in vitro properties, in U.S.-China Joint Seminar on Virus Diseases of Fruit and Vegetable Crops, Homestead, Fla., August 15 to 17, 1979.
125. **Nielsen, L. W.,** Internal cork of sweet potatoes. Production of corky lesions in core-grafted roots, *Plant Dis. Rep.*, 36, 132, 1952.
126. **Doolittle, S. P. and Harter, L. L.,** A graft-transmissible virus of sweet potato, *Phytopathology*, 35, 695, 1945.
127. **Cali, B. B. and Moyer, J. W.,** Characterization of sweet potato feathery mottle virus RNA, *Phytopathology*, 71, 864, 1981.
128. **McLean, D. L.,** Some aphid vector-plant virus relationships of the feathery mottle virus of sweet potato, *J. Econ. Entomol.*, 52, 1057, 1959.
129. **Lawson, R. H., Hearon, S. S., and Smith, F. F.,** Development of pinwheel inclusions associated with sweet potato russet crack virus, *Virology*, 46, 453, 1971.
130. **Atcham, T., Lockhart, B., and Bantarri, E.,** Identification and characterization of a virus in sweet potato found in southeast Zaire, Central Africa, *Phytopathology*, 73, 787, 1983.
131. **Salama, F. M., Lyman, C. M., and Whitehouse, U. G.,** Isolation of the internal cork virus from sweet potato leaves having the purple ring structure by use of differential filtration, *Phytopathology*, 57, 89, 1966.
132. **Nome, S. F.,** Sweet potato vein mosaic virus in Argentina, *Phytopathology*, 77, 44, 1973.
133. **Loebenstein, G. and Harpaz, I.,** Virus diseases of sweet potatoes in Israel, *Phytopathology*, 50, 100, 1959.
134. **Costa, A. S., Kitujima, E. W., and Normanhu, E. S.,** Sweet potato mosaic induced by a virus of the potato y group, *Fitopatologica*, 8, 7, 1973.
135. **Liao, C. H., Chien, I. C., Chung, M. L., Chiu, R. J., and Han, Y. H.,** A study of sweet potato virus disease in Taiwan. I. Yellow spots, *J. Agric. Res. China*, 28, 127, 1979.
136. **Sheffield, F. M. L.,** Virus diseases of sweet potato in East Africa. I. Identification of the viruses and their insect vectors, *Phytopathology*, 47, 582, 1957.
137. **Terry, E. R.,** Sweet potato *(Ipomoea batatas)* virus diseases and their control, in, *Sweet potato, Proc. First International Symposium*, Villareal, R. L. and Griggs, T. D., Eds., Asian Vegetable Research and Development Center, Shanhua, Tainan, Taiwan, 1982, 481.
138. **Hildebrand, E. M.,** A whitefly, *Trialeurodes abutilonea*, an insect vector of sweet potato feathery mottle in Maryland, *Plant Dis. Rep.*, 43, 712, 1959.
139. **Girardeau, J. H., Jr. and Ratcliff, T. J.,** The vector-virus relationship of the sweet potato whitefly and a mosaic of sweet potatoes in South Georgia, *Plant Dis. Rep.*, 44, 48, 1960.
140. **Sheffield, F. M. L.,** Virus diseases of sweet potato in East Africa. II. Transmission to alternative hosts, *Phytopathology*, 48, 1, 1957.
141. **Moyer, J. W.,** unpublished.

Chapter 3

MAJOR INSECT PESTS*

James M. Schalk and Alfred Jones

TABLE OF CONTENTS

I.	Introduction	60
II.	Coleoptera	60
	A. *Alcidodes dentipes* (Oliver)	60
	B. *Aspidomorpha* spp.	60
	C. *Cylas* spp.	61
	D. *Diabrotica* spp.	65
	E. *Euscepes postfasciatus* (Fairm.)	65
	F. *Plectris aliena* Chapin	68
	G. *Typophorus* spp.	70
III.	Homoptera	71
	A. *Aphis gossypii* Glov.	71
	B. *Bemisia tabaci* (Genn.)	71
	C. *Myzus persicae* (Sulzer)	72
IV.	Lepidoptera	73
	A. *Herse convolvuli* L.	73
	B. *Omphisa anastomosalis* (Guernée)	73
References		75

---

* This chapter was prepared by U.S. Government employees as a part of their official duties and legally cannot be copyrighted. Mention of a trademark or proprietary product does not constitute a guarantee or warranty of the product by the U.S. Department of Agriculture. The authors wish to thank Mr. Robert Hamalle for his technical expertise in arranging and photographing the drawings for publication.

I. INTRODUCTION

This chapter is on the major known insect problems of sweet potato that occur in tropical and temperate regions of the world. The insects are listed alphabetically by order and genus. Illustrations were drawn by the senior author from specimens, photographs, and publications listed in the bibliography.

Chemical control recommendations have been limited primarily to organophosphorus, carbamate, and synthetic pyrethroids because of environmental concerns. Generally these compounds have had no long-term harmful effect on the environment. However, many are highly toxic to warm-blooded animals and should be applied only by a trained applicator wearing protective equipment.

Research in the breeding of sweet potatoes has resulted in the development of insect-resistant lines and cultivars. The text stresses the use of these resistant plants as an effective alternative to insecticides. Additional advantages of plant resistance are its compatibility with biological, chemical, and cultural insect control methods.

II. COLEOPTERA

A. *Alcidodes dentipes* (Oliver) (Figure 1)

Common name — striped sweet potato weevil.

Family — Curculionidae.

Host plants — This insect feeds mainly on sweet potato but will also feed on ground nut and legumes.[1]

Distribution and injury — *A. dentipes* is found in the tropical areas of Africa where it girdles sweet potato stems above the ground level. Injured plants usually wilt and die. Larvae bore into the stem causing galls.[1]

Appearance and biology — Larvae are thought to feed inside the stem on the pith. Adults are 14 mm long and have longitudinal stripes on the elytra. Pronounced spines are found on the inner edge of all tibia and on the front femur of adults.[1] Two other species have been taken from sweet potato: *A. congeanus* Fáust (Central Africa) and *A. convexus* Oliver (Madagascar), however, verification of their pest status is required.[2]

Control — Because of sporadic damage, routine controls are not recommended. However, if severe injury results stomach poisons should prove effective. Consult recommendations for the best insecticides and rates of application.

B. *Aspidomorpha* spp. (Figure 2)

Common name — tortoise beetle.

Family — Chrysomelidae.

Host plants — Primarily sweet potato. Other hosts are wild *Ipomoea* spp., coffee, beet, and potato.[1,3-5]

Distribution and injury — This species complex is found in Africa, China, and India. Both larval and adult stages feed on the foliage eating small holes in the leaves. When newly set plants are attacked the injury may be severe.[1,3,5]

Appearance and biology — There are over 12 spp. of *Aspidomorpha* beetles. Adults oviposit eggs singly in rows on the underside of the foliage that hatch in 3 to 9 days. The larval period is 8 to 15 days with 5 instars. Mature larvae (5 mm long) are light green with a fringe of spines along their margins. The exuvia (cast skin) remain attached to the dorsal area of the body. The pupae are green (6 mm long) with the exuvia held dorsally. Duration of the pupal stage is 4 to 6 days and occurs on the foliage. The life cycle from egg to adult requires 28 to 37 days. The adult is oval and shield-like (6 to 8 mm long) with broad flat elytra. Newly emerged adults are white and darken in 2 to 3 hr. It takes 7 to 9 days for the

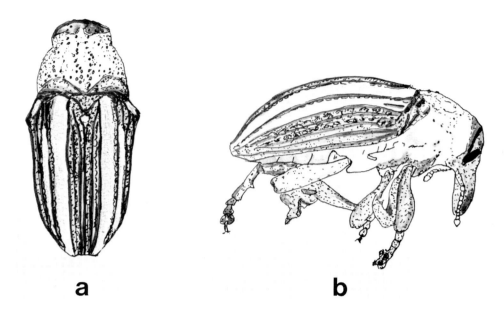

FIGURE 1. *Alcidodes dentipes* (Oliver), adult; dorsal (a); lateral (b).

beetle to develop a golden irridescence. Adults can live between 58 to 76 days. Several generations occur per year in the tropics and only one or two in temperate regions.[1,3-5]

Control — Larvae and adults are controlled by applying carbaryl 1 to 1.8 kg ai/ha to the foliage as needed in the seedbed or production fields. Care should be taken to insure complete coverage of the leaf undersides.[6]

C. *Cylas* spp. (Figure 3)

Common name — sweet potato weevil.

Family — Brentidae.

Host plants — These pests primarily feed on sweet potato and have also been reported feeding on *Cuscuta* sp., *Daucus carota* L., *Dichondra carolinensis* Michx., *Ipomoea* sp., maize, *Manihot palmata*, *M. utelissima* Pohl and *Thunbergia* sp.[1,7,8]

Distribution and injury — On a world basis the genus *Cylas* constitutes the most important insect threat to sweet potato. There are 27 described species of *Cylas* and *Protocylas*, but their taxonomy is plagued with problems. The species complex most important are *puncticollis* F., *femoralis* Fáust which is African and *formicarius* F. which is worldwide (India, Asia, Pacific, New World, and parts of Africa).[2] *C. f. elegantulus* (Summers) is the only recognized member of this genus in the New World.[9,10] Larvae and adults of these species feed on roots in the field and in storage. In storage the insects can build up to very high populations. Oviposition is in small cavities in the root skin or at the crown of the plant. The larvae migrate into the tissue feeding and growing. Repulsive terpene odors produced by larval feeding make the roots unfit for human or animal consumption.[1,7,11-15]

Appearance and biology — *Cylas* breeds continually and throughout the year all stages can be found. Females lay eggs (0.71 mm long) singly in small cavities eaten out of the stem crown or roots. Average reproduction per female is about 100 eggs *(C. f. elegantulus)*. After each oviposition she covers the eggs with whitish excretion. In less than a week eggs hatch and the larvae eat into the root tissue with the tunnels enlarging as the larvae grow. Larval feeding in the crown, in severe infestations, can cause vine death. Larval feeding lasts 2 to 3 weeks with the roots developing an offensive terpene odor. Larvae are legless

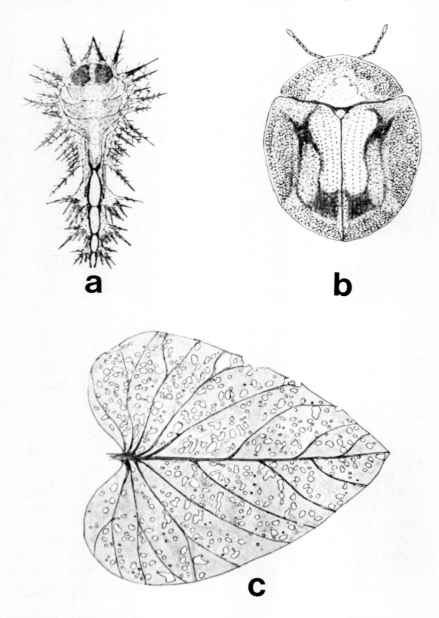

FIGURE 2. *Aspidomorpha* spp.; larvae (a); Adult (b); feeding damage by larvae and adult to sweet potato foliage (c).

and white (5 to 8 mm long) with a light brownish head and darker brown mandibles. Pupal (6 mm long) duration lasts one week and occurs in a cavity within the root. The emerging white adult eats its way out of the root and gradually matures developing a normal coloration. Mature adults vary from all black *(C. puncticollis* Boh) to bluish elytra and a red pronotum *(C. formicarius)*. The adults can be from 6 to 8 mm long and feed on stems, foliage and roots near the surface of the ground. Adults can also reach developing roots through cracks in the soil. Generation completion can occur in 4 to 6 weeks. Adults can live for as long as 8 months but are not capable of living more than 5 to 8 days in summer without food.

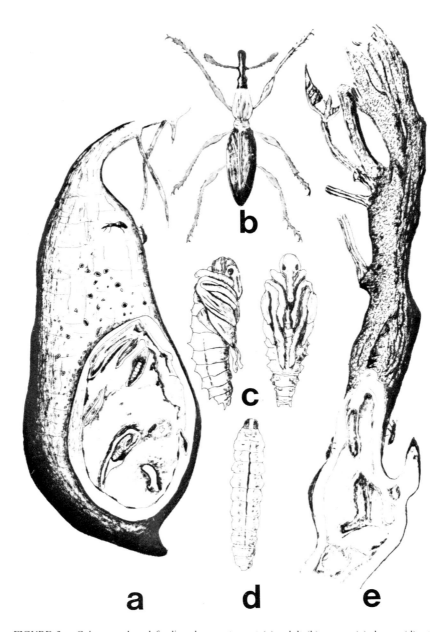

FIGURE 3. *Cylas* spp. larval feeding damage to root (a); adult (b); pupae (c); larvae (d); stem damage (e).

In storage, adults must feed on enlarged roots as dry fibrous roots do not offer sustenance. Adults can fly up to 2 km, but the main way of disseminating into new fields is through the planting of infested cuttings or slips. New fields can become infested when they are planted next to fields having wild hosts or volunteer plants which harbor adult insects.[1,7-9,14-18]

Control — The weevil is difficult to control in the tropics because of continuous reproduction throughout the year. Pest management programs must be as thorough as possible, concentrating on insecticide usage, crop rotation, sanitation, and the use of the available resistant lines and cultivars.

Table 1
INSECTICIDES FOR EFFECTIVE SWEET POTATO WEEVIL PROTECTION IN SWEET POTATO PLANT BEDS AND PRODUCTION FIELDS

Insecticides[a]	Other name	Rate/ha	Species controlled	Ref
Aldicarb G	Temik	1.5 kg	C. formicarius	24
Ambush 2 EE	Permethrin	1 ℓ	C. formicarius elegantulus	25
Carbaryl WP	Sevin	0.1%	C. formicarius	26
Carbofuran 3 G	Furadan	2 kg	C. formicarius formicarius	14
Carbofuran 75 WP	Furadan	0.05%	C. formicarius formicarius	14
Dimecron 100 EC	Phosphamidon	0.03%	C. formicarius	27
Disulfoton G	Disyston	1.5 kg	C. formicarius	24
Dyfonate 4 E	Fonofs	1.7 kg	C. formicarius elegantulus	16
Endosulfan 50 WP	Thiodan	1.12 kg	C. formicarius elegantulus	28
Fenitrothion EC	Sumithion	0.05%	C. formicarius	29
Fenthion EC	Lebaycid	0.05%	C. formicarius	29
Hostathion 40 EC	Triazophos	1.0 kg	C. formicarius elegantulus	30
Lannate 90 WP	Methomyl	1.0 kg	C. formicarius	30
Methyldemeton EC	Metasystox	0.05%	C. formicarius	29
Parathion EC	Paraphos	0.05%	C. formicarius	27
Permethrin 25 WP	Ambush	0.22 kg	C. formicarius elegantulus	28
Phorate G	Thimet	1.5 kg	C. formicarius	24
Sumithion EC	Fenitrothion	0.1%	C. formicarius	29
Surecide 25 EC	Cyanofenphos	1.0 kg	C. formicarius	30
Thimet 50 WP	Phorate	0.56 kg	C. formicarius elegantulus	16
Trithion 20% EC	Carbophenothion	0.1%	C. formicarius	27

[a] G = Granular formulation, E = Emulsifiable formulation, WP = Wettable powder formulation, and EC = Emulsifiable concentration.

The most efficient method to propagate sweet potatoes is through the use of plant beds. Bedding (seed) roots should be weevil free but if infested roots must be used they should be treated with granular dasnit (15 g at 2.2 kg ai/ha) at planting. Roots should be thoroughly covered with insecticide and the insecticide should extend to both sides of the bed in order for it to be well mixed with the soil used to cover the roots.[16] Cuttings taken from bedded roots should be made 2.5 cm from the soil surface. The procedure of cutting above the soil surface insures a 98% chance of selecting weevil-free plants. In regions where sweet potato cuttings are obtained directly from production fields only new tip growth (20 to 45 cm) should be taken. No roots should be included in cuttings.[16,19,20]

All cuttings should be dipped in a carbofuran (0.05% ai) solution for 20 min prior to planting. This will kill all stages of the weevil and provide some residual protection for the new transplants.[14,20]

Crop rotation, with nonhost plants (beans, rice, tobacco, sugar cane) is another method to reduce weevil infestations. Important consideration should be given in selecting a planting site. Fields and adjacent areas should be free of wild host plants and volunteer sweet potatoes. Irrigation at 7-day intervals will also reduce weevil infestation.[14,21-23]

Weevil infestation in production fields can be reduced with effective insecticides. Treatments are applied once at planting (granular formulation) and during root enlargement (spray formulation) at 2 to 4 week intervals until harvest. In applying insecticide sprays it is extremely important to get deep foliage penetration so that weevils feeding in the crown area are controlled (Table 1).[16,24-30]

After harvest, roots can be protected by treating them in storage crates with 5% imidan dust (84 g to 23 kg of roots). The interior of the storage area should be thoroughly cleaned and then sprayed with malathion (280 g in 9.5 ℓ of water). Malathion offers an effective residual treatment for the adult weevils in their search for food.[16]

There should be no residual crop left in the field after harvest because this provides an ideal breeding ground for the weevil. The old vines and roots should be destroyed by burning or by feeding to livestock.[7]

The use of resistant sweet potatoes is compatible with chemical and nonchemical control measures and also offers built-in protection against insect attack.[31] Researchers have shown that the use of moderately resistant sweet potato lines increased the effectiveness of soil insecticides in reducing soil insect injury.[32] The synergistic effect of resistant lines and insecticide may also apply in sweet potato weevil control. There are many lines and cultivars exhibiting levels of weevil resistance.[33-43] The resistance can be identified as antibiosis, nonpreference, and escape (Table 2). The latter is where roots have long necks or develop deep in the soil where they are protected from the weevils.[11,43,44]

D. *Diabrotica* spp. (*D. adelpha* Harold, *D. balteata* LeConte, *D. speciosa* Germar, *D. undecimpunctata howardi* Barber) (Figure 4)

Common name — rootworms.

Family — Chrysomelidae.

Host plants — Members of this genera are capable of feeding on more than 200 common weeds, grasses, and cultivated crops including sweet potato.[3,45]

Distribution and injury — The tropical species of *Diabrotica* are very diverse and only limited knowledge is available on their biology. *D. adelpha* has been reported in Costa Rica while *D. speciosa* has been found in Argentina.[46,47] The distribution of *D. balteata* is from Colombia, S.A. to the southern U.S. *D. undecimpunctata howardi* can be found from Central Mexico to east of the Rocky Mountains in the U.S.[45]

The larvae of *Diabrotica* eat small round holes through the root periderm (skin) and form irregular enlarged cavities just under the skin. Feeding holes are found in groups. Holes may become enlarged as the roots develop. In the U.S. sweet potatoes are often attacked during early development, which results in many unsightly healed holes at harvest. Adult feeding on sweet potato foliage produces irregular holes.[48]

Appearance and biology — Adult *Diabrotica* are about 6 mm long and in *D. balteata* the elytra are marked with alternating green and yellow bands while in *D. undecimpunctata howardi* the elytra have 11 spots on a yellowish green background. *Diabrotica* spp. which feed on the plant family Convolvulaceae lay their eggs in the soil where they hatch in 1 to 2 weeks depending on temperature. Larval species are nearly indistinguishable and are 6 to 12 mm long, creamy white with brown heads and a dark brown spot on the last abdominal body segment. Larval duration is from 8 to 30 days, which is dependent on temperature and food supply. Pupae (6 mm long) are found in cells just below the soil surface and emerge as adults (6 mm long) in 1 week. In warm climates of the U.S. the insect can overwinter as adults.[48]

Control — There are no effective insecticides which provide adequate control of the larvae. However, resistant sweet potatoes have been developed recently which provide protection against *D. balteata* and *D. undecimpunctata howardi*.[37,38,49] This resistance may also provide protection against other *Diabrotica* species.

E. *Euscepes postfasciatus* (Fairm) (Figure 5)

Common name — scarabee of the sweet potato.

Family — Curculionidae.

Host plants — In addition to the sweet potato, this insect can be found in the wild hosts *Ipomoea horsfalliae* Hook, *I. pes-caprae* (L.) R. Br., *I. pentaphylla* Jacq., and *I. triloba* L.[50]

Table 2
SUMMARY OF RESISTANCE IN SWEET POTATO LINES AND CULTIVARS TO INFESTATIONS AND DAMAGE BY THE SWEET POTATO WEEVIL (*CYLAS* SPP.)

Entries	Resistant[a] Shoots	Resistant[a] Roots	Nature of resistance[b]	Insect species	Flesh color	Ref.
L3-64		+	Nonpreference	*C. f. elegantulus*	Orange	33
W19-2		+	Nonpreference	*C. f. elegantulus*	Orange	33
N-2A		+	Nonpreference	*C. f. elegantulus*	Orange	34
L2-175		+	Nonpreference	*C. f. elegantulus*	Orange	34
L-187		+	Nonpreference	*C. f. elegantulus*	Unknown	35
TIS 2532	+		Antibiosis, nonpreference	*C. puncticollis*	White	36
TIS 3052		+	Antibiosis	*C. puncticollis*	White	36
TIS 3030		+	Antibiosis	*C. puncticollis*	White	36
TIS 3017	+		Antibiosis, nonpreference	*C. puncticollis*	White	36
W-71		+	Nonpreference	*C. f. elegantulus*	Orange	37
W-115		+	Nonpreference	*C. f. elegantulus*	Orange	37
W-119		+	Nonpreference	*C. f. elegantulus*	Orange	37,42
Resisto (W-125)		+	Antibiosis nonpreference	*C. f. elegantulus*	Orange	37
W-149		+	Nonpreference	*C. f. elegantulus*	Orange	37,41
W-154		+	Nonpreference	*C. f. elegantulus*	Orange	37
Regal		+	Antibiosis nonpreference	*C. f. elegantulus*	Orange	38
Picadito		+	Antibiosis	*C. f. elegantulus*	White	39
W-109		+	Nonpreference	*C. f. elegantulus*	Orange	40
W-113		+	Nonpreference	*C. f. elegantulus*	Orange	40
W-132		+	Nonpreference	*C. f. elegantulus*	Orange	41
W-141		+	Nonpreference	*C. f. elegantulus*	Orange	41
W-182		+	Nonpreference	*C. f. elegantulus*	Orange	42
W-191		+	Nonpreference	*C. f. elegantulus*	Orange	42
W-193		+	Nonpreference	*C. f. elegantulus*	Orange	42
S3		+	Escape	*C. formicarius*	Unknown	43
S13		+	Escape	*C. formicarius*	Unknown	43
S234		+	Escape	*C. formicarius*	Unknown	43
S238		+	Escape	*C. formicarius*	Unknown	43
S248		+	Escape	*C. formicarius*	Unknown	43
S324		+	Escape	*C. formicarius*	Unknown	43
S327		+	Escape	*C. formicarius*	Unknown	43
S332		+	Escape	*C. formicarius*	Unknown	43
S334		+	Escape	*C. formicarius*	Unknown	43
S335		+	Escape	*C. formicarius*	Unknown	43
S336		+	Escape	*C. formicarius*	Unknown	43
S360		+	Escape	*C. formicarius*	Unknown	43

[a] + = positive response of resistance in shoots or roots.
[b] Nonpreference = insect responses that lead them away from a plant for oviposition, food and shelter. Antibiosis = plants that have an adverse effect on insect life as reduced fecundity, decreased size, abnormal length of life, increased mortality. Escape = roots grow deeper in the soil and have long necks making weevil penetration difficult.

Distribution and injury — The scarabee is a sweet potato pest in South America, Caribbean, and Pacific. Larvae and adults feed on roots and stems of sweet potato. Eggs are laid in small chambers in the skin and covered with a fecal plug. After hatching, the larvae feed and produce narrow tunnels in the roots. Larval injury is similar to that of the sweet potato weevil and causes the roots to produce terpene compounds which make them inedible.[9,50-52]

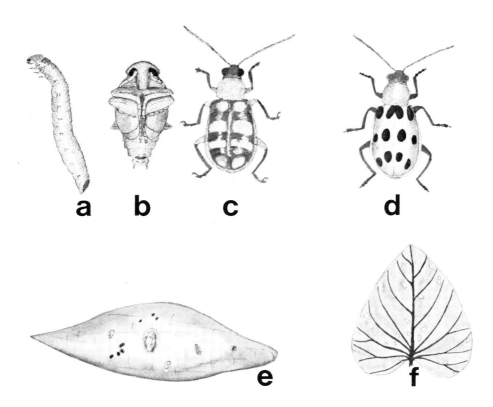

FIGURE 4. *Diabrotica balteata* LeConte, larvae (a); female pupae (b); adult (c). *D. undecimpunctata howardi* Barber; adult (d); larval root injury (e); adult foliage injury (f).

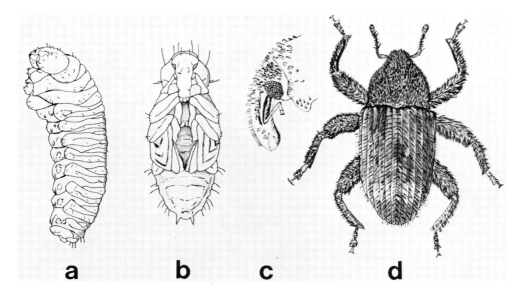

FIGURE 5. *Euscepes postfasciatus* (Fairmaire) larvae (a); pupae (b); adult snout (c); adult (d).

Appearance and biology — The scarabee breeds throughout the year and has overlapping life cycles. Females oviposit their eggs in small cavities eaten out of the stem or root. A female can lay up to 100 eggs per month. Eggs hatch within 6 to 7 days and the larvae eat into the tissue with the tunnels enlarging as the larvae grow. Severe stem infestations can result in foliage wilting. Larval feeding can last 20 to 30 days and the roots, on which they feed, develop offensive terpene odors. Larvae are legless, white, and 5 mm long with yellowish heads and reddish brown mandibles tipped with black. The pupae (4 mm long) are white. Newly emerged adults (4 mm long) are yellowish and later turn brown mottled with lighter areas on their elytra. The life cycle is completed in 33 to 46 days. Adults can live, with food, from 4 to 10 months. Starved adults can survive up to 1.5 months. Adults feed on the roots by eating shallow pits on the surface. Although the adults have wings they have not been observed in flight. Therefore, the main way of dissemination is through the planting of infested roots or cuttings.[9,50-52]

Control — Insect damage can be reduced by crop rotation, sanitation, irrigation, and effective use of insecticides.

Sweet potatoes should be followed with nonhost plants such as beans, rice, sugar cane, and tobacco. The particular rotation is dependent on local preferences.[53]

Important consideration should be given in selecting a planting site free of wild host plants and volunteer plants.[20] Immediately after harvest all fields should be cleared of remaining foliage and roots by feeding to livestock or burning.[23]

Cuttings taken from plant beds or production fields should be done carefully so that the stems are cut 2.5 cm above the soil or crown. This produces cuttings free of insect eggs which are normally oviposited near the crown. Cuttings can be dipped in an insecticide such as carbofuran (0.05% ai) for 20 min prior to planting. Although carbofuran has not been tested against the scarabee it has proven to be effective against the sweet potato weevil.[14]

In production fields the scarabee can be controlled with soil applications of ethyl parathion at 2 kg ai/ha granular.[54]

Irrigation that prevents soil cracking reduces scarabee infestation.[19]

Roots should be protected in storage with an effective insecticide. Imidan dust 5% (84 g to 23 kg of roots) has provided effective control in storage of the sweet potato weevil and may provide the same protection against the scarabee.[16]

The interior of the storage area should be thoroughly clean of all plant debris and treated with a contact insecticide. Malathion is recommended for contact control of sweet potato weevil in storage (280 g ai/9.5 ℓ water) and may also provide protection against the scarabee.[16]

Lines and cultivars resistant to the sweet potato weevil may also prove resistant to the scarabee and should be tried (Table 2).

F. *Plectris aliena* Chapin (Figure 6)

Common name — South American pasture beetle, white grub.

Family — Scarabaeidae.

Host plants — This insect feeds on sweet potato and pasture grasses.[48,55]

Distribution and injury — The species is found in Argentina, Brazil, and Uruguay. It was introduced into the U.S. and Australia probably in the early 1900s. The larvae feed mainly on pasture sod.[55] However, severe damage to sweet potatoes may result when they are planted in fields previously grown to pastures. The larvae cause severe root damage by gouging out broad shallow areas in the sweet potato.[48]

Appearance and biology — Eggs (1 mm diameter) are laid in the soil during the spring and summer and soon hatch into C-shaped larvae. Larvae have three instars and vary in color from white to cream with light-tan heads and grayish areas on the tip of the abdomen. Mature larvae (19 mm long) overwinter in the soil. In the spring they pupate and emerge as adults. Adults (10 mm long) are light tan with typical scarab-like characters.[48] Adult beetles live mainly in the soil but emerge about 20 minutes after sunset from late spring until mid-summer. Males are capable of flight and move in random directions (7 to 14 cm above the ground) in search of females. When a female emerges several males are attracted

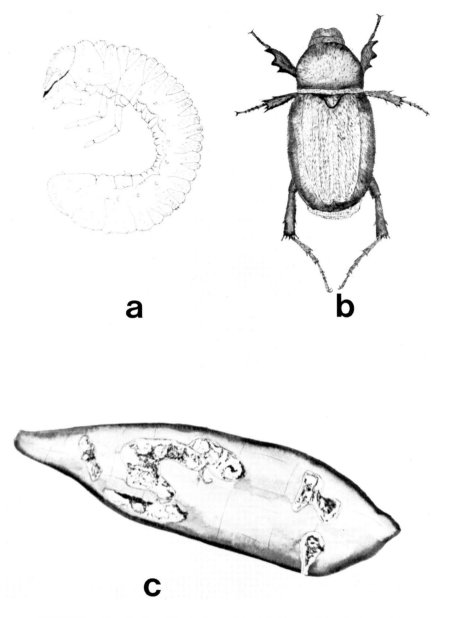

FIGURE 6. *Plectris aliena* Chapin; larvae (a); adult (b); root injury by larvae (c).

to her. After mating, all above ground activity ceases (15 to 20 minutes after sunset) and both sexes burrow back into the soil. Adult beetles do not feed and have poorly developed mouth parts. Their digestive tract is devoid of food. A mated female has been observed laying 33 eggs and she survived 21 days without food. A life cycle takes 1 to 2 years depending on the environment. The species appear to be restricted to light sandy soils which may limit distribution.[55]

Control — The use of insecticides to control grubs has given variable and often ineffective results. However, one of the most effective and economical ways to prevent grub damage in sweet potatoes is to grow resistant lines or cultivars. Several lines and cultivars are available having high levels of grub resistance.[37,38,49]

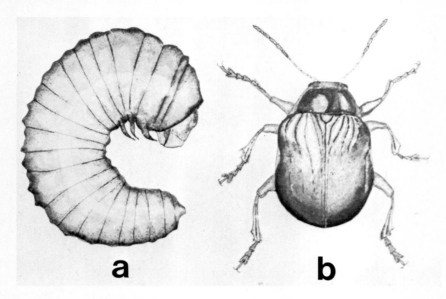

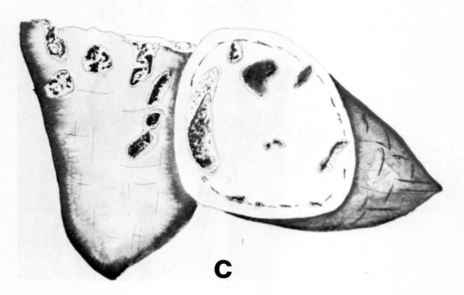

FIGURE 7. *Typophorus* spp., larvae (a); adult (b); root injury by larvae (c).

G. *Typophorus* spp. (*T. nigritus viridicyaneus* (Crotch), *T. nigritus nitidulus* (F) (Figure 7)

Common name — sweet potato leaf beetle.

Family — Chrysomelidae.

Host plants — These spp. feed on sweet potato, wild *Ipomoea* spp. and *Convolvulus arvensis* L.[48]

Distribution and injury — These species have been reported in the southern U.S. and Brazil.[48,56] In the U.S. they are widely distributed pests but usually cause little loss of production. The adults begin feeding on the margins of the foliage and work inward. Larvae tunnel into vines and roots. Root damage is easily recognized by the large excrement-filled tunnels that penetrate deeply into the roots. Larvae can be found in the tunnels sometimes a month or more after harvest.[48]

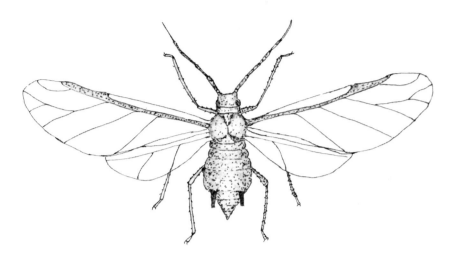

FIGURE 8. *Aphis gossypii* Glov., adult winged (alatae) form.

Appearance and biology — Adults are metallic bluish-green (7 mm long). Eggs are laid during the spring and summer in clusters on the underside of leaves or just beneath the surface of the soil under the sweet potato plants. Larvae are pale yellow or yellow with brown heads. Mature larvae are 13 mm long. The insect overwinters as larvae but pupates as the weather becomes warmer. Adults emerge in the spring.[48]

Control — Effective control of adults can be obtained with carbaryl. Consult labels for application rates.[6]

III. HOMOPTERA

A. *Aphis gossypii* Glov. (Figure 8)
Common name — cotton aphid.
Family — Aphididae.
Host plants — A polyphagous insect which has a wide host range.[1]
Distribution and injury — This aphid is found throughout the world. The damage consists of leaf distortion and leaves having a cupped appearance. Aphids are predominantly found on the underside of the leaves. The insect is a very important vector of virus diseases.[1,57,58]
Appearance and biology — Adult females exist in winged (alatae) or wingless (aptera) forms, 1 to 2 mm long, live for 2 to 3 weeks and give birth to living young. Females are blackish green with red eyes and antennae about half the body length. The wingless females are generally larger and paler. Young nymphs are greenish or brownish.[1]
Control — Aphid species often develop resistance to repeatedly used insecticides. However, alternating treatments of two or more of the following insecticides should control them: diazinon, disulfoton, endosulfan, malathion, mevinphos, naled, oxydemeton-methyl, parathion, and pirimor. Consult labels for dosage recommendations.[59,67] Naturally occurring predators and parasites can be nurtured to provide an effective means of aphid suppression through careful use of insecticides.

B. *Bemisia tabaci* (Genn.) (Figure 9)
Common name — tobacco whitefly, sweet potato whitefly.
Family — Aleyrodidae.
Host plants — Whiteflies attack over 100 spp. of plants. Many of the host plants belong to the families Compositae, Convolvulaceae, Cucurbitaceae, Euphorbiaceae, Leguminoseae, Malavaceae, and Solanaceae.[71]

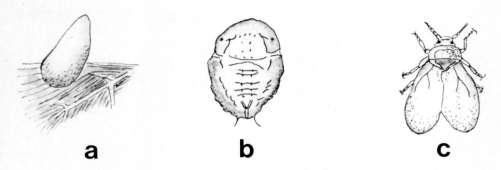

FIGURE 9. *Bemisia tabaci* (Genn.); egg (a); nymph (b); adult (c).

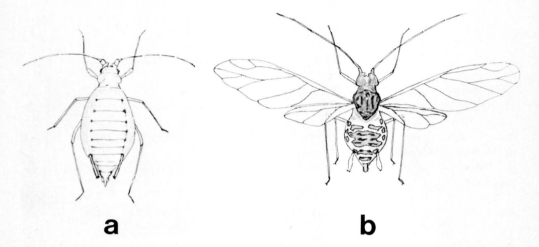

FIGURE 10. *Myzus persicae* (Sulzer); adult wingless (aptera) form (a); adult winged (alatae) form (b).

Distribution and injury — This insect is found throughout the world and when infestations are high severe plant damage results. Dry conditions are favorable for its development. Of significant concern is the ability of this insect to transmit virus diseases.[69,71-76]

Appearance and biology — The eggs are 0.2 mm long and oviposited on the under surface of the leaf. Newly laid eggs are white but eventually turn brown. Eggs are held onto the leaf by a small spine attached to the egg base which is inserted into the leaf stoma. The nymphs emerge in 3 to 7 days in warm weather and are active for 1 to 2 days after which they become fixed to the leaf surface. Nymphs undergo three instars in about 10 days. The late instar nymphs are greenish white (1.5 mm long). The red eyes of the adults can be seen through the integument of the pupae (1 mm long). Adults are 1 mm long with four wings and covered with a white waxy bloom. Females can lay an average of 160 eggs over a 16 day (mean) oviposition period. The duration of the pupal stage is 1 to 3 weeks. The complete life cycle is 2 to 4 weeks in warm environments.[72]

Control — Suppression of this insect is difficult because it rapidly becomes resistant to organic insecticides. The new synthetic pyrethroids show promise in controlling whiteflies. Consult labels for recommendations and dosages.[1,59-67]

C. *Myzus persicae* (Sulzer) (Figure 10)
Common name — green peach aphid.
Family — Aphididae.

Host plants — A polyphagous insect that feeds on peach, plum, apricot, cherry, many crops, ornamentals, and flowering plants.[1,3]

Distribution and injury — Leaf damage is characterized by distortion or leaf curling and other virus symptoms for which the aphid is the vector. The insect is mostly present on the undersides of the leaves.[58,68,69]

Appearance and biology — In temperate climates the insect passes the winter as an egg (shiny black) on the bark of peach, plum, apricot, and cherry. The young (yellowish-green) hatch in the spring about when peach trees bloom. When mature they begin giving birth to living young and remain on peach trees for 2 to 3 generations. Adults are 1 to 2 mm long. Adults acquire wings (aptera) and migrate to alternate host plants (crops) in late spring where reproduction continues until the fall. In the fall females fly back to the peach trees, where they give birth to the true sexual females. These females mate with males and lay eggs on the bark. In warm climates continuous reproduction occurs and there probably is no sexual generation.[3,70]

Control — Chemical control is not always practical because the aphid is often resistant to insecticides. Populations of naturally occurring parasites and predators can be encouraged through effective pest management practices where insecticides are applied only when needed. The following insecticides are used for control: diazinon, disulfoton, endosulfan, mevinphos, naled, oxydemeton-methyl, parathion, and pirimor. Consult labels for application rates.[59,67]

IV. LEPIDOPTERA

A. *Herse convolvuli* L. (Figure 11)

Common name — sweet potato moth.

Family — Sphingidae.

Host plants — This insect feeds on sweet potato, other Convolvulaceae, Leguminosae, and taro.[1,23]

Distribution and injury — The species has a wide distribution and is found in Africa, Europe, Middle and Far East, Australia, New Zealand, New Guinea, South China, and the Pacific Islands. Larvae severely defoliate production fields and plant beds.[1,11,19,23]

Appearance and biology — Adult females oviposit anywhere on the host plant. Eggs are laid singly and are 1 mm in diameter (subspherical). Larvae have 5 instars which take 3 to 4 weeks to develop. Mature larvae are large (95 mm × 14 mm) and body color varies from green to dark brown. Larvae have 7 oblique brown stripes laterally and a wavy white lateral line extending the full body length. A conspicuous horn is found on the last abdominal segment. Pupation is in the soil (8 to 10 cm deep). The duration of the pupal stage can be 17 to 26 days in warm climates and 4 to 6 months in colder climates. Adult moths are grey with black lines on the wings and broad incomplete bands on the abdomen with a wing span of 80 to 120 mm. Adults can be seen feeding on flowers at dusk *(Hibiscus, Ipomoea, Begonia)*.[1,23]

Control — Effective control can be obtained with trichlorphon and parathion.[1,20] Other economical insecticides that effectively control lepidoptera are carbaryl and *Bacillus thuringiensis* var. *kurstaki*.[77,78] Consult labels for correct application rates.

B. *Omphisa anastomosalis* (Guernée) (Figure 12)

Common name — sweet potato stem borer, sweet potato vine borer.

Family — Pyralidae.

Host plants — Sweet potato and wild *Ipomoea* spp. are the plants fed upon.[11]

Distribution and injury — This insect has been reported in India, Malaysia, and China. Where it occurs this pest is considered to be as destructive as the sweet potato weevil (*Cylas* spp.). Losses have been estimated to be up to 5 metric tons/ha. Plants are attacked by larvae

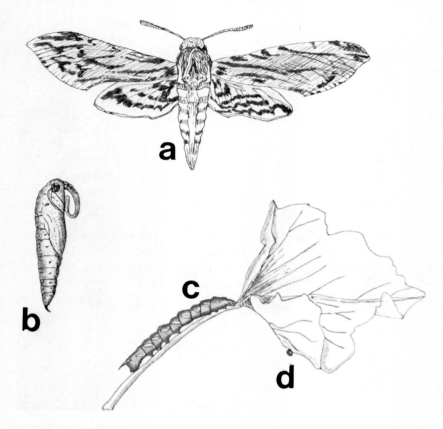

FIGURE 11. *Herse convolvuli* L.; adult (a); pupae (b); larvae (c); egg (d).

boring into the main stems leading to the roots. Severely damaged stems during early growth results in weak plant development. Larval feeding signs are lumps of granular or powdery fecal matter that accumulate around the stem base.[11,79]

Appearance and biology — Eggs are laid singly on exposed main stems attached to the roots. Eggs are flat, yellow-white and hatch in 5 to 8 days. Larvae eventually bore into the stems through the leaf axils and feed on stem tissue for 28 to 50 days. Mature larvae are 18 mm long and pinkish brown with a brown head. Conspicuous wart-like outgrowths arranged dorsolaterally and semicircular pale black patches dorsally are characteristic of the larvae. Mature larvae are found just above the roots where they construct cocoons for pupation. Pupation lasts 14 days. The emerging adults (28 mm wing span) mate and lay eggs within 3 days and live for only 7 days. The cycle from egg to adult emergence is 57 days.[79]

Control — Carbaryl provides effective protection against this pest. A total of 6 treatments, starting 2 weeks after planting and continued every 2 weeks is recommended.[79] Consult labels for dosage rates.

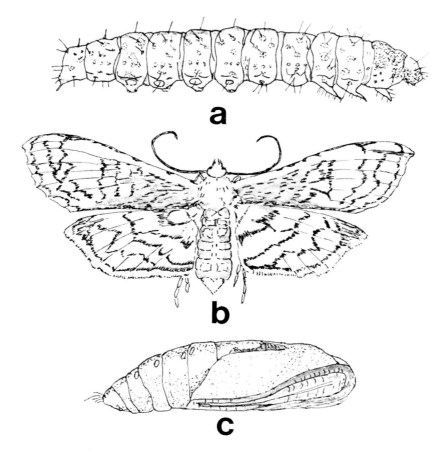

FIGURE 12. *Omphisa anastomosalis* (Guernée); larvae (a); adult (b); pupae (c).

REFERENCES

1. **Hill, D. S.**, *Agricultural Insect Pests of the Tropics and Their Control*, Cambridge University Press, New York, 1983, 198, 201, 202, 354, 457, 469, 472.
2. **Wolf, W. G.,** personal communication, 1983.
3. **Metcalf, C. L., Flint, W. P., and Metcalf, R. L.,** *Destructive and Useful Insects*, McGraw-Hill, New York, 1962, 511, 651, 754.
4. **Trehan, K. N. and Bagal, S. R.,** Life history, bionomics and control of sweet potato weevil *(Cylas formicarius* F.) with short notes on some other pests of sweet potato in Bombay State, *Indian J. Entomol.*, 19, 245, 1957.
5. **Visalakshi, A., Kashy, S. G., and Nair, M. R. G.,** Biological studies on *Aspidomorpha furcata* Thumb. (Chrysomelidae:Cassidinae:Coleoptera), *J. Entomol.*, 5, 167, 1980.
6. **Anonymous,** Agricultural Products, Chemical Guide, Union Carbide, Research Triangle Park, N. C., 1983.
7. **Cockerham, K. L., Deen, O. T., Christian, M. B., and Newsome, L. D.,** The biology of the sweet potato weevil, *La. Agric. Exp. Stn. Bull.*, No. 483, 1, 1954.
8. **Jayaramaiah, M.,** Bionomics of sweet potato weevil *Cylas formicarius* (Fabricius) (Coleoptera:Curculionidae), *Mysore J. Agric. Sci.,* 9, 99, 1975.
9. **Pierce, W. D.,** Weevils which affect Irish potato, sweet potato and yams, *J. Agric. Res.*, 12, 601, 1918.
10. **Wolf, W. G.,** personal communication, 1984.

11. **Ayyar, T. V. R.**, *Handbook of Economic Entomology for South India*, Government Press, Madras, 243, 1963.
12. **Akazawa, T. L., Uritani, I., and Kubata, H.**, Isolation of Ipomoeamarone and two coumarin derivatives from sweet potato roots injured by the weevil, *Cylas formicarius elegantulus*, *Agric. Biol. Chem. Biophy.*, 88, 150, 1960.
13. **Talekar, N. S.**, Effects of a sweet potato weevil (Coleoptera:Curculionidae) infestation on sweet potato root yields, *J. Econ. Entomol.*, 75, 1042, 1982.
14. **Talekar, N. S.**, Infestation of a sweet potato weevil (Coleoptera:Cucurlionidae) as influenced by pest management techniques, *J. Econ. Entomol.*, 76, 342, 1983.
15. **Uritani, I., Saito, T., Honda, H., and Kim, W. K.**, Induction of furanoterpenoids in sweet potato roots by the larval components of the sweet potato weevil, *Agric. Biol. Chem.*, 39, 1857, 1975.
16. **Rolston, L. H., Barlow, T. B., and Riley, E. G.**, Control of the sweet potato weevil in planting material, *La. Agric. Exp. Stn. Bull.*, No. 752, 1983.
17. **Reinhard, H. J.**, The sweet potato weevil, *Texas Agric. Exp. Stn. Bull.*, 308, 1, 1923.
18. **Mullen, M. A.**, Sweet potato weevil, *Cylas formicarius elegantulus* (Summers), development, fecundity and longevity, *Ann. Entomol. Soc. Am.*, 74, 479, 1981.
19. **Holdaway, F. G.**, Insects of sweet potato and their control, *Hawaii Agric. Exp. Stn. Prog. Notes*, 26, 7, 1941.
20. **Anonymous**, Sweet Potato Weevil Control in Centerpoint, Asian Vegetable Research and Development Center, Shanhua, Tainan, Taiwan, 1983.
21. **Trehan, K. N.**, Life history bionomics and control of sweet potato weevil *(Cylas formicarius* F) with short notes on some other pests of sweet potato in Bombay State, *Indian J. Entomol.*, 19, 240, 1957.
22. **Moreno, R. A.**, Intercropping with sweet potato *(Ipomoea batatas)* in Central America, in *Sweet Potato Proc. First Int. Symp.*, Villareal, R. L. and Griggs, T. D., Eds., Asian Vegetable Research and Development Center, Shanhua, Tainan, Taiwan, 1982, 25.
23. **Smee, L.**, Insect pests of sweet potato and Taro in the Territory of Papua and New Guinea, their habits and control, *Papua New Guinea Agric. J.*, 17, 99, 1965.
24. **Singh, R., Sinka, P. K., and Yadav, R. P.**, Evaluation of some granular insecticides in the control of sweet potato weevil *Cylas formicarius* F., *Proc. Bihar Acad. Agric. Sci.*, 25, 32, 1977.
25. **Cruz, C.**, Chemical control of the sweet potato weevil *Cylas formicarius*. in Puerto Rico, Breeding new sweet potatoes for the tropics in *Proc. Am. Soc. Hortic. Sci.*, Tropical Region, Martin, F. W., Ed., Mayaguez, Puerto Rico, 1983, 27, part B, 95.
26. **Subramaniam, T. R. and Gopalaswamy, P. V. S. R.**, Evaluation of certain new insecticides in the control of the sweet potato weevil, *Cylas formicarius* F., *Madrasque J.*, 60, 621, 1973.
27. **Das, N. M. and Nair, M. R. G. K.**, Control of the sweet potato weevil *Cylas formicarius* F. with some of the newer synthetic insecticides, *Agric. J. Kerola*, 4, 78, 1966.
28. **Waddill, V. H.**, Control of the sweet potato weevil *Cylas formicarius elegantulus* by foliar applications of insecticides, in *Sweet Potato Proc. First Int. Symp.*, Villareal, R. L. and Griggs, T. D., Eds., Asian Vegetable Research and Development Center, Shanhua, Tainan, Taiwan, 1982, Chap. 15.
29. **Pillai, K. S. and Magoon, M. L.**, Studies on chemical control measures for sweet potato weevil *Cylas formicarius* F., *Indian J. Hortic.*, 26, 201, 1969.
30. **Kung, S. P., Su, C. Y., and Rose, R. I.**, Sweet potato weevil: *Cylas formicarius elegantulus* (Sum), Control of sweet potato weevil, *Insecticide Acaracide Test* 1, Entomological Society of America, Philadelphia, 1975, 75.
31. **Schalk, J. M. and Ratcliffe, R. H.**, Evaluation of the USDA program on alternative methods of insect control. Host plant resistance to insects, *FAO Plant Prot. Bull.*, 25, 9, 1977.
32. **Cuthbert, F. P. and Jones, A.**, Insect resistance as an adjunct or alternative to insecticides for control of sweet potato soil insects, *J. Am. Soc. Hortic. Sci.*, 103, 443, 1978.
33. **Rolston, L. H., Barlow, T. B., Hernandez, T., Nilakhe, S. S., and Jones, A.**, Field evaluation of breeding lines and cultivars of sweet potato for resistance to the sweet potato weevil, *HortScience*, 14, 634, 1979.
34. **Barlow, T. B. and Rolston, L. H.**, Types of host plant resistance to the sweet potato weevil found in sweet potato roots, *J. Kans. Entomol. Soc.*, 54, 649, 1981.
35. **Cockerham, K. L. and Harrison, P. K.**, New sweet potato seedlings that appear resistant to sweet potato weevil attack, *J. Econ. Entomol.*, 45, 132, 1952.
36. **Hahn, S. K. and Leuschner, K.**, Resistance of sweet potato cultivars to African sweet potato weevil, *Crop Sci.*, 21, 499, 1981.
37. **Jones, A., Dukes, P. D., Schalk, J. M., Mullen, M. A., Hamilton, M. G., Paterson, D. R., and Boswell, T. E.**, W-71, W-115, W-119, W-125, W-149 and W-154 sweet potato germplasm with multiple insect and disease resistance, *HortScience*, 15, 835, 1980.

38. **Jones, A., Dukes, P. D., Schalk, J. M., Hamilton, M. G., Mullen, M. A., Baumgardner, R. A., Paterson, D. R., and Boswell, T. E.,** Recent release of "Regal", personal communication, 1984.
39. **Waddill, V. H. and Conover, R. A.,** Resistance of white fleshed sweet potato cultivars to the sweet potato weevil, *HortScience*, 13, 476, 1978.
40. **Mullen, M. A., Jones, A., Arborgast, R. T., Schalk, J. M., Paterson, D. R., Boswell, T. E., and Earhart, D. R.,** Field selection of sweet potato lines and cultivars for resistance to the sweet potato weevil, *J. Econ. Entomol.*, 74, 288, 1980.
41. **Mullen, M. A., Jones, A., and Arborgast, R. T.,** Resistance of sweet potato lines to infestations of sweet potato weevil *Cylas formicarius elegantulus* (Summer), *HortScience*, 16, 539, 1981.
42. **Mullen, M. A., Jones, A., Paterson, D. R., and Boswell, T. E.,** Resistance of sweet potato lines to the sweet potato weevil, *HortScience*, 17, 931, 1982.
43. **Pillai, K. S. and Kamalam, P.,** Screening sweet potato germplasm for weevil resistance, *J. Root Crops*, 3, 65, 1977.
44. **Franssen, C. J. H.,** *Insect Pests of Sweet Potato Crop in Java*, Korte Meded, Inst. Pezilkt, Buitenzorg, 1934, 20.
45. **Kryson, J. L. and Branson, T. F.,** Biology, ecology and distribution of *Diabrotica*, *Proc. Inter. Maize Virus Disease Colloquium and Workshop*, The Ohio State University, Ohio Agricultural Research and Development Center, Wooster, August 2 to 6, 1983, 1-7 pp.
46. **Risch, S. J.,** A comparison by sweep sampling of the insect fauna from corn and sweet potato monocultures and dicultures in Costa Rica, *Oecologia*, 42, 195, 1979.
47. **Christensen, J. R.,** Study of the genus *Diabrotica* Chev. in Argentina, *Rev. Fac. Agron. Vet.*, 10, 464, 1943.
48. **Cuthbert, F. P.,** Insects Affecting Sweet Potatoes, Handbook No. 329, U.S. Department of Agriculture, Washington, D.C., 1967, 28.
49. **Jones, A., Dukes, P. D., Schalk, J. M., Hamilton, M. G., Mullen, M. A., Baumgardner, R. A., Paterson, D. R., and Boswell, T. E.,** "Resisto" sweet potato, *HortScience*, 18, 251, 1983.
50. **Sherman, M. and Tamashiro, M.,** The sweet potato weevil in Hawaii, their biology and control, *Hawaii Agric. Exp. Stn. Tech. Bull.*, 23, 36, 1954.
51. **Parasram, S.,** The scarabee a major pest of sweet potato, *Caribbean Farming*, 2, 18, 1970.
52. **Tucker, R. W. E.,** The control of scarabee *(Euscapes batatas* Waterh.) in Barbados, *Agric. J. (Bridgetown Barbados)*, 6, 133, 1937.
53. **Wan, H.,** Cropping systems involving sweet potato in Taiwan, in *Sweet Potato Proc. First Int. Symp.*, Villareal, R. L. and Griggs, T. D., Eds., Asian Vegetable Research Development Center, Shanhua, Tainan, Taiwan, 1982, chap. 23.
54. **Monterio, D. A., Parra, J. R. P., Cavalcante, R. D., and Jque, T.,** Controle da broca da batata - doce *Euscepes postfasciatus* (Fairmaire 1849) (Coleoptera, Curculionidae), com insecticides modernos, *Biologica*, 38, 204, 1972.
55. **Roberts, R. J.,** An introduced pasture beetle *Plectris aliena* Chapin (Scarabeidae: Melolonthinae), *J. Aust. Entomol. Soc.*, 7, 15, 1968.
56. **Santoro, F. H., Bezzi, A., Vigevanoy, A., and Cantos, F.,** Biologia del Negrito de la Batata *Typophorus nigritus nitidulus* (F) Y Ensayo Preliminar Sobre Control Quimico de Adutos (Coleoptera-Chrysomelidae-Eumolpinae), Instituto Nacional de Tecnologia Agropecuario Centro de Investigaciones de Recursos Naturales, de Idia N 373-378 Enero 1979.
57. **Arena, O. B. and Nwankiti, A. O.,** Sweet Potato Disease in Nigeria, *PANS*, 24, 294, 1978.
58. **Mahto, D. N. and Sinba, D. C.,** Mosaic disease of sweet potato and its transmission, *Indian J. Entomol.*, 40, 443, 1978.
59. **Anonymous,** Sample Labels, Agricultural Division, Ciba-Geigy Corp., Greensboro, N.C., 1981, 139 pp.
60. **Anonymous,** Product Guide, Mobay Chemicals, Kansas City, Mo., 1981, 441 pp.
61. **Anonymous,** Product Label Guide, FMC Corp., Agricultural Chemical Group, Philadelphia, 1983, 55 pp.
62. **Anonymous,** Pesticide Products Guide, American Cyanimide, Princeton, N.J., 1980, 187 pp.
63. **Anonymous,** Agricultural Chemicals Label Book, Shell Chemical Co., Houston, 1984, 102 pp.
64. **Anonymous,** Product Guide, Ortho-Chevron Chemical Co., San Francisco, 1982.
65. **Anonymous,** Product Guide, Mobay Chemicals, Kansas City, Mo., 1984, 441 pp.
66. **Anonymous,** Crop Chemical Sample Label Guide, Monsanto Co., St. Louis, Mo., 1981, 64 pp.
67. **Anonymous,** Product Label Guide, ICI Protection Division, Wilmington, Del., 1983.
68. **Nome, S. F.,** Sweet potato vein mosaic virus in Argentina, *Phytopathol. Z.*, 77, 44, 1972.
69. **Sheffield, F. M. L.,** Virus diseases of sweet potato in East Africa. I. Infestation of the virus and their insect vectors, *Phytopathology*, 47, 582, 1957.
70. **Bodenkeimer, F. S. and Swirski, E.,** *The Aphidoidea of the Middle East*, The Weizmann Press of Israel, Jerusalem, 1957, chap. 4.

71. **Azab, A. K., Megahed, M. M., and El-Mirsawi, D. H.**, On the range of host-plants of Bemisia tabaci (Genn), *Bull. Entomol. Soc. Egypt*, LIV, 319, 1970.
72. **Azab, A. K., Megahed, M. M., and El-Mirsawi, D. H.**, On the biology of *Bemisia tabaci* (Genn), *Bull. Entomol. Soc. Egypt*, 55, 305, 1971.
73. **Hollings, M. and Stone, O. M.**, Purification and properties of sweet potato mild mottle, a whitefly borne virus from sweet potato *(Ipomoea batatas)* in East Africa, *Ann. Appl. Biol.*, 82, 511, 1976.
74. **Liao, C., Chien, I., Chung, M., Chiu, R., and Han, Y.**, A study of sweet potato virus diseases in Taiwan. I. Sweet potato yellow spot virus diseases, J. Agric. Res. China, 28, 127, 1979.
75. **Schaefers, G. A. and Terry, E. R.**, Insect transmission of sweet potato disease agents in Nigeria, *Phytopathology*, 66, 642, 1976.
76. **Steinbauer, C. and Kushman, L. J.**, Sweet Potato Culture and Diseases, *Handbook* No. 388, U.S. Department of Agriculture, Washington, D.C., 1971.
77. **Anonymous,** Chemical and Agricultural Products Division, Abbott Laboratories, Chicago, 1983.
78. **Subramanian, R. R., David, B. V., Thangavel, P., and Abraham, E. V.**, Insect pest problems of tuber crops in Tamil Nadu, *J. Root Crops*, 3, 43, 1977.
79. **Hua, H. T.**, Studies on some major pests of sweet potatoes and their control, *Malaysian Agric. J.*, 47, 437, 1970.

Chapter 4

THE PHYSIOLOGY OF YIELD IN THE SWEET POTATO

Stanley J. Kays

TABLE OF CONTENTS

I.	Reproductive Biology of the Sweet Potato			80
II.	Plant Architecture			81
	A.	The Root System		82
		1.	Root Types	82
			a. Storage Roots	83
			b. Primary Fibrous Roots	85
			c. Pencil Roots	85
			d. Lateral Roots	86
		2.	Root Distribution and Architecture	86
		3.	Carbon Allocation Within the Root System	88
	B.	Aboveground Plant Parts		88
		1.	Vine and Internode Length	89
		2.	Branching	89
		3.	Petiole Length	89
		4.	Leaf Number, Area, and Shape	90
III.	Acquisition of Energy			90
	A.	Plant Factors		95
		1.	Leaf Chlorophyll Concentration	95
		2.	Leaf Stomata	95
		3.	Leaf Geometry	96
		4.	Leaf Age	96
		5.	Leaf Starch and Sugar Concentration	96
	B.	Environmental Factors		97
IV.	Transport of Energy			98
	A.	Phloem Loading		99
	B.	Phloem Transport		99
	C.	Phloem Unloading		100
	D.	Modeling Transport		102
V.	Allocation of Energy			103
	A.	Allocation Within the Plant		103
	B.	Sink Strength in Relation to Allocation		103
	C.	Recycling of Carbon		104
VI.	Respiratory Losses of Energy			104
	A.	Photorespiration		104
	B.	Dark Respiration		105
	C.	Respiratory Rate as a Function of Plant Part and Age		107
	D.	Respiratory Quotients and Plant Energy Input/Loss Ratios		109

VII. Storage of Energy...110
    A. Storage Root Initiation 110
    B. Storage Root Induction 111
    C. Storage Root Development 111

VIII. Growth Analyses..113
    A. Plant Area and Volume Measurements 113
        1. Leaf Area..113
        2. Leaf Area Index 114
        3. Leaf Area Duration 114
        4. Storage Root Volume 115
    B. Plant Weight Measurements...116
        1. Biological Yield 116
        2. Harvest Index...116
    C. Rate Measurements..116
        1. Crop Growth Rate 119
        2. Relative Growth Rate 119
        3. Net Assimilation Rate...121

IX. Environmental Factors Affecting Yield 122
    A. Radiation 122
    B. Temperature..122
    C. Length of Growing Season 122
    D. Rainfall..123

X. Production Factors Affecting Yield 123
    A. Soils..123
        1. Soil Fertility 123
        2. Soil pH 124
        3. Soil Type 124
        4. Soil Oxygen Concentration 124
        5. Ridging or Mounding...124
    B. Irrigation 124
    C. Plant Population...125
    D. Length of Growing Period...125
    E. Insects, Diseases, and Nematodes 126
    F. Plant Growth Regulators 126

XI. Conclusions..126

References..126

I. REPRODUCTIVE BIOLOGY OF THE SWEET POTATO

In nature, plants have evolved energy allocation strategies that tend to maximize their survival potential. The energy acquired through photosynthesis can be partitioned into two general directions, that needed for maintenance of the plant and that allocated to production (foliage and reproductive parts). Through the process of natural selection, wild species have developed favorable energy allocation cost-benefit ratios for the environmental conditions under which they evolved. With the onset of agriculture man began to domesticate a number of plant species that greatly altered this balance. Thus the process of artificial selection resulted in significant changes in the energy allocation patterns within these domesticated species.

The sweet potato represents an excellent example of the effect of the intervention of man on the reproductive (i.e., energy allocation) strategy of a species. While there are numerous wild species of *Ipomoea* (estimated at over 400), *Ipomoea batatas* (L.) Lam., a "cultigen", is not found in the wild. In fact, with the exception of short-term escapes from cultivation, the species does not appear to effectively compete and survive in the wild. Nevertheless, viewing the sweet potato as one would a wild species, i.e., as a reproductive unit with a specific energy allocation strategy for a given set of environmental conditions, allows us to more thoroughly understand the plant as it currently exists.

The sweet potato can reproduce by one of three primary means. The plant can reproduce and colonize an area by allocation of carbon (energy) acquired through photosynthesis into storage roots, which subsequently sprout to produce new plants. It can also reproduce by allocating a major proportion of its nonmaintenance energy into vines, which under suitable conditions form roots readily at the nodes, producing daughter plants. Lastly, and only of minor importance in a numerical sense, is sexual reproduction via the formation of seed. Very little energy is allocated for sexual reproduction in the sweet potato. This contrasts sharply with species such as *I. aquatica* Forsk. where significantly more of the carbon budget of the plant is allocated for seed production.

II. PLANT ARCHITECTURE

When reduced to its most basic roles, a sweet potato plant can be divided into three component parts each with distinctly different functions. These include the photosynthetic canopy which absorbs light energy and converts it into a manageable chemical form (carbon compounds), the petioles and vines which are responsible for the transport of this energy and of resources acquired by the root system throughout the plant, and the root system which absorbs water and nutrients and anchors the plant. In addition, the root system through the formation of storage roots also serves as a storage site for the energy accrued above and beyond that required for maintenance and structural development. Each part of the plant and their respective functions are closely integrated. Changes in one can result in significant and often quite rapid alterations in the others. These interrelationships and the precise mechanisms controlling them are not yet fully understood, however, gross effects have been well documented in the literature on sweet potatoes.

It is evident, based on the net energy allocation patterns in the plant at harvest, that the interrelationship between the aboveground plant parts and the root system is distinctly different between the sweet potato, and that found in most domesticated food plants. Root and tuber crops, endemic to the tropics and subtropics, developed survival and reproductive strategies that involved storage of accrued energy in fleshy storage organs. This was further exaggerated under domestication. This strategy is in sharp contrast to the majority of domesticates evolving in the temperate zone (cereals, legumes). There, energy was sequestered into seeds that were capable of surviving both dry and/or cold conditions during the nongrowing season. This difference in reproductive strategy greatly affects the relationship between the above and below ground portions of the plant. Distinctly different energy allocation requirements and, therefore, patterns emerged.

It is important to note that the allocation pattern for energy within the plant is not static, but is in a relatively constant state of flux. For example, diurnal changes are marked. A different set of conditions is operative at night when carbon dioxide is not being fixed and carbon temporarily stored in the leaf as starch during the day is being remobilized, than during peak periods of photosynthesis during the day. These allocation patterns are also subject to change as the plant itself changes during the growing season. In addition, energy allocation patterns are altered in response to both short and long term environmental factors — e.g., water stress, insufficient oxygen in the root zone, excess soil nitrogen, and so on. What we measure as yield represents the integration of these changing patterns over the entire production season.

It is important then, to first ascertain where the plant allocates its resources. To do this, one must have some understanding of the general architecture of the plant, both above and below ground.

A. The Root System

While there have been a number of anatomical studies on the sweet potato root system, it is surprising in light of the fact that a portion of the root system, the storage roots, comprises the yield component of agronomic interest, that a more precise understanding of the general architecture has not been developed. Perhaps the most detailed study to date was by Weaver and Bruner[1] in 1927.

Roots can be classified based upon a number of morphological, physiological, or biochemical parameters. For example, the root system of the sweet potato could be separated into components based on function, geotropic response, presence or absence of pigmentation, point or time of origin, specific morphological characteristics and so on. The following description of the root system is based upon information from several studies.

1. Root Types

Several approaches have been utilized to describe the types of roots comprising the root system of the sweet potato. This has resulted in a less than clear picture of the actual structural architecture of the below ground plant parts. The following description divides the root system initially into two basic groups based on origin. These are (a) the adventitious roots — those arising from the underground stem portion of a vine cutting or transplant or from a root piece when used for propagation and (b) lateral roots — those arising from existing roots. These two general groups are then subdivided into the following classes.

Adventitious roots
  1. storage roots
  2. primary fibrous roots
  3. pencil roots
Lateral roots
  1. primary
  2. secondary
  3. tertiary

During the early ontogeny of young adventitious roots emerging from the stem, they may be separated into two classes: "thin" roots and "thick" roots.[2] The former arise primarily from internodal regions of the underground stem while the latter develop from the nodal area. "Thin" roots are typically tetrarch in the arrangement of their primary vascular tissue, i.e., four xylem and phloem poles found within the vascular cylinder (Figure 1). The "thick" roots, however, are pentarch or hexarch in structure (Figure 1). Both penetrate out from the underground stem either horizontally or obliquely in the soil.

The development of young adventitious roots into either "thick" or "thin" roots depends upon both the above and below ground environment at the time of their development. Research by Togari[2] indicates that young adventitious roots represent the precursors for the

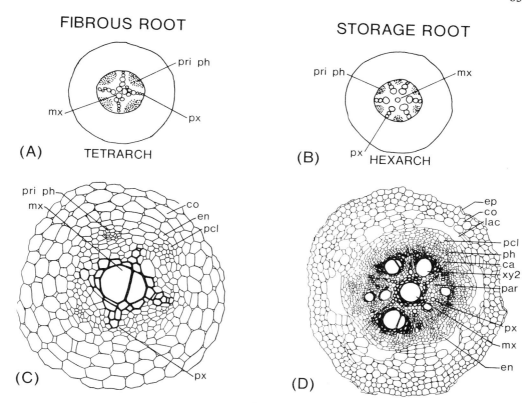

FIGURE 1. Cross-sections of a sweet potato fibrous root showing (A) a tetrarch stele, and (C) components parts; and a young storage root showing (B) a hexarch stele and (D) component parts: ca, primary cambium; co, cortex; en, endodermis; ep, epidermis; lac, lacuna; mx, metaxylem; PAR, parenchyma; PCL, pericycle; ph, phloem; PX, protoxylem and XYZ, secondary xylem. (C and D after Artschwager, E., *J. Agric. Res.*, 23, 157, 1924.)

development of storage, pencil, and primary fibrous roots (Figure 2). If the environment is conducive, young "thick" roots develop into storage roots. Under conditions of high nitrogen or low soil oxygen these roots develop into primary fibrous roots, i.e., the principal conductive system for the fibrous lateral roots that emerge from them. The effect of low oxygen (e.g., excess soil moisture) on this shift is quite pronounced. Plants grown for 81 days at 2.5% root zone oxygen concentration had 88.7% of their root dry weight as nonstorage roots (principally fibrous) while plants grown at 21% root zone oxygen had only 10.9% of their total root dry weight partitioned into nonstorage roots.[3] Hence the root zone environment is a major modulator in this shift between young adventitious roots and "thick" and "thin" roots and subsequently to storage, primary fibrous, or pencil roots.

Young "thin" adventitious roots appear to develop into either primary fibrous roots or in some cases pencil roots. Under dry compacted soil conditions young "thick" adventitious roots begin to enlarge, however, this is terminated by rapid lignification of the stele.[2] The result is the formation of pencil roots which are larger in diameter than primary fibrous roots, but much smaller than storage roots.

a. Storage Roots

Storage roots arise from pentarch or hexarch thick young roots if the cells between the protoxylem points and the central metaxylem cell do not become lignified,[4,5] or if only a slight proportion of these cells are lignified.[2]

The increase in size of the storage root is attributed to the activity of the vascular cambium

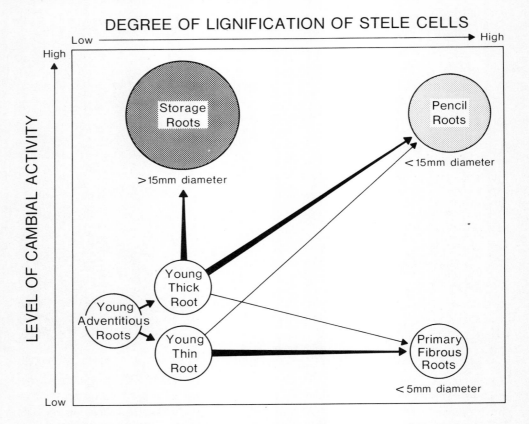

FIGURE 2. The development of young adventitious roots into the three basic adventitious root types found in the sweet potato. The eventual fate appears to be related to the level of cambial activity and the degree of lignification of the stele cells. (Adapted from Togari, Y., *Nat. Agric. Exp. Stn. Bull. (Tokyo)*, 68, 1950.)

as well as the activity of the anomalous cambia.[4,5] Phellogen activity on the outer surface of the storage root gives rise to the periderm,[4,6] although the periderm contributes little to the overall bulk of the storage root at harvest.

The appearance of anomalous cambia is a general phenomenon. In addition to appearing around the protoxylem groups and the central metaxylem cell, anomalous cambia will also arise in close proximity to xylem differentiated from the vascular cambium. New cambia may also arise from the original phloem groups. Later in the development of the storage root, cambia may arise independently of vascular groups. New cambia may also develop around xylem elements produced from anomalous cambial activity.[4]

The contribution of the vascular cambium relative to the anomalous cambia to increases in the volume of the storage root is a cultivar characteristic.[5] This is based on the observation that xylem elements differentiated from the vascular cambium will show a regular arrangement in radial rows.[4]

The information available on longitudinal growth of the storage root[5] indicates that most of the longitudinal growth occurs towards the distal end in the initial stages; later growth at this end is delimited with subsequent longitudinal development in the direction of the proximal end.[7]

The proximal end of the storage root is connected to the plant by a storage root "stalk". Generally the stalk is only 10 to 15 cm in length and the storage roots are found in the upper portion of the bed clustered around the stem. Occasionally, under certain conditions,

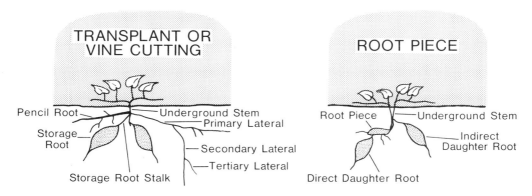

FIGURE 3. The primary components of the sweet potato root system developed from transplants or vine cuttings and from root pieces.

some cultivars will form storage roots on extended stalks as much as 90 to 150 cm away from the underground stem or quite deep in the soil.[8,9]

At the distal end of the storage root the stelar arrangement is often found to be tetrarch.[5] This portion of the root continues to grow with a similar geotropic orientation downward into the soil forming a root system similar to the primary fibrous roots. Thus, a considerable volume of fibrous roots occurs below the storage root. In addition, a relatively large number of lateral roots emerge directly from the storage root *per se*.

When root pieces (sections of storage roots or small storage roots) are used as a means of propagation instead of vines or transplants, a slightly different pattern of storage root development occurs (Figure 3). Here the storage root may emerge directly from the root piece (termed direct daughter root), or indirectly through an adventitious root emerging from the below ground portion of the stem (termed indirect daughter root) as in vine cuttings and transplants. In some cases, the root piece enlarges with little or no carbon allocation into new storage roots.[10,11] When root pieces are used they retain their proximal dominance, i.e., sprouts emerge from the end of the root piece that was originally closest to the stem, regardless of how small the root piece is sectioned.[12]

It should be noted at this point that the storage organ of the sweet potato is in fact a root and not a tuber.[4] This fact appears to be largely disregarded, even by some sweet potato anatomists. Storage roots and tubers differ substantially in both anatomy, the former being a root and the latter a modified stem, and in general physiology. It follows then that the expression "tuberization" is quite incorrect when applied to the sweet potato. Rather the correct terminology would be storage root enlargement or development.

b. Primary Fibrous Roots

Primary fibrous roots emerge largely from tetrarch "thin" adventitious roots (Figure 2) although under adverse conditions they may be from pentarch, hexarch, and even septarch "thick" roots. Generally, these are less than 5 mm in diameter and are branched and rebranched with lateral roots forming a dense network throughout the root zone. Primary fibrous roots have heavily lignified steles and very low levels of vascular cambial activity. Their formation is favored by high nitrogen and low oxygen within the root zone.[3,13]

c. Pencil Roots

Pencil roots, generally between 5 and 15 mm diameter, are the least well defined of the adventitious roots emerging from the underground stem of the plant. They develop primarily from young thick adventitious roots under conditions which are not conducive for the development of storage roots,[2] for example dry, compacted soils. In some cases, they may also apparently be derived from thin roots. In pencil roots, lignification of the stele is not total, so that limited meristematic activity can arise with the remaining stelar parenchyma, concurrent with secondary thickening due to the vascular cambium. Like the primary fibrous roots, these roots also give rise to numerous lateral roots.

d. Lateral Roots

The lateral roots of the sweet potato emerge from existing roots, thus each type of adventitious root (fibrous, storage, and pencil) have a profusion of lateral roots at varying densities along their axis. The primary lateral roots emerge from adventitious roots at approximately 90° angles. Many of these primary laterals grow profusely downward into the soil forming secondary laterals (laterals emerging from primary laterals) also emerging at approximately a 90° orientation to the axis of the primary lateral. In some cases, tertiary laterals form (laterals emerging from secondary laterals). Thus, the lateral roots of the sweet potato fill much of the soil volume and form the water and nutrient absorbing system of the plant.

2. Root Distribution and Architecture

The study of plant root systems is difficult and laborious even under the most favorable of conditions. As a consequence, our understanding of the general morphology and physiology of the root system of most crop plants has lagged behind that of the aboveground plant parts. Perhaps the most thorough study of the general architecture and distribution of the root system of the sweet potato was published in 1927. Weaver and Bruner[1] excavated the roots of plants of the cultivar Yellow Jersey planted at a field spacing of 38 × 122 cm. Measurements were made at three stages during the development of the plants: early — 23 days after transplanting; midsummer — 55 days after transplanting; and fall — 125 days after transplanting. The following is a brief summary of their findings.

By the 23rd day after transplanting the sweet potato slips, the plants had 71 adventitious roots, generally 1 mm in diameter, that had developed from the underground portion of the stem (~ 12.7 cm in length). These extended outward obliquely from the stem at angles of 45° or less. Many were very short (~ 2.5 cm), however, others extended to a maximum spread of 10 cm and reached a depth of 43 cm in the soil. Lateral roots arose at 90° angles from these roots at a rate of about 2.4/cm. Some of the longest of these laterals had begun to branch. The aboveground portion of the plant had an average of 30 to 35 leaves.

Fifty-five days after planting, the sweet potato plants had started to develop storage roots and had a root system with a radius of 50 to 76 cm and reached a maximum depth of 104 cm (Figure 4). Twelve to 16 adventitious roots from the underground portion of the stem were 1.5 to 4 mm in diameter. Storage roots 5 to 10 cm long and 5 to 13 mm in diameter were developing on part of the adventitious roots. Many roots ran horizontally out from the plant 60 cm or more, branching freely but terminating in the top 15 to 20 cm of soil. Others had developed obliquely outward and downward, however, remaining in the top 30 cm of soil (Figure 4). From these arose numerous long lateral roots penetrating vertically to a depth of 60 to 75 cm. These were branched with short secondary laterals, and occasionally tertiary laterals. Laterals typically formed at a rate of 1.5 to 3.2/cm. The aerial portion of the plant had 12 to 19 stems ranging from 0.3 to 1.4 m with as many as 400 to 450 leaves per plant.

By the final sampling date, 125 days after planting, individual plants had approximately 12 storage roots found within the upper 15 to 23 cm of soil, the longest being 5 cm thick and 13 cm long. The storage root stalks were only 8 cm long, but contained a number (8 to 15) of lateral roots extending horizontally outward. The entire root system by this time extended outward from the plant 1.8 m and downward to a working depth of 1.3 m (1.8 m maximum depth).

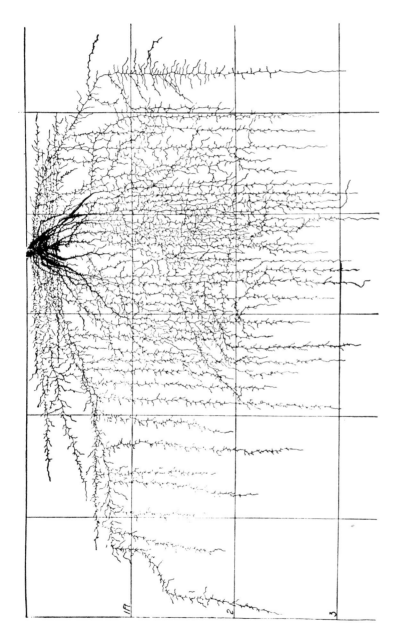

FIGURE 4. The root architecture of the sweet potato (cultivar Yellow Jersey) 55 days after planting. (From Weaver, J. E. and Bruner, W. E., *Root Development of Vegetable Crops*, McGraw-Hill, New York, 1927, 351. With permission.)

The storage roots had 1.2 to 3.2 laterals per centimeter which extended horizontally outward for 20 to 30 cm, having 1.2 to 2.4 secondary laterals per centimeter. In addition, typically about 3 large laterals ($\sim$ 3 mm in diameter) extended horizontally outward from the storage roots 30 to 45 cm, then penetrated vertically to 1.2 m or more. At the distal end of the storage roots, the root was 3 to 4 mm in diameter and continued downward. Emerging from this root, near the base of the storage root were 1 or 2 large lateral roots which had developed horizontally 25 to 45 cm then penetrated downward to a depth of 1.2 m or more. Secondary laterals from these were 2.5 to 15 cm long and had short tertiary laterals.

Adding to the root volume of the plant by this date were roots developing at the nodes of the aboveground vines. As many as 10 to 15 roots, usually in 2 rows had developed from single nodes, with 3 to 4 such clumps per 30 cm of stem. Some nodes, however, had only a single root. Most nodal roots were only 1 mm in diameter although occasionally they were up to 5 mm. These extended vertically downward, as much as 1 m into the soil. Roots arising in clumps at the nodes were profusely branched (4.8 to 7.2/cm) in the surface 15 to 20 cm of soil. This greatly extended the plant's effective water and nutrient absorbing zone. one cultivar at one spacing on one soil type. Additional research altering agronomic (cultivar, fertility, spacing, irrigation, soils) and climactic (temperature, light intensity, photoperiod) parameters would greatly expand our limited present understanding of the root system.

3. Carbon Allocation Within the Root System

Most studies of the sweet potato that indicate root dry matter reflect only the storage roots. Some data are available from studies where the plants were grown in an artificial media within containers or in a solution culture. This, however, probably restricts the actual fibrous root system due in part to the limited soil (media) volume and the abundance of water and nutrients within the restricted zone. Therefore, these data can only be used as an approximation of the possible carbon allocation in plants grown under field conditions. It probably underestimates the total allocation of carbon to the fibrous root system (pencil, primary fibrous, and lateral roots).

Field measurements by Yoshida et al.[14] indicated that at harvest the fibrous roots accounted for only 1.1% of the total root dry weight and 0.7% of the total plant dry weight. As the plants developed during the growing season, the dry weight of fibrous roots increased while its percent of the total root and plant dry weight decreased markedly.

B. Aboveground Plant Parts

The aboveground portion of the sweet potato plant is comprised of the leaves, which absorb light energy converting it into carbohydrates through the fixation of atmospheric carbon, and the leaf petioles and stems which form the conduits for transport of this carbon throughout the plant. The stems and petioles also determine the spacial arrangement of the leaves within the canopy. The general architecture of the canopy is critical in maximizing the light reception by the plant. Sweet potato genotypes are predominantly prostrate vining plants, and in contrast to most agricultural crop plants, they establish a largely horizontal and shallow canopy. Other species, like *Zea mays,* exhibit a more vertical canopy development. This affords the sweet potato, as well as other vining species, rapid horizontal expansion and a highly plastic canopy. As a consequence, the sweet potato is able to maximize its growth in areas where the sunlight has not been exploited, thus rapidly maximizing the plant's reception of incoming radiation. This advantage, however, is partially negated by the lack of canopy depth, seen in the low leaf area index of the plant (area of leaves per area of land). When the vines are tied up to wire mesh (1.2 to 1.5 m) increasing the light exposure of the canopy, yield is increased.[15]

A number of studies have related integrated measurement of the aboveground plant parts (e.g., plant part weight or leaf area/acre of land)[16,19] to growth and yield. None, to my knowledge, have precisely determined the spatial-weight relationships within the canopy. Critical components which determine the spatial arrangement or architecture of the above-ground plant parts with reference to light interception are vine and internode length, branching, petiole length, and leaf number, area, orientation, and shape.

1. Vine and Internode Length

Sweet potato genotypes are generally classified as either bunch (bush), intermediate, or vining types based on the length of their vines at harvest. Yen[20] expanded this to six classifications, however, his criteria were based more on general habit than vine length alone. Classes I and II exhibited a compact bunch or bush habit, classes III and IV had a greater vine spread and classes V and VI exhibited extended nodes and long vines (criteria for each class were not given).

Yen[20] found that vine length varied widely in a large population of sweet potato selections collected in South America and Oceania. The length of vines ranged from approximately 0.5 to 4 m and varied to some degree with the geographical area in which the plant was collected. For example, selections from western South America typically had shorter vines than those from the island of Timor. Internode lengths were also highly variable ranging from several centimeters up to 10 cm in length. Somda and Kays[21] determined the precise internode and vine length in the cultivar Jewel grown at three plant densities (15, 30 or 45 cm × 98 cm). Initially, internode lengths were relatively short, however, as the vines began to develop, internode length increased substantially (Figure 5). Planting density had a significant effect on the internode length as well as a pronounced effect on vine mean length. At lower plant densities the vines were longer. Hence plant populations would be anticipated to have a subsequent effect on the spatial relationships within the canopy.

2. Branching

Branching of the sweet potato vines is known to vary widely with genotype. This includes not only the number, but also the distances of the branches outward from the crown of the plant. Most plants exhibit considerable primary and secondary branching, however, generally tertiary branching is limited. In the cultivar Jewel, the number of branches increased dramatically as the plant population per hectare decreased.[21] The branching pattern and vine lengths of representative plants after 126 growing days are displayed in Figure 6.

3. Petiole Length

The length of an individual petiole determines the relative level of the leaf within the canopy's light hierarchy. Leaves with short petioles within a canopy of long petioles (on the same plant or neighboring plants) are at a distinct competitive disadvantage. Petiole length varies widely with genotype. These have been shown to range from approximately 9 up to 33 cm in length.[20] In the early stages of development of the canopy of a genotype, the petiole length is at its minimum.[21] Toward the middle and latter part of the growing season, the petiole length increases substantially, with the height of the overall canopy increasing correspondingly (Figure 5). Late in the growing season, leaves at the base of the vines begin to shed. These represent the oldest leaves within the canopy, but also typically the lowest physically in the light reception hierarchy. On a single branch are apical petioles that are still elongating, petioles of maximum length, shorter basipetal petioles, and those that have already been shed. Occasionally (e.g., Figure 5, 15 cm spacing) a short petiole is found within a series of long petioles. This typically supports a smaller leaf and is adjacent to an internode that is likewise repressed in length. The stem and petiole geometry appears to present a fairly accurate permanent record of the growth conditions experienced by the vine at each chronological stage of development.

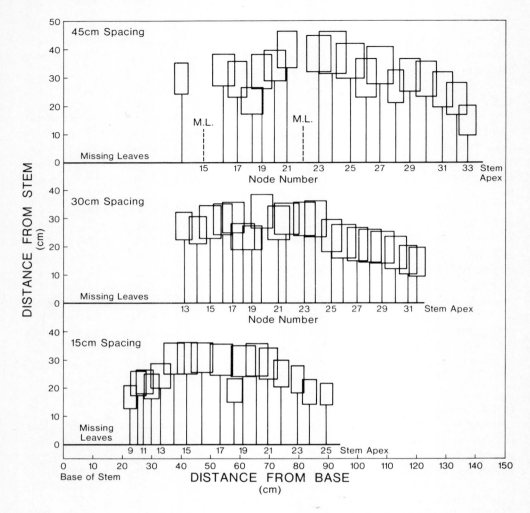

FIGURE 5. Architecture of representative vine of the sweet potato (cultivar Jewel) 126 days after planting at three in-row plant spacings (15, 30, or 45 cm × 96 cm). (From Somda, Z. C. and Kays, S. J., manuscript in preparation, 1985.)

4. Leaf Number, Area, and Shape

Leaf area and shape vary widely among genotypes. Yen[20] distinguished 15 different leaf types. Their shape ranged from broad entire to highly loded (Figure 7). As a plant develops during the growing season, leaf width, length and area increase (Figure 5)[21] and the leaf area/unit dry weight decreases.[22] Therefore, the greater leaf area found as the season progresses represents not only more leaves, but more surface area per leaf and thicker leaves.

III. ACQUISITION OF ENERGY

In the photosynthetic process, plants acquire energy from the sun and convert it into a usable and storable form, carbohydrates. The process of photosynthesis can be divided into three primary steps, two light reactions which occur in the grana of the chloroplasts and the dark fixation and reduction of carbon dioxide from the atmosphere. The light reactions collect energy and transport it in the form of electrons from water through a series of

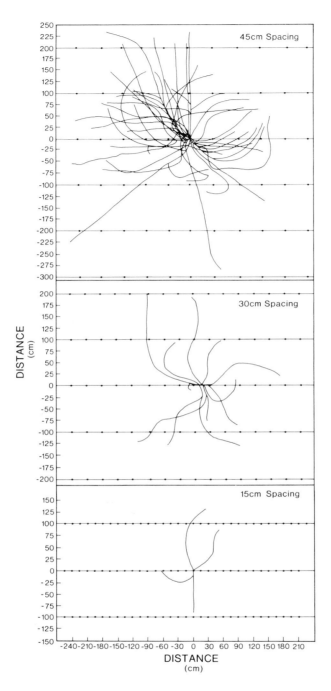

FIGURE 6. Vine and branch pattern of representative plants of the sweet potato (cultivar Jewel) 126 days after planting at three in-row plant spacings (15, 30 or 45 cm × 96 cm). (From Somda, Z. C. and Kays, S. J., manuscript in preparation, 1985.)

intermediates to NADP* where it can be temporarily stored. The effectiveness of light in these photoreactions is known to depend upon its spectral composition. In most plants,

---

* Nicotinamide adenine dinucleotide phosphate.

FIGURE 7. Genotypic variation in the leaf shape of the sweet potato. (From Yen, D. E., *The Sweet Potato and Oceania*© 1974, Bishop Museum Press, Honolulu, Hawaii, 389. With permission.)

photosystem I is excited by radiant energy of wavelengths of less than 680 nm while photosystem II is excited by wavelengths of greater than 685 nm.

Both light quality, quantity, and duration are important in maximizing the photosynthetic process. The effect of wavelength on the magnitude of photosynthesis in the sweet potato (action spectrum) is illustrated in Figure 8. There is a broad peak at 600 to 680 nm and a lower peak at 435 nm, typical of many herbaceous species.[23] Part of the light striking the

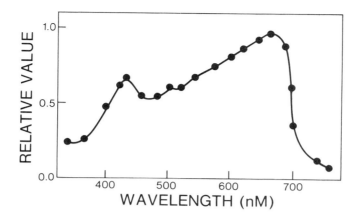

FIGURE 8. The relationship between light wavelength and photosynthesis in sweet potato leaves. (From Inada, K., *Plant Cell Physiol.*, 17, 355, 1976. With permission.)

leaves in the upper portion of sweet potato canopy is not absorbed, but is transmitted through the leaf, striking either lower leaves in the canopy or the ground. Individual leaves tend to saturate at a light intensity of 750 $\mu E\ m^{-2}s^{-1}$ [24] or around 30 klux,[25] which is approximately one-third of full sunlight. Because of partial absorption, the spectral composition of the transmitted light is significantly altered making it photosynthetically less active. Hence, lower leaves within the canopy or light reception hierarchy tend to have progressively lower photosynthetic rates. When the surface area of leaves per unit area of land is sufficiently high (i.e., a leaf area index greater than 4.0), the respiratory utilization of energy by these lower leaves can be greater than the net energy fixed. Such conditions may account for much of the leaf shedding found in the latter part of the growing season.

The energy obtained in the light reactions is subsequently utilized in the dark reactions of photosynthesis. The sweet potato, along with a majority of the plant species on earth, fixes carbon from atmospheric carbon dioxide by way of the reductive pentose phosphate pathway. Carbon dioxide reacts with ribulose-1,5-biphosphate, a 5-carbon sugar, forming two 3-carbon phosphoglyceric acid molecules. Since the first products are 3-carbon compounds, this portion of the photosynthetic process is referred to as the C_3 pathway. Part of the phosphoglyceric acid is cycled back to ribulose-1,5-biphosphate. This cycling back of phosphoglyceric acid represents the primary sink for the energy derived in the light reactions. Each molecule of carbon dioxide fixed requires 9 ATP equivalents of energy.

The phosphoglyceric acid which is not cycled back into ribulose-1,5-biphosphate, is converted through several intermediate steps to fructose-6-phosphate which can be utilized via the glycolytic, tricarboxylic acid, and electron transport system pathways of the respiratory system as an energy and/or carbon source for maintenance and synthetic reactions within the cell. It may also be converted to starch for short-term storage within the leaf or sucrose for transport out of the leaf to other parts of the plant. These represent what are known as the mobile pool of photosynthates (primarily sucrose) and the nonmobile pool (primarily starch) within the sweet potato leaf. It is possible to measure, using the unstable isotope ^{11}C, the percent export of photosynthate and the size of the mobile pool.[26] Under normal conditions (cultivar Centennial) approximately one-third of the fixed carbon is temporarily stored in the nonmobile pool within the leaf, with the remaining two-thirds being translocated out of the leaf (Figure 9). A measure of how rapidly the carbon is moving through this mobile pool is indicated by the pool turnover time, which is approximately 21 min. Conditions which block or impede transport, but not photosynthesis greatly increase

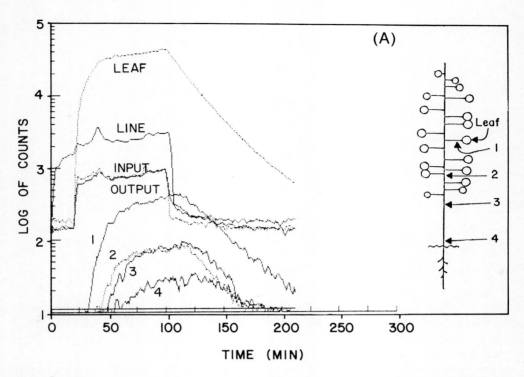

FIGURE 9. The use of the unstable isotope of carbon, ^{11}C, to study transport of photosynthate within the plant. (A) Raw data as seen in real time on the computer screen for a sweet potato plant treated with $^{11}CO_2$ on a single leaf. Data represent the log of the counts in the leaf, line from the accelerator, input and output from the leaf and individual points on the plant. Detector 1 is position on the petiole of the treated leaf; detector 2 is six internodes below the treated leaf; detector 3 is 9 internodes below the treated leaf and detector 4 is at the soil surface. (B) The percent export of photosynthate and the export pool turnover time is calculated using the "washout" curve of ^{11}C in the leaf (Figure A) following cessation of steady-state labelling of the leaf. The analysis uses a 2 compartment model where activity from the storage compartment is lost only by isotope decay and activity in the export compartment is lost by export. Data are corrected for a number of variables, e.g., ^{11}C half-life. (From Kays, S. J., Chua, L. K., Goeschl, J. D., Magnuson, C. E., and Fares, Y., in *Sweet Potato Proc. 1st Int. Symp.*, Villareal, R. L. and Griggs, T. D., Eds., AVRDC, Tainan, Taiwan, 1982, 95. With permission.)

the percent of carbon incorporated into the nonmobile pool and markedly increase the turnover time of the mobile pool.[27] In addition, there is evidence now that much of the acropetal transport of carbon into vine growth comes from this nonmobile pool which is hydrolyzed to sucrose during the night and translocated to the apical tips of the growing branches.

The maximum rate of photosynthesis of the sweet potato reported in most studies falls within the 18 to 22 mg $CO_2 dm^{-2} hr^{-1}$ range.[24,28,29] Recent work[30,31] suggests that this may be low. The maximum rates for some cultivars are now known to be between 35 to 39 mg $CO_2 dm^{-2} hr^{-1}$ and the rate of photosynthesis is known to vary widely between cultivars. Differences between studies may represent differences among cultivars, variation in experimental conditions and the condition of the plants. For example, the gas flow rate through photosynthetic chamber, the thickness of the leaves, leaf age and other parameters may have a pronounced effect. These maximum rates, however, are well within the range reported for C_3 species.

The rate of photosynthesis is modulated by a number of internal plant factors and external environmental factors. When these are integrated over an entire growing season, they have a pronounced effect on yield.

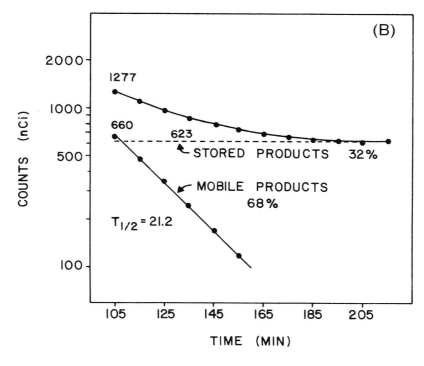

FIGURE 9B.

A. Plant Factors

The relationship between a number of plant factors and the rate of photosynthesis has been explored by various researchers. If rate limiting steps can be identified, then integrating improvements into new lines can be greatly facilitated.

1. Leaf Chlorophyll Concentration

The concentrations of α- and β-chlorophyll, carotene, and xanthophyll in sweet potato leaves are known to vary among cultivars,[32] however, with the exception of β-chlorophyll this variation does not appear to be significant in relation to photosynthetic rates. Chlorophyll concentrations between 7.6 and 10.6 mg/g leaf dry weight are common for cultivated selections.[30]

2. Leaf Stomata

Since leaf stomata largely control the exchange of gases moving both into and out of the leaf, the possible relationship between either stomatal density and/or stomatal responsiveness to environmental factors and net photosynthetic rate has been explored. The density of stomata on the leaves of 16 cultivars varied from 47 to 87/mm^2 on the adaxial (upper) surface and 163 to 253/mm^2 on the abaxial (lower) surface.[30] Hence, the stomatal density on the lower surface was 2 to 3 times that of the upper. While these differences among cultivars were significant, they were not correlated with the net photosynthetic rate of each cultivar.

Photosynthetic rate may be reduced by a water deficit within the leaf and corresponding increases in stomatal resistance. As the water vapor conductance (the reciprocal of the water vapor diffusion resistance) decreases, photosynthesis decreases.[33] This resistance to diffusion appears to be mediated largely by the resistance of the stomata (r_s) rather than by the boundary layer around the leaf (r_a). When the leaf moisture status is satisfactory, photosynthesis tends to be high even at high ambient temperatures. The conversion of water from a liquid to a gaseous phase has a tremendous effect on leaf temperature. When the actual leaf temperatures were measured over two entire growing seasons under severe environmental conditions (high ambient temperatures and solar radiation) the maximum leaf temperature attained by the sweet potato was only 32.7°C, while the soil surface temperature exceeded 50°C.[34]

3. Leaf Geometry

The amount of light absorbed by the leaves in a sweet potato canopy depends to a large extent on the size of the leaves, their orientation to the incoming radiation, and the leaf area index. It is known from other species that small erect leaves have lower light extinction coefficients. Hence, less of the total incoming radiant energy is intercepted by the uppermost portion of the canopy. This results in a more equal distribution of incident light over the entire leaf area of the canopy giving a higher net assimilation rate. When the leaf area index is high, i.e., sufficient leaves in the lower canopy to intercept the penetrating light, these differences can be quite significant. Tsunoda[22] found that the highest yielding sweet potato cultivars in response to high light intensity, had relatively thick, small leaves which allowed good light penetration. Although leaf size, angle, and shape are known to vary widely, these parameters are seldom intentionally selected for in ongoing breeding programs. As a consequence, most cultivars have relatively horizontal leaves with high extinction coefficients.

4. Leaf Age

Leaf age also has a significant effect on the rate of photosynthesis of individual leaves.[19,31,35] Relatively young, fully expanded leaves tend to have the highest photosynthetic rates. Using a specific leaf blade length for a morphological stage of development (e.g., Lo = 20 mm), a leaf plastochron index number is assigned sequentially to each leaf below the reference leaf (Figure 10A). Thus, the photosynthesis of leaves of an equivalent stage of development can be compared at different times of the growing season. As the age of the leaf increases (Figure 10B, i.e., increasing plastochron index number), there is typically a decline in the rate of photosynthesis. In addition, the rate of photosynthesis varies during the season (Figure 10B). Highest photosynthetic rates for the plant were found at mid-season (early August). This suggests that the rate of photosynthesis is being modulated in response to additional plant factor(s).

5. Leaf Starch and Sugar Concentration

The rate of leaf photosynthesis has been shown to be correlated with the concentration of nonstructural carbohydrates (starch and sugars) within the leaf blade (r = 0.924).[29,36] In addition, the leaf carbohydrate concentration appears to decline after the storage roots begin to develop.[37] The majority of these carbohydrates are in the form of starch. When conditions are artificially created to increase the carbohydrate concentration within the leaf, photosynthesis tends to be depressed; however, the precise relationship has not been sufficiently clarified. The sweet potato has the capacity for significant increases in the concentration of starch within the leaf without affecting the rate of photosynthesis. This occurs daily in the diurnal photosynthetic cycle of the leaf. It is anticipated from studies with other species, that some conditions (e.g., a high carbon dioxide atmosphere for a prolonged period) could increase the starch concentration to a point that would impede photosynthesis. There is presently no indication that this occurs under normal growing conditions in the sweet potato. Therefore, it is thought that the export rate of photosynthates out of the leaf may be of greater importance than the concentration within the leaf.[29] The sweet potato, like several high yielding C_4 species, has a high percent export of photosynthate (i.e., 72%, cultivar Jewel).[38] This is substantially higher than C_3 species such as cotton (64%) and velvet bean (67%).[39]

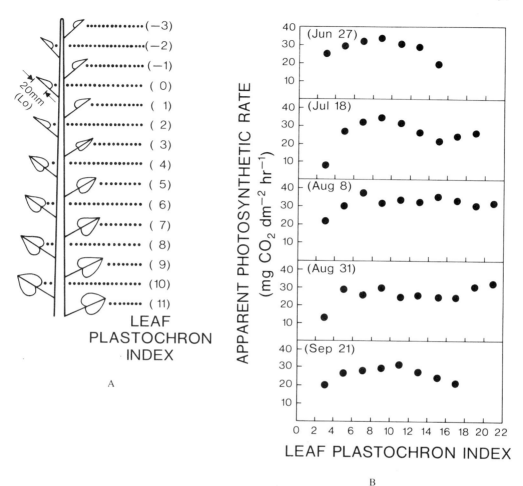

FIGURE 10. (A) Expression of leaf age by leaf plastochron index (leaf reference length = 20 mm). (From Kato, S., Hozyo, Y., and Shimotsuko, K., *Jpn. J. Crop Sci.*, 48, 254, 1979. With permission.) (B) Changes in the apparent photosynthetic rate with leaf plastochron index and date during the growing season. (From Kato, S., Hozyo, Y., and Shimotsuko, K., *Jpn. J. Crop Sci.*, 48, 254, 1979. With permission.)

B. Environmental Factors

The plant's environment has a pronounced effect on the rate of photosynthesis of individual leaves and the rate within the canopy. The major modulating factors are light intensity, temperature, and water balance. While individual leaves exhibit typical light saturation kinetics, the response of the canopy is more complex. Early in the season when the leaf area index is small (0.5 to 1.5) the canopy is saturated at a relatively low light intensity. As the leaf area increases (3.7 to 5.5), increases in gross photosynthesis are realized with progressively higher light intensity.[40] The rate of photosynthesis, therefore, was found to be closely correlated with the level of solar radiation intercepted by the plant.

Air temperature also has a significant effect on the rate of photosynthesis of the sweet potato; however, this response is closely tied to the water balance of the leaf.[33] In addition, the photosynthetic rate response of the plant to increasing temperature appears to vary with the stage of development of the plant. At early stages of growth, apparent photosynthesis was not strongly affected by temperature increases from 30 to 40°C. Later, however, the photosynthetic rate was significantly depressed by the higher temperatures. Leaf temperature is greatly affected by the level of solar radiation.[41] As the temperature increased, leaf moisture content decreased and the stomatal resistance increased,[42] causing a decline in the rate of photosynthesis.[33] It has also been shown that as the water vapor pressure gradient between the leaf and the air increases, leaf stomatal resistance increases.[43] The response of the stomata to high water pressure deficit appears to be less when the roots are high in carbohydrates.[33]

Stomatal resistance was found to be lower than mesophyll resistance;[44] however, both increase with age of the leaf. High stomatal resistance has a greater effect on the rate of transpiration of the leaf than the rate of photosynthesis. Hozyo[44] suggests that carbon dioxide diffusion resistance in the leaf is a rate limiting process at a saturating light intensity and atmospheric carbon dioxide concentration. Significant increases in photosynthetic rate result from short-term increases in the ambient carbon dioxide concentration.[44] This response increases as the light intensity increases.[45]

Wind speed also affects the movement of gases into and out of the leaf. This effect is primarily on the boundary layer around the leaf. As the wind speed increases, boundary layer resistance decreases exponentially.[46] Leaf angle and flutter also affect the boundary layer resistance. At wind speeds of less than 50 cm s^{-1}, boundary resistance may become a critical factor affecting photosynthetic rate.[46]

IV. TRANSPORT OF ENERGY

After carbon is assimilated, part of it is translocated in the form of sucrose out of the leaf into the stem and eventually to the site of utilization or deposition. Transport within the stem may be either acropetal or basipetal in direction. When transported acropetally, the carbon not used for maintenance reactions is utilized to expand the photosynthesizing surface area of the plant. Carbon transported to the roots may be used for maintenance, expansion of the fibrous root system or deposition in the specialized storage roots of the sweet potato. Thus, both the absolute flux of photosynthate and the directional allocation are important in determining yield.

The carbon fixed by the leaf is moved out of the leaf blade via the phloem tissue. This is composed of four types of cells: the sieve tube elements and companion, fiber, and parenchyma cells. The sieve elements, connected end to end, form the actual conduits (sieve tubes) for sucrose being transported out of the leaf. At the end of each sieve element, between connected elements, is a sieve plate permeated with small pores, through which the photosynthates must pass.

The actual movement of photosynthate through the system is thought to be mediated by a pressure flow mechanism described by Münch in 1930.[47] When the concentration of sucrose is high in the source (Figure 11) water will move into the phloem transporting the sucrose toward the sink which is low in sucrose. The water then recycles back to the leaf via the xylem. This concentration differential in photosynthates between the source and the sink results in the buildup of a hydrostatic pressure head that drives the system. For the system to work, there has to be water returned from the sink to the source (xylem). If sucrose is removed from the sink, i.e., hydrolyzed and then converted to starch, the concentration remains low in the storage root. With continued synthesis of sucrose in the leaf and removal by the storage root, the gradient is maintained. The flux or mass transport of carbon through the system, therefore, is a direct function of the difference in pressure between the source and the sink and the radius of the conduits to the fourth power. It is inversely proportional to the length of the conduits and the viscosity of the liquid.

A precise understanding of the transport system is essential since there is considerable evidence which suggests that phloem transport may play a limiting role in productivity. In its simplest form, phloem transport can be broken down into three basic components: (1) loading, (2) transport, and (3) unloading.

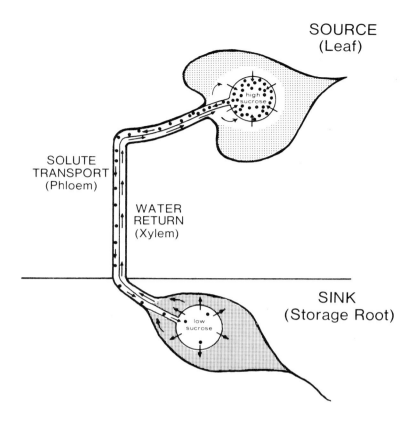

FIGURE 11. A model illustrating the Münch pressure flow theory of solute transport in the sweet potato. When the concentration of sucrose is high in the source, water moves into the phloem transporting the sucrose toward sinks which are low in sucrose. The water then recycles back through the xylem. Sucrose at the sink is hydrolyzed and converted to starch, keeping the concentration low, thus maintaining the hydrostatic pressure head that drives the system.

A. Phloem Loading

After synthesis in the cytoplasm of leaf chlorenchyma cells, sucrose in the leaf must pass through a series of steps prior to transport. The initial step involves transversing the plasmalemma of the cell in which it originates and movement into the apoplast. The molecules of sucrose diffuse in the apoplastic solution to phloem cells at which time they are transported across the plasmalemma against a concentration gradient into the phloem cells (termed phloem loading). Loading is an active process requiring energy and is thought to function in many plants through a sucrose-proton cotransport system.[48]

Little is presently known about this process in the sweet potato. The turnover time (T) of the mobile or export pool of photosynthates, however, gives us some measure of the rate at which photosynthates are being loaded into the phloem and moved out of the leaf. These rates have been measured for two cultivars grown under similar conditions (Centennial: $T^1/_2$ = 21.2 min, Jewel: $T^1/_2$ = 22.6). Conditions that impede transport greatly increase (e.g., double) the length of time required for turnover of the pool in the sweet potato.[27,38] In addition, the pool size is small relative to cotton, another C_3 species (i.e., ~ 50%).[39]

B. Phloem Transport

After molecules of sucrose enter the sieve tubes, they are moved toward dominant sink sites. The direction of flow of photosynthate from individual leaves appears to be controlled by both the individual leaf's position on the plant and the hierarchy of demand for photosynthates within the plant at any given moment. Typically fibrous roots and lateral buds are much lower in the hierarchical chain of control of photosynthate partitioning than are storage roots and shoots. In addition, neither leaf position relative to the apex nor photosynthate demand are static. As a consequence, directional flow of carbon may change radically in response to changing conditions.

Kato et al.[31] found that sweet potato leaves ranging in position on the plant from LPI 4 to 19 display bidirectional transport. The lower leaves (LPI 19) transport a greater percentage of their fixed carbon in a basipetal direction. Increasing storage root sink strength through grafting the same scion on different stocks had a distinct effect on the directional flow of carbon. Transport direction of carbon from leaves of LPI 4 and LPI 19 were not affected by stock type, that is, storage root sink potential; however, leaves of LPI 7 had a substantial increase in the proportion of their translocated carbon which was directed toward the stock. Therefore, sink strength appeared to affect only certain leaves in the canopy.

This allocation of photosynthates within the plant changes during the day. We have found that very little of the photosynthate in the cultivar Jewel, regardless of leaf position, is translocated during daylight to the apical growing tip. Rather this carbon appears to be moved acropetally during the night and is derived from the temporary nonmobile pool (starch) within the leaf. Sucrose which is being translocated down the stem is also not directed during the day into young lateral branches after they reach several centimeters in length. The sweet potato, therefore, appears to have a diurnally controlled scheme for the bidirectional allocation of its photosynthates.

The importance of phloem transport velocities in relation to leaf photosynthetic rate has been stressed in several studies.[16,49] Based on field dry matter production rates, Austin and Aung[16] also suggest that the rate by which assimilates move appears to be more important for storage root growth than increases in leaf area. Kato and Hozyo[49] have described the transport velocities and coefficients of transport of ^{14}C-photosynthates in both the acropetal and basipetal directions using reciprocal grafts to alter both source and sink potential. Transport velocities of carbon toward the apex were not as substantially affected as were the velocities toward the root system when the storage root sink potential was altered. Using the cultivar Okinawa No. 100 (0-100) which has a strong storage root sink potential and the wild type, *Ipomoea trifida* HBK (T-15) which has a limited storage root sink potential, 0-100/0-100 (scion/stock) grafts had a mean apical transport velocity of 0.80 cm/min while 0-100/T-15 grafts were 0.70 cm/min and T-15/0-100 and T-15/T-15 grafts were 0.77 and 0.52 cm/min, respectively. Basipetal transport, in contrast, was very strongly affected by storage root sink strength. Grafts of 0-100/0-100 had a mean velocity of 2.13 cm/min while 0-100/T-15 grafts had a velocity of 0.62 cm/min. It was concluded by Kato and Hozyo[49] that sink activity affects the speed of translocation and the coefficient of translocation. It should be noted, however, based on the close rates of basipetal transport between 0-100/T-15 (0.62 cm/min) and T-15/0-100 (0.52 cm/min) grafts, that factors other than storage root sink strength have a pronounced effect on transport velocity.

It has been suggested that the development of phloem bundles and secondary phloem in the portion of the root between the storage root and the stem may limit the rate of translocation of carbon into the storage root.[5] Based on the direct relationship between radius of the sieve tubes (r^4) and mass transport as described by the Münch hypothesis, small differences could have a pronounced effect. However, correlations of measurements of the phloem cross-sectional area with storage root dry weight, although from a very limited number of cultivars, did not support this contention.[50]

C. Phloem Unloading

If one accepts, based on substantial correlative evidence in the sweet potato[51-54] and other plants,[55] that sink potential is of major importance in the net rate at which carbon is fixed by the plant, then processes dealing with the deposition of carbon at active sink sites become of critical interest. These processes can be grouped under several very general headings: (a) phloem unloading, (b) hydrolysis of sucrose, (c) transfer of the carbon to the site of polymerization, and (d) starch synthesis. Each of these general processes is composed of secondary and tertiary subprocesses or steps. When contrasted with phloem loading and transport in most crop plants, relatively little is known about the pathway of photosynthates upon reaching a major sink such as a storage root. This is especially true in the sweet potato.

Based on the apparent importance of sink strength in the sweet potato, it is probable that either the unloading process *per se* or one of the early steps in sink metabolism regulates the concentration of solutes in the phloem. Reducing the rate at which this step proceeds (i.e., decreasing sink strength) would, therefore, subsequently lead to an increase in the concentration of carbon within the phloem which in turn would decrease the net rate of loading and eventually photosynthesis. This effect can be seen in grafting experiments where stocks from plants which are poor sinks decrease the rate of photosynthesis of the scion.[56-58] It is apparent, therefore, that there is a strong interactive relationship between loading, transport, and unloading within the plant.

Are there other lines of evidence that lend support to the idea that the unloading process or subsequent steps may be a critical modulator? One might anticipate a sweet potato storage root to function in some ways similar to wheat grains where the rate of starch synthesis in the endosperm is closely related to the concentration of sucrose in the sink tissue. In this case, regulation is not predominantly due to the supply of sucrose to the sink,[59,60] but rather to processes operating within the sink itself. Factors regulating the movement of sucrose into and through the grain and influencing its conversion to starch appear to be major controlling steps.[61] In addition, declining deposition of carbon as the plant approaches maturity may more accurately reflect a decline in the sink's capacity to accept or metabolize sucrose rather than a decline in supply potential from the source.[62]

While there are obvious parallels between the deposition of carbon in cereal grains and sweet potato storage roots, there are numerous points of divergence. Perhaps one of the most important differences and advantages of the sweet potato in accumulating stored carbon is that the storage organ is relatively plastic. That is, both its volume and length of time for deposition are not rigidly fixed. In contrast, cereal grains have a relatively short period in which photosynthates are deposited and a fixed potential volume. If the environmental conditions are not favorable during this period, yield is often greatly reduced. The sweet potato, however, can withstand periods of low carbon deposition since deposition occurs over much of the growing season. In addition, deposition does not appear to cease once the root attains a certain size. A substantial portion of the sweet potatoes grown are harvested before the plant reaches its maximum yield potential in that relatively small storage roots are desired. As a consequence, reported sweet potato yields are often much lower than they would be if the plant was allowed to accumulate carbon until the end of the growing season.

The actual phloem unloading process may be either apoplastic or symplastic in nature. It's probable that the hydrolysis of sucrose may represent the primary rate limiting step. Hydrolysis would decrease the concentration of sucrose, thus facilitating a diffusion mediated unloading process. Generally, roots contain three enzymes which are capable of hydrolyzing sucrose: sucrose synthetase, acid invertase and alkaline invertase.[63] Hydrolysis could be regulated through the concentration or activity of the enzyme(s) controlling this step. Hirai[64] has looked at the relative activity of starch phosphorylase in sweet potato roots in relation to plant moisture status. When water absorption and transpiration were approximately equal, the enzymatic activity was at its highest. When uptake exceeded transpiration or transpiration exceeded uptake, activity was repressed. It is much more likely, however, that a metabolic step preceding starch synthesis is rate limiting.

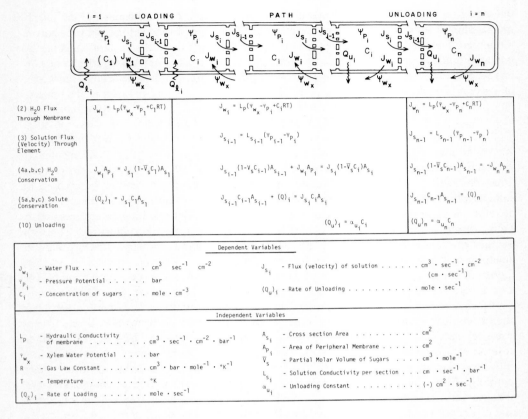

FIGURE 12. The minimal set of equations sufficient to express the Münch hypothesis of phloem transport. These equations take into account differences in the five types of elements which might be found along a sieve tube. From left to right these are (1) an initial loading element, (a) a typical loading element, (b) a typical path element, (c) a typical unloading element, and (n) a terminal unloading element. Dependent variables pertaining to each element are depicted within the element, and independent variables are depicted outside. (From Magnuson, C. E., Goeschl, J. D., Sharpe, P. J. H., and De Michele, D. W., *Plant Cell Environ.*, 2, 181, 1979. With permission.)

D. Modeling Transport

The photosynthesis, transport, and storage system in the sweet potato can be seen as a highly integrated series of processes. With transport alone, the interactive relationship of loading, unloading, membrane permeability, sieve tube geometry, and solute viscosity on profiles of concentration, pressure, and velocity within the sieve tubes are complex and nonlinear in behavior. As a consequence, it is often difficult to interpret the effect of an environmental alteration on the system in other than a very superficial way.

One extremely useful approach has been to model transport. Assuming the model is mechanistic in nature and contains a sufficient number of biophysical considerations, it allows a relatively rapid analysis of how the transport system will react under various conditions (e.g., water stress or altering sink strength). A model, therefore, will allow us to better understand the interactive effects within the system and more precisely select critical experiments and conditions.

A biophysical mathematical model of an osmotically driven pressure flow system in the phloem has been described[65,66] which, when applied to sweet potatoes will allow exploring this relationship between transport and productivity. Figure 12 gives the minimal number of equations needed to express the Münch hypothesis of phloem transport. These equations can be separated into three classes, those which describe phloem loading, transport, and unloading. It is of interest that the results derived by a number of researchers[51,58,67] on the relationship of sink strength to the rate of photosynthesis and for that matter the results of a number of other experiments on the sweet potato, are predicted by the model. Thus, the potential effects of altering the soil water potential, sink location, sink strength, and so on, can be explored rapidly and easily. Interesting relationships can then be tested with appropriate experiments and further explored.

V. ALLOCATION OF ENERGY

A. Allocation Within the Plant

The photosynthetically fixed carbon resources are allocated between maintenance reactions, production of additional structural components and deposition within the specialized storage roots of the plant. Various growth centers (points of high metabolic activity and, therefore, high demand for photosynthates) within the plant compete for available resources. Thus, within the plant at any particular time during the growing season, there exists a dominance hierarchy for photosynthates. Typically the fibrous roots and lateral buds are much lower in the heirarchical chain of photosynthate distribution than are storage roots or shoots. The change in dry weight of a particular sink, relative to alternative sinks, is seen as a measure of the strength of its competitive ability within the system.

The environmental and cultural conditions, as well as differences among cultivars, can have a pronounced effect on how the energy (carbon) accrued by the plant is allocated. For example, under conditions of high rainfall or excess nitrogen, the allocation pattern is shifted in favor of the vines.

Even within a specific type of sink, e.g., storage roots, there appears to also be a dominance hierarchy. The control of the allocation of carbon to these competitive sinks is not well understood. A number of mechanisms can influence the preferential movement of carbon to one sink over another. Variations in the ability of one sink to deplete the supply of photosynthates in the phloem (unloading coefficient) causes a concentration gradient which enhances subsequent movement of carbon to the sink. Variation in the chronological order of inception of the sink is also a factor. Hence, large storage roots will have more cells which increases their potential to deplete the level of carbon in the phloem. The position of a sink relative to the source of assimilate production can also be significant. Using mass flow models, sinks with identical unloading coefficients (strength) but at different distances from the source loading zone can be shown to have differences in the absolute amount and percentage of the total photosynthate unloaded into each; sinks closer to the source have the advantage.[39] Finally, vascular connections and lateral transport potential, may also be important in the establishment of a hierarchy of carbon allocation between competitive storage root sinks. Current breeding programs for edible sweet potatoes have selected toward minimizing variation in the patterns of carbon allocation to the roots. As a consequence, desirable new cultivars tend to have a number of roots of uniform size.

B. Sink Strength in Relation to Allocation

It has been over a century since Boussingault[68] presented the hypothesis that "... the accumulation of assimilates in an illuminated leaf may be responsible for a reduction in the net photosynthesis rate of that leaf". While the hypothesis has been supported by considerable experimental evidence, it has not been strictly proven (see review by Neales and Incoll[55]). In the sweet potato considerable correlative evidence, based primarily on altering sink strength or source potential, points in favor of the hypothesis. In addition, we have seen the relationship between sink strength and transport. According to the Münch hypothesis for phloem transport, the greater the sink strength, the greater the depression in solute concentration will be in the phloem at the sink. This increases the concentration differential between the source and sink or the hydrostatic pressure head that drives the system.

A number of studies have explored the relationship between source potential and sink strength within the sweet potato. Through the use of reciprocal grafts between a high yielding cultivar and *I. trifida,* a related species that does not produce storage roots, it has been shown that plants with stocks that are good sinks have a much higher dry matter accumulation than plants with stocks that are poor sinks.[49,56,58,67,69,70] It has been inferred,[52,54] therefore, that yield is determined primarily by the rootstock rather than the foliar portion of the plant. However, both photosynthetic source potential and storage root sink capacity can be rate limiting.[51]

C. Recycling of Carbon

This subject has not been adequately addressed in the sweet potato. We do know that starch within the storage roots is hydrolyzed and recycled during sprout formation.[71] Both field and storage conditions can induce this remobilization of stored carbon. Sprouts originate in the vascular cambium region[178] or in the cortical region adjacent to it.[5] Autoradiographs of sprouts from labeled storage roots indicate that the sprouts continue to utilize carbon recycled from the roots even after they have a number of actively photosynthesizing leaves.[27] By comparing the pattern of distribution of recycled [^{14}C]-carbon in the new leaves between storage roots with differing internal patterns of label deposition, it appears that hydrolysis and remobilization is a general phenomenon throughout the root rather than occurring in a localized region.[27]

Recycling of carbon out of senescing leaves may be important in the overall carbon balance within the plant. When the leaf area index of the sweet potato rises above three to four, the plants begin to shed the older and lower leaves within the canopy. Nutrients are remobilized out of these organs and moved to alternate sinks. The extent and relative importance of the recycling of carbohydrates and minerals has not been ascertained in the sweet potato. Data that appear to illustrate this phenomena, however, can be found in several studies.[72,73]

VI. RESPIRATORY LOSSES OF ENERGY

Sweet potato plants allocate their carbon (energy) resources in several general directions, each of which represents a form of utilization (i.e., not or at least not readily recycled). For example, carbon directed toward the formation of new leaves and stems represents a form of utilization in which, for the most part, much of the carbon is irreversibly tied up. Respiratory utilization of carbon differs, however, in that it results in a terminal transfer of the energy of the carbon bond into short term energy transfer molecules and the liberation of the carbon atom as carbon dioxide. Thus, it represents a direct irreversible form of utilization of the plants' energy resources.

Respiration is a central process in the cells of plants which mediates the release of energy through the breakdown of carbon compounds and the formation of carbon skeletons that are necessary for maintenance and synthetic reactions within the plant. Respiration, therefore, in a very broad sense, can be thought of as a generally essential but opposing force to the acquisition of energy. There are two general types of respiratory processes in plants, those which occur only in the light (light or photorespiration) and those that occur at all times, both in the presence or absence of light (dark respiration).

A. Photorespiration

A significant portion of the carbon fixed during photosynthesis is lost due to both photorespiration and dark respiration. Species, such as the sweet potato, that have the C_3 photosynthetic pathway for carbon fixation have distinctly higher levels of photorespiration than do C_4 species. C_3 plants lose an estimated 30 to 50% of their photosynthetically assimilated carbon due to photorespiration. In the sweet potato, the respiratory rate of the leaf as measured by the loss of carbon dioxide from the tissue, proceeds at a substantially higher rate in the light than in the dark. This light-stimulated loss of carbon is in addition to and superimposed upon normal dark respiration.

Photorespiration is found in the photosynthesizing tissue of the plant, principally the leaves and is stimulated by high (i.e., ambient, 21%) oxygen levels. As the oxygen concentration decreases (21 → 2%) photorespiration decreases and the net acquisition of carbon through photosynthesis increases. This inhibition of photosynthesis by oxygen, the Warburg effect, involves the competition between molecules of carbon dioxide and oxygen for ribulose bisphosphate carboxylase, the primary photosynthetic carboxylating enzyme. At higher oxygen levels the oxygenation reaction is favored and more glycolic acid, the substrate for photorespiration, is produced. This compound moves out of the chloroplasts and through a cyclic series of reactions in which it is partially oxidized and carbon dioxide is liberated. The oxidation step, however, is not linked to ATP* formation, thus photorespiration results in both a loss of energy and photosynthetic carbon from the plant. Since there is no net energy gain, a classical requirement for respiration, photorespiration actually is more accurately termed oxidative photosynthesis.

Accurately measuring photorespiration is difficult since carbon dioxide is being fixed by the leaf and simultaneously evolved by both dark respiration and photorespiration. One technique used to measure the relative amount of photorespiration is to determine the increase in the photosynthetic rate upon lowering the oxygen concentration to 2%. The sweet potato has about a 30% inhibition of photosynthesis in the ambient environment (Table 1) which is comparable to many other C_3 species but markedly greater than C_4 species.

B. Dark Respiration

The living cells of plants respire continuously utilizing photosynthetic carbon and oxygen from the surrounding environment and liberating carbon dioxide. Dark respiration represents a series of oxidation-reduction reactions in which energy is released and carbon skeletons needed for synthetic and maintenance reactions are formed. It involves several pathways and a series of individual reactions. The glycolytic, tricarboxylic acid, and electron transport system pathways are each involved in the breakdown of many of the common substrates utilized by the cells. It is through these respiratory pathways that the carbon from photosynthesis begins its transformation into the majority of the other compounds in the plant.

The series of steps in the respiratory oxidation of sugar or starch involves three major interacting pathways. The initial pathway is that of glycolysis where sugar is broken down into pyruvic acid, a three carbon compound. This takes place in the cytoplasm and does not require oxygen. The second pathway is the tricarboxylic acid or Krebs cycle. Here pyruvic acid is oxidized to carbon dioxide, the reactions occurring in the mitochondria. Oxygen, although not reacting directly in these steps, is required for the pathway to proceed as are several organic acids. The third pathway, the electron transport system, transfers hydrogen atoms which have been removed from organic acids in the tricarboxylic acid cycle and from 3-phosphoglyceraldehyde during glycolysis. They are moved through a series of oxidation-reduction steps terminating upon uniting with oxygen in the formation of water. Energy is trapped chemically in the form of ATP, which can be utilized to drive various energy requiring reactions within the cell. A fourth respiratory pathway, the pentose phosphate system, while not essential for the complete oxidation of sugars, functions by providing

---

* Adenosine triphosphate.

Table 1
THE INFLUENCE OF OXYGEN CONCENTRATION ON LEAF PHOTOSYNTHESIS

Species	Photosynthesis ($mgCO_2 dm^{-2} hr^{-1}$)		Photosynthesis inhibition at 21% oxygen (%)
	21% O_2	2% O_2	
C_3 Plants			
Sweet potato	12.7[a]	18.1	30
Cotton	17.1	23.3	27
Potato	14.5	21.5	33
Broad bean	17.7	25.6	30
Soybean	13.0	18.8	31
C_4 Plants			
Corn	22.0	22.5	2
Crabgrass	30.4	30.4	0

[a] Irradiance intensity 600 $\mu Em^{-2}s^{-1}$.

From Vines, H. M., Tu, Z.-P., Armitage, A. M., Chen, S.-S., and Black, C. C., *Plant Physiol.*, 73, 25, 1983.

carbon skeletons, reduced NADP* required for synthetic reactions and ribose-5-phosphate for nucleic acid synthesis. It appears to be operative in varying degrees in all respiring cells.

The storage roots of the sweet potato,[74] along with a number of other plants and plant parts, possess an alternate or cyanide-resistant electron transport pathway. This is in addition to the normal electron transport system. When operative, electrons move from the normal pathway into the alternate pathway at coenzyme Q. In the cyanide resistant pathway, the electrons are transferred to a flavoprotein of intermediate potential, then to an alternate cytochrome oxidase and subsequently to oxygen forming water. It is important to note that only one ATP is formed from each reduced NAD** compared to the formation of three in the normal pathway. Thus, a major portion of the energy derived from the substrate is lost as heat. The precise role of the alternate electron transport pathway has not been ascertained nor has the significance of its presence in the sweet potato.

A number of environmental[40] and internal[75,76] factors affect the rate at which respiration proceeds. Perhaps the most important of which is temperature. The respiratory rate of the entire plant increases approximately 1.6 times for every 10°C increase in temperature between 15 and 30°C.[40] Not all of the plant parts respond similarly to temperature. For example, the storage roots have a greater response to temperature changes than the entire plant.

Oxygen also has a pronounced effect on the rate of respiration. While oxygen is not required for the operation of the glycolytic pathway, it is essential for the tricarboxylic acid cycle, the pentose phosphate pathway and the electron transport system. Glycolysis can proceed, therefore, under anaerobic conditions, i.e., in the absence of oxygen. The occurrence of anaerobic conditions often poses a serious problem in the production of sweet potatoes (for a recent review see Collins and Pharr[77]).

Low oxygen conditions occur in the root zone of the sweet potato when the plants are exposed for a period of time to excessive soil moisture, a condition occurring with surprising frequency due to high rainfall. During the early stages of growth of the plant, storage root formation and/or enlargement is repressed and top growth increased.[3] Since the inhibition

---

* Nicotinamide adenine dinucleotide phosphate.
** Nicotinamide adenine dinucleotide.

can be reversed upon returning the plant root zone to aerobic conditions and nonstorage roots (i.e., fibrous) grow readily in solution culture, it appears that the storage roots are substantially more sensitive to oxygen deficiency than are the fibrous roots.

In the later growth stages of the plant, after storage roots have formed, anaerobic conditions within the root zone are extremely detrimental. Losses are realized either directly due to the roots rotting in the field or indirectly due to rotting during curing and storage[78] and decreased quality. Quality losses include decreases in baking quality, carotenoid pigments, and dry matter[79] and an increase in pithiness.[80]

When the oxygen concentration within the root zone and subsequently within the root tissue falls below a threshold level, pyruvic acid can no longer proceed through the tricarboxylic acid cycle and instead forms either lactate or ethanol which may accumulate in toxic levels. Prolonged exposure to anaerobic conditions results in death of the cells and loss of the storage roots. Exposure for short periods often results in the formation of off-flavors which may or may not, depending on the length of exposure, be eliminated upon returning to aerobic conditions.

Under anaerobic conditions the actual rate of respiration by the tissue increases (termed the Pasteur effect) although the yield of energy decreases markedly. When ethanol is formed each free glucose molecule gives a net yield of 2 ATPs, thus the energy yield is only 1/19 that derived when glucose is fully oxidized under aerobic conditions (38 ATPs). Therefore, much more glucose must be oxidized to meet the energy requirements of the cells. The additional energy remains in the form of alcohol. This has disastrous consequences for live tissue in terms of losses of stored photosynthetic carbon and the accumulation of undesirable compounds.

The susceptibility of storage roots to low oxygen conditions is known to vary widely due to cultivar.[81-86] Neither the reason for these differences nor the genetics of resistance to flooding have been established. The concentration of ethanol within the roots was not correlated with damage symptoms. In addition, the activity of the ethanol-forming enzymes (pyruvate decarboxylase and alcohol dehydrogenase) were not correlated with the ethanol concentration developed under simulated waterlogging conditions.[83,87,88] Resistant clones do appear, however, to recover normal metabolic patterns more quickly than susceptible lines.

The ambient temperature when waterlogging occurs also affects the extent of storage root damage. More direct losses (field) occur when the temperatures are warm than when cool.[81]

The actual critical ambient oxygen concentration for the shift to anaerobic conditions appears to be between 5 to 7% (12 to 13°C),[87] however, it is the internal concentration of oxygen that is of primary importance. The internal oxygen concentration is modulated by (1) the differential in partial pressures for the molecule between the interior and exterior of the root, (2) the rate of utilization of oxygen by the tissue, and (3) the diffusion resistance of the tissue. Elevating the temperature increases the rate of use of oxygen, and in turn, the rate at which the extinction point is reached upon flooding.

Root size is also important. The rate of diffusion of oxygen molecules into the storage root and conversely carbon dioxide molecules outward, is affected by the diffusion resistance of the tissue and the length of the path of the molecule. As the root size increases, there is a progressive decrease in surface area relative to its volume. This is because volume (assuming the shape is not significantly changed) increases as the cube of length while surface area increases only as the square of length. As a consequence, as the root increases in size the rate at which oxygen can diffuse into it decreases and under undesirable conditions (e.g., excess soil moisture) the extinction point is reached more rapidly and at higher ambient oxygen concentrations.

C. Respiratory Rate as a Function of Plant Part and Age

The rate of dark respiration varies considerably between the individual plant parts of the

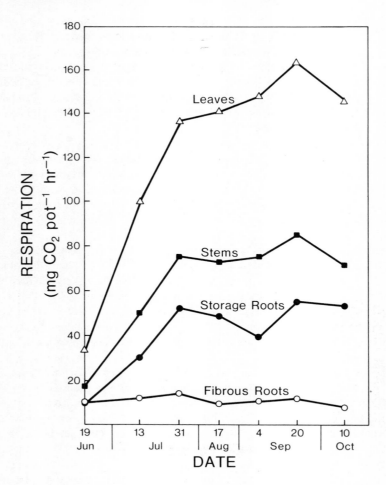

FIGURE 13. Changes in the respiratory rate of individual sweet potato parts during the growing season. (From Tsuno, Y. and Fujise, K., *Bull. Nat. Agric. Sci. (Jpn.)*, D13, 1, 1965. With permission.)

sweet potato. When expressed on a whole plant basis (Figure 13)[19] the leaves have the highest rate, followed by the stems, storage roots and fibrous roots. As the mass of the leaves, stems, and storage roots increased, so did their contribution to the total respiratory rate of the plant. The fibrous roots, however, maintained a relatively constant rate per plant.

When respiration is expressed on a weight basis, the leaf blades (minus petiole) were found to have the highest rates of dark respiration (Table 2),[40] followed by the petioles + stem, fibrous roots, and storage roots. The rate of respiration per unit of plant weight decreased markedly in the individual plant parts with increasing age. The storage roots displayed the greatest change. Toward harvest, the rate of respiration had declined to only 10% of its original value. The leaf blades displayed the lowest percent change with time (31%). This change in respiration with age is illustrated most dramatically when expressed on a whole plant weight basis (Table 2). By October 1 the respiratory rate of the entire plant (0.48 mg CO_2 $g^{-1}hr^{-1}$ @ 25°C) had decreased to 12% of its rate on July 18 (4.10 mg CO_2 $g^{-1}hr^{-1}$).

Cultivar also has a significant effect on storage root respiration[19] although a similar effect was not found for the leaves. This appears to be, in part, due to differences in the moisture content of the storage roots which is correlated with the rate of root respiration (Figure 14).

Table 2
RESPIRATORY RATES OF EACH PLANT PART AT DIFFERENT GROWTH STAGES (IN mg CO_2 $g^{-1}hr^{-1}$ AT 25°C)

Date	Leaf	Petiole + stem	Storage roots	Root	Entire plant
July 18	4.49	3.36	—	5.09	4.10
Aug. 4	4.68	3.07	2.09	2.43	3.27
Aug. 18	5.66	2.19	1.09	1.03	2.34
Sept. 3	1.12	0.67	0.24	1.42	0.48
Oct. 1	1.38	0.66	0.20	1.27	0.48

From Agata, T. and Takeda, T., *J. Fac. Agric. Kyushu Univ.*, 27, 75, 1982. With permission.

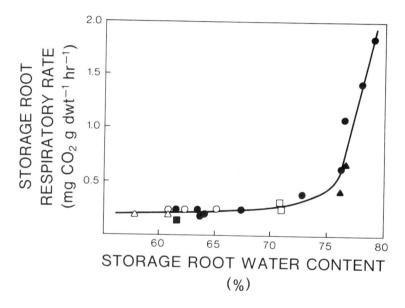

FIGURE 14. The relationship between sweet potato storage root water content and respiratory rate. Different symbols represent six cultivars. (From Tsuno, Y. and Fujise, K., *Bull. Nat. Agric. Sci. (Jpn.)*, D13, 1, 1965. With permission.)

The actual rate changed little between 58 to 75% moisture; however, above 75% respiration increased tremendously. This may, in part, account for the shorter storage life of many low dry matter cultivars.

After harvest there is a substantial decline in the rate of respiration of the storage roots.[87] Uncured roots of the cultivar Jewel had a rate of respiration of approximately 70 mg CO_2 $kg^{-1}hr^{-1}$ immediately after harvest (Figure 15), but decreased rapidly, stabilizing at about 25 mg CO_2 $kg^{-1}hr^{-1}$ after 38 to 40 hr (25°C).

D. Respiratory Quotients and Plant Energy Input/Loss Ratios

The respiratory quotient (RQ_{10}) is a measure of the change in rate of respiration due to temperature. It is the ratio of the rate at one temperature (T_1) vs. the rate at that temperature + 10°C (i.e., RQ_{10} = mg CO_2 $kg^{-1}hr^{-1}$ @ T_1 + 10°C/mg CO_2 $kg^{-1}hr^{-1}$ @ T_1) when

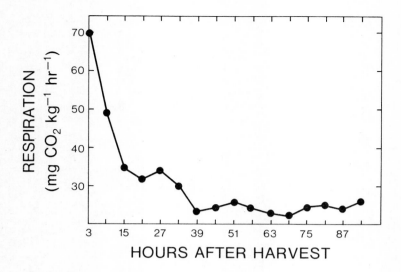

FIGURE 15. Changes in the rate of respiration of sweet potato storage roots after harvest. (From Chang, L. A. and Kays, S. J., *J. Am. Soc. Hortic. Sci.*, 106, 481, 1981. With permission.)

temperature is measured in degrees centigrade. The RQ_{10} for dark respiration of entire sweet potato plants was found to be 1.6 between 15° and 30°C,[40] while the RQ_{10} of the storage roots was approximately 2.5,[19] varying slightly with cultivar. With a higher RQ_{10} for storage roots, the entire plant's RQ_{10} would be expected to shift during the growing season based on the increased contribution (i.e., weight) of this part of the plant.

When the respiration rate of the entire plant was compared with the net photosynthetic rate, respiration was found to have a mean value of 3.79 g m^{-2} day^{-1} over the entire growing season, while net photosynthesis had an average rate of 7.94 g m^{-2} day^{-1} (Figure 16).[40] Net photosynthesis tended to increase until the middle of September and then gradually decrease as harvest approached. This decline was due to a decline in gross photosynthesis rather than an increase in respiration.

VII. STORAGE OF ENERGY

The development of storage roots as a repository for photosynthates can be divided into three basic processes. These include: (1) the initiation of a root capable of becoming a storage root; (2) the induction or triggering of the development of the root as a storage organ; and (3) the actual development of the storage organ. Most of the information currently available comes from anatomical studies.

A. Storage Root Initiation

Young roots that have the potential to become storage roots are morphologically distinct at the time of initiation. These have pentarch or hexarch steles and a central pith.[2,4] The majority of the nonstorage roots are tetrarch, although some may be pentarch, hexarch, or septarch. Some of the roots that do not develop, for one reason or another, into storage roots appear to form pencil roots.[2] These tend to be small in diameter (~ 15 mm) and relatively uniform in thickness over their entire length. They do not form the vascular bundles and secondary phloem parenchyma as do the roots that subsequently develop into normal storage roots.

Potential storage roots generally arise from the nodes of the below ground portion of the

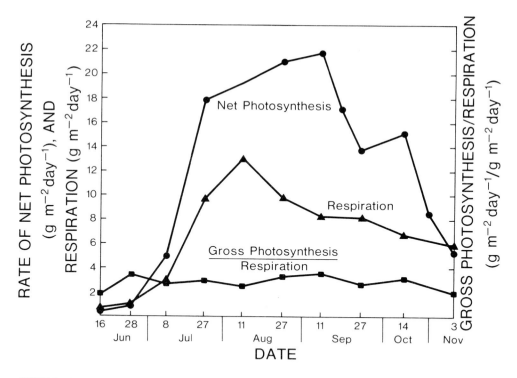

FIGURE 16. Changes in the rate of net photosynthesis and dark respiration of a sweet potato plant (cultivar Koganesenga) and the ratio of gross photosynthesis to respiration during the growing season. (From Agata, T. and Takeda, T., *J. Fac. Agric. Kyushu Univ.*, 27, 75, 1982. With permission.)

stem and frequently from the more apical area of the stem.[2] In contrast, fibrous roots are typically of internodal origin.

B. Storage Root Induction

The induction phase is the beginning of the period of development of the storage root. At this time, a sufficient number of the pentarch and hexarch roots begin to enlarge to accommodate the photosynthate moving into the root zone. It is evident that the plant modulates the number of storage roots that are induced in response to the prevailing conditions. How this assessment is made is not presently clear. It may simply be in response to the concentration of photosynthates in the phloem.

Not all of the potential storage roots which are initiated are subsequently induced, however. A number appear to become pentarch and hexarch fibrous roots.[5] Some begin to develop but terminate their radial enlargement after a brief period of growth, becoming pencil roots. Togari[2] found the relationship between the rate of cambial activity within the root and the rate of lignification of the stele to be an accurate index of the eventual fate of the young root.

The length of the induction period and the number of roots induced appear to be cultivar characteristics. Lu et al.[89] found the secondary cambial growth to begin about 15 to 25 days after planting with secondary and tertiary cambium formed between 30 days and maturity.

C. Storage Root Development

Relatively little is known about the actual process of carbon deposition in the storage roots. Starch is the major storage form of carbon, accounting for approximately 20% of the

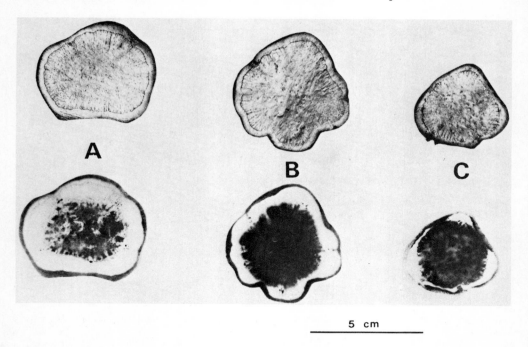

FIGURE 17. Distribution of assimilated $^{14}CO_2$ in cross sections of a sweet potato storage root obtained from a plant treated 44 days after transplanting and harvested for autoradiography after 106 days. (A) Section from the proximal portion of the root; (B) section from the medial portion; and (C) section from the distal portion. Mounted specimens on top, autoradiographs below. (From Chua, L. K. and Kays, S. J., *J. Am. Soc. Hortic. Sci.*, 107, 866, 1982. With permission.)

fresh weight of the roots. Starch grains are located in storage parenchyma cells of which Artschwager[4] discerns two types. There is the normal bundle parenchyma which are high in starch and are the product of the vasuclar cambium. These cells represent the major type of storage parenchyma. The other type is the interstitial parenchyma derived from the parenchymatous sheath of the young rootlet, i.e., parenchyma cells found between the primary xylem and phloem. These cells apparently contain considerably less starch than the bundle parenchyma.

Storage roots pulsed with [^{14}C]-photosynthate initially accumulate radioactivity in the vascular cambial region,[7] the zone of maximum meristematic activity. It is probable that the labeled carbon at this point is in a form other than starch, in that starch grains are generally not found in the vascular cambial region. This carbon appears to move into adjacent parenchyma cells where it is polymerized into starch grains.[90] Discernible activity can be found throughout the root 24 hr after labeling. This probably represents carbon directed toward maintenance reactions within the broad variety of cell types in the root.

Based on autoradiographs of labeled roots, the deposition of [^{14}C]-photosynthate within the storage roots appears to occur for more than 24 hr after an initial one hour pulse label. Hozyo[53] found some redistribution of [^{14}C]-photosynthate within the plant between the 1st and 7th day after labeling. An alternative possibility which would support this hypothesis would be the outward redistribution of stored carbon from within the root as new cells were added. This, however, is less likely since the patterns of demarcation between labeled and unlabeled areas in roots labeled when young, would not be as distinct as they were found to be when harvested at the end of the season (Figure 17).

Roots labeled early in their development do not have a uniform distribution of [^{14}C] in the central stele (Figure 17) when reaching a harvestable size.[7] This appears to be due to the later development of anomalous cambium within this region. The contribution of the anomalous cambia to the overall volume of the storage root is a cultivar characteristic. Wilson and Lowe[5] found that with three of the West Indian cultivars they studied, the anomalous cambia accounted for approximately 50% of the storage tissue, while with one cultivar it accounted for virtually none.

The degree of contribution of the anomalous cambia to the development of the roots total volume also appears to vary with location within the root.[7] [^{14}C]-photosynthate deposited early in the development of the storage root was concentrated at the distal (root) end with longitudinal bulking proceeding toward the proximal end. This is seen in the amount of unlabeled tissue in the proximal end of the root at harvest when labeled as a young storage root (Figure 17). It is also supported by anatomical studies.[5] Specialized tissues at the distal end of the storage root allow for the longitudinal extension in very young roots. Later these tissues exhibit the structure and presumably the function of normal secondary thickened roots, thus delimiting the storage root at the distal end. Therefore, longitudinal growth appears to progress toward the proximal end.

VIII. GROWTH ANALYSES

The development of new more productive sweet potato cultivars is facilitated by a thorough understanding of the operative mechanisms affecting crop growth and yield. These can be studied at an integrated level where gross area and weight relationships of specific plant parts are analyzed relative to some variable of interest (e.g., time, fertility, and so on). Another approach is to study the plant at a general process level (e.g., photosynthetic or respiratory rate.) One may also focus on specific steps within a process (e.g., the activity of a particular enzyme). The general objectives of each type of analysis are to understand the factors controlling plant growth, and in so doing, help to both maximize the productivity of existing cultivars and facilitate the development of new higher yielding cultivars.

A. Plant Area and Volume Measurements

As with most plant species, the analysis of plant area, or in some cases, volume relationships has focused largely upon the aboveground plant parts, specifically the leaves. Leaf area has been utilized as a general measure of the amount of photosynthetic machinery, thus a measure of the energy trapping and carbon dioxide fixing potential of the plant. Leaf area has been studied on the basis of: (1) the individual plant [leaf area (LA), leaf area/plant]; (2) unit of land covered [leaf area index (LAI), leaf area/unit area of land]; and (3) longevity of leaf area [leaf area duration (LAD), leaf area integrated over time]. In nearly all cases, the leaves are treated as two-dimensional units. Sweet potato storage roots, however, are distinctly three-dimensional and do not lend themselves to this type of surface area analysis. As a consequence, root measurements have been made largely on a volume or length and diameter basis.

1. Leaf Area

Leaf area is a function of the total number of leaves and the size of the leaves per plant. Both are modulated by the environmental conditions under which the plant is grown, however, leaf number has a much greater plasticity in response. For example, the mean number of leaves on plants of the cultivar Jewel at the end of the growing season ranged from a high of 373 at the widest plant spacing (45 × 96 cm) to a low of 117 at the closest spacing (15 × 96 cm).[21] The mean area per leaf, however, only varied slightly (72.9 vs. 66.1 cm^2, respectively). Leaf number is a function of the number of growing points (branches), the general rate of growth of the plant and the length of time the leaves remain on the plant. Differences in leaf size arise due to effects on cell division, thus number, and/or on cellular expansion.

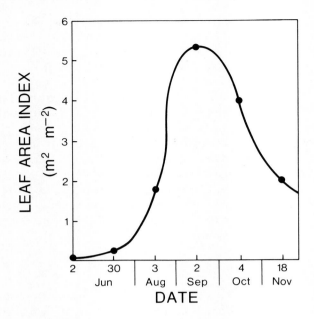

FIGURE 18. Changes in the leaf area index of the sweet potato during the growing season. (From Kotama, S., Chuman, K., and Tanoue, M., *Bull. Kyushu Agric. Exp. Stn.*, 15, 493, 1970.)

The sweet potato continues to produce leaves throughout its entire growth period. However, under some conditions the leaf area per plant may plateau and/or decline. This is due, in part, to an increase in the number of leaves that are shed by the plant late in the season.

2. Leaf Area Index

The area of leaves per unit area of land for terrestrial plants ranges from 2 up to 15. The optimum leaf area index is generally lower in the sweet potato than in many other crop plants. For example, rice has an optimum leaf area index of 6 to 12,[91] pineapple 9 to 10,[92] while the sweet potato has an optimum leaf area index of only 3 to 4. This is due largely to the essentially horizontal development of the canopy and the plants poor leaf orientation, which accentuates the shading of leaves within the canopy.

The leaf area index for the sweet potato changes with light intensity[40] and during the development of the plant (Figure 18). Tsuno and Fujise[93] found the optimum leaf area index at an average solar radiation of 430 g cal cm^{-2} day^{-1} to be 3.2 m^{2}m^{-2}. This gave a maximum dry matter production rate of 120 g m^{-2} week^{-1}. Generally, as the leaf area index rose above 3 to 4, the crop growth rate (g m^{-2} week^{-1}) declined.

Numerous factors are known to alter the leaf area index. For example, soil fertility has a substantial effect (Figure 19).[19] Under high soil fertility conditions the maximum leaf area index are 1 to 2 units higher than under low fertility conditions. Soil nitrogen concentration is known to have a pronounced effect on leaf area index; under conditions of excessively high nitrogen the leaf area index may increase to above the optimum resulting in a decrease in total dry matter accumulation rate.[94]

Leaf area index also varies due to cultivar. Cultivars that are adapted to high altitudes (Africa) and poor soils tend to have lower leaf area indexes.[95]

3. Leaf Area Duration

In opposition to the development of progressively greater leaf areas within a sweet potato

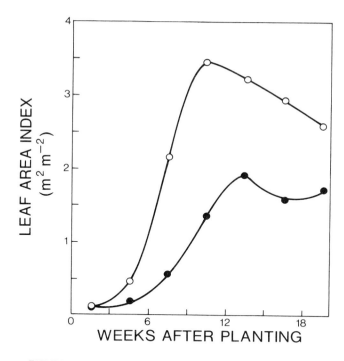

FIGURE 19. Changes in the leaf area index of the sweet potato during the growing season at two soil fertility levels: -O- high fertility, -●- low fertility. (From Tsuno, Y. and Fujise, K., *Bull. Nat. Agric. Sci. (Japan)*, D13, 1, 1965. With permission.)

canopy are a number of forces. External factors such as insect or animal predation depresses the leaf area. Perhaps most importantly, however, is the shedding of leaves by the plant. Because of the physical nature of the canopy, the light intensity within the lower most portions often becomes insufficient to maintain the leaves found there. These tend to be the oldest leaves and have the shortest petioles (Figure 5). As a consequence, these leaves are often the first to abscise.

Leaf area duration, leaf area integrated over time*, gives a measure of the longevity of the photosynthetic surface of the plant. Leaf area duration values of up to 88 weeks have been reported.[96] There is an inverse linear relationship found between the leaf area duration and the dry weight of plant parts (leaf blades and petiole) shed from the plant.[19] Plant nitrogen status has a pronounced effect on leaf area duration. As the amount of nitrogen absorbed by the plant increased from 5 to 11 g m^{-2} of field, the leaf area duration increased from 20 to 50. Under conditions of excessively high fertilization, however, an overabundance of leaf growth occurs accelerating leaf shedding and reducing the leaf area duration.[19]

Leaf area duration varies substantially among cultivars grown under similar conditions. For example, the cultivar "Fadenawena" had a leaf area duration of 36.4 weeks while for Laloki No. 2 it was 87.9 weeks.[96] The leaf area duration also varies widely from year to year for the same cultivar. The percent difference between two years for the same cultivar ranged from 5.8 to 45.4.[96]

4. Storage Root Volume

Storage root volume can be ascertained by water displacement or by determining the

---

* $LAD = 1/2 \Sigma (LAI_n + LAI_{n+1})(time_{n+1} - time_n)$.

relationship between length and width. Lowe et al.[97] used the following equation to determine storage root volume in six West Indian cultivars:

$$v = 4/3\ \pi ab^2 \qquad (1)$$

where v = volume, a = 0.5 × storage root length (cm), and b = 0.5 × storage root diameter (cm). The single most important root characteristic indicative of high yield was root diameter. This varied widely between cultivars as did the final shape of the storage roots. Shape established by the length to diameter ratio at final harvest ranged from globular (1.5) to narrow fusiform (4.4).[97] Thus, root shape is a cultivar characteristic[98] which is modulated by time, environmental, and cultural conditions.[2,99]

The importance of root diameter is seen in the chronological sequence of shape changes as the storage root enlarges. Storage root length and number are more or less established in the first two-thirds of the growing season. Thus, increases in yield have to come largely from increases in the radial diameter of the existing roots.

B. Plant Weight Measurements

Weight measurements of plant growth and component parts represent the most common form of information gathered for growth analyses. Generally, these measurements fall into one of two classes, either measures of biological yield (w) and its individual components (e.g., vines, leaves, storage roots, and fibrous roots) or the relationship between biological yield and one of its components. For example, harvest index (HI), the relationship between economical yield and biological yield (economic yield dry weight/total plant dry weight × 100), has been examined in a number of studies.

1. Biological Yield

Dry matter yields of the sweet potato vary widely, with yields of up to 26.5 t/ha reported.[96] Typically, biological yields are expressed on both a per plant and per unit of land area basis. While there have been numerous reports of "total" plant dry weight, seldom, however, are the fibrous roots included. Yoshida et al.[14] monitored the development of each major plant part (storage roots, fibrous roots, vines, and leaf blades) during growth (Figure 20). The storage roots constituted the largest percentage of the final plant weight/unit area of land (66.8%), followed by the vines (25.3%), leaf blades (7.1%), and fibrous roots (0.7%) (Figure 21).

2. Harvest Index

The harvest index gives an indication of the relative distribution of photosynthates between the storage roots and the remainder of the plant. This varies widely due to cultivar, plant nutrition, and other cultural and environmental factors. Huett[100] found that the harvest index ranged from 9.5 to 66.3% for 16 cultivars evaluated in subtropical Australia. The relationship of the distribution of photosynthate (HI) and yield to photosynthetic efficiency between diallel crosses between 2 lines with high and 2 lines with low photosynthetic efficiency was studied by Hahn[101] (Figure 22). It was found that to a certain point photosynthetic efficiency may not be a limiting factor in high yield, but rather the distribution of photosynthates may be of greater importance.

C. Rate Measurements

The integration of weight or area measurements over time provides rate values that are highly useful for studying the growth of the sweet potato. Three of the plant measurements that are most commonly expressed on a rate basis are plant weight/unit area of land, weight/

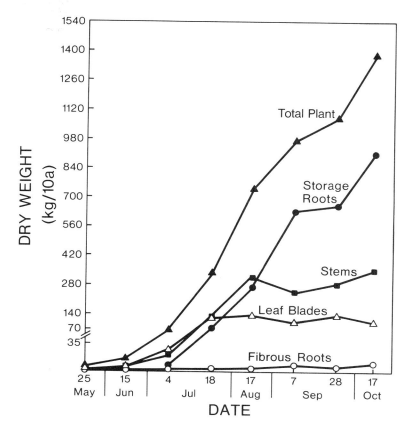

FIGURE 20. Changes in the dry weight of various plant parts per 10 a (0.1 ha) of land during the growing season. (From Yoshida, T., Hozyo, Y., and Murata, T., *Proc. Crop Sci. Soc. Jpn.*, 39, 105, 1970. With permission.)

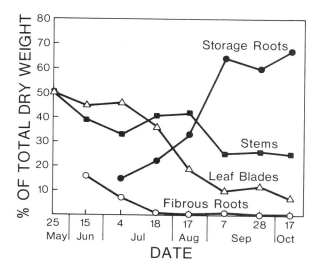

FIGURE 21. Changes in the percent of the total dry weight accounted for by each plant part during the growing season. (From Yoshida, T., Hozyo, Y., and Murata, T., *Proc. Crop Sci. Soc. Jpn.*, 39, 105, 1970. With permission.)

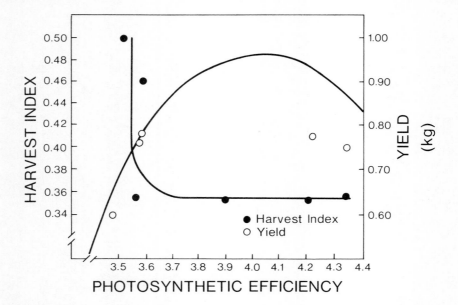

FIGURE 22. Relationship between photosynthetic efficiency and the yield components: harvest index and fresh yield. (From Int. Inst. of Tropical Agriculture, *Annu. Rep. Root and Tuber Improvement Program* 1974, Ibadan, Nigeria, 1974. With permission.)

plant and plant weight/leaf area. These are referred to in the literature as crop growth rate (CGR*, dry weight accumulated/unit of land/unit of time), relative growth rate (RGR**, dry weight/unit plant/unit time), and net assimilation rate (NAR***, dry weight/unit leaf area/unit time). Likewise, the component parts of the plant can be similarly treated. For example, the storage root or yield growth rate (YGR†) is a measure of the storage root weight/unit plant/unit time.

$$* \text{CGR} = \frac{(w_2 - w_1)}{(t_2 - t_1)} \tag{2}$$

where w_1 and w_2 are initial and final yields (g dwt m^{-2}) and $t_2 - t_1$, the length of the time interval (weeks or days) between each yield measure.

$$** \text{RGR} = \frac{\ln w_2 - \ln w_1}{t_2 - t_1} \tag{3}$$

where w_1 and w_2 are initial and final yields (g dwt plant^{-1}) and $t_2 - t_1$ the length of the interval (weeks or days) between each yield measure.

$$*** \text{NAR} = \frac{(w_2 - w_1)(\ln L_2 - \ln L_1)}{(t_2 - t_1)(L_2 - L_1)} \tag{4}$$

where w_1 and w_2 are initial and final yields (g dwt plant^{-1}), L_1 and L_2 are the leaf areas per plant at those times (dm^2 plant^{-1}) and $t_2 - t_1$ the length of the interval (weeks or days) between each yield measure.

$$† \text{YGR} = \frac{\text{SRW}_2 - \text{SRW}_1}{t_2 - t_1} \tag{5}$$

where SRW_1 and SRW_2 are the initial and final storage root weight/plant (g dwt plant^{-1}) and $t_2 - t_1$ the length of the interval (weeks or days) between each yield measure.

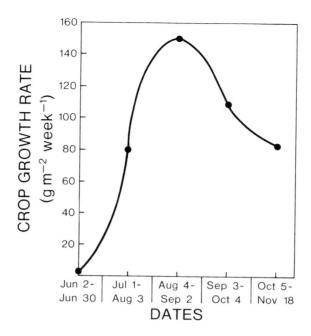

FIGURE 23. Changes in the crop growth rate during the growing season. (From Kotama, S., Chuman, K., and Tanoue, M., *Bull. Kyushu Agric. Exp. Stn.*, 15, 493, 1970. With permission.)

1. Crop Growth Rate

The rate at which dry matter is accumulated by the sweet potato is considerably lower than for many crops. For example, maize (245 g m^{-2} week^{-1}), sugar beet (230), paddy rice (220), and soybeans (160) are all substantially higher than the sweet potato (120).[102] The rate of growth of the crop is not static but changes as the plant and its environment changes during the growing season. The highest CGR (150 g m^{-2} week^{-1}) for a cultivar grown in Japan was 10 to 14 weeks after planting (Figure 23).[18] Early in the season the CGR is relatively low. Since rate of growth of the plant is, in part, a function of the leaf area available for photosynthesis, Tsuno[102] determined the relationship between leaf area and growth rate (Figure 24). Maximum growth rate of the plant was found to be when the leaf area index was 3.2. As the leaf area index increased or decreased from this point the crop growth rate declined.

The length of time the plant maintains a high crop growth rate is also an important factor in determining final yield. Compared to paddy rice which has a maximum crop growth rate substantially higher than the sweet potato, but similar yields/unit area, Tsuno found that the sweet potato compensates for its lower rate by functioning at its maximum for a longer period of time (Figure 25).

2. Relative Growth Rate

Most analyses of the relative growth rate of sweet potatoes have focused upon the harvested yield component of the plant (yield growth rate) rather than the total plant dry weight. Occasionally yield growth rate (YGR) is referred to as relative growth rate.

When the yield growth rate is plotted against the crop growth rate, the plants are seen to proportion the bulk of their dry matter into structures other than storage roots early in the season (Figure 26).[72] As the season progresses, the g dwt m^{-2} day^{-1} proportioned into the storage roots increases reaching a peak value of approximately 23 g m^{-2} day^{-1}. At the latest

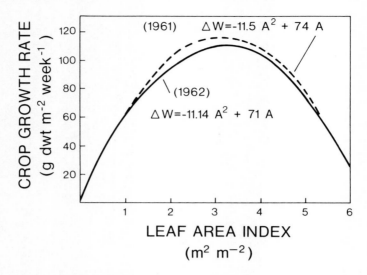

FIGURE 24. The relationship between leaf area index and crop growth rate. (From Tsuno, Y. and Fujise, K., *Bull. Nat. Agric. Sci. (Jpn.)*, D13, 1, 1965. With permission.)

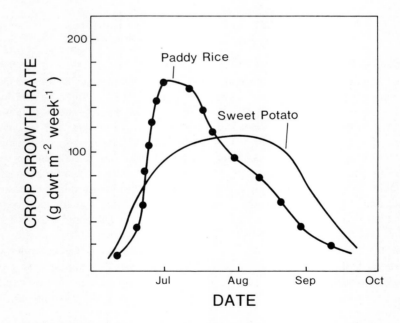

FIGURE 25. A comparison between the sweet potato and paddy rice with respect to their crop growth rates during an entire growing season. (From Tsuno, Y., *Fertilite*, 38, 3, 1970. With permission.)

sampling date, the yield growth rate actually superseded the crop growth rate indicating that there was a transport of carbon out of other plant parts into the storage roots and/or a loss of leaves. Yield growth rate was found to be highly correlated with crop growth rate (0.95★), net assimilation rate (0.95★) and mean solar radiation (0.94★), however, the correlations between crop growth and leaf area index (0.56) and mean air temperature (0.52) were not significant.[72]

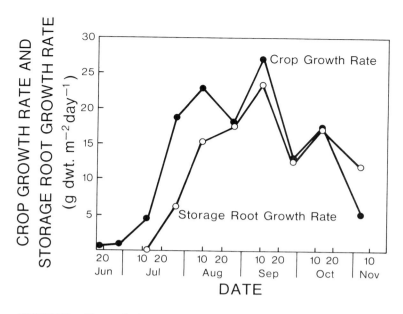

FIGURE 26. Changes in the storage root growth rate and the crop growth rate during the growing season. (From Agata, T. and Takeda, T., *J. Fac. Agric. Kyushu Univ.*, 27, 65, 1982. With permission.)

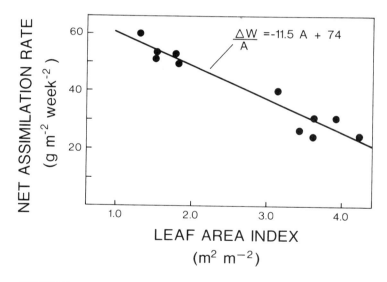

FIGURE 27. The relationship between the net assimilation rate and the leaf area index of the sweet potato. (From Tsuno, Y. and Fujise, K., *Bull. Nat. Agric. Sci. (Jpn.)*, D13, 1, 1965. With permission.)

3. Net Assimilation Rate

The increase in dry weight of the plant is, in part, a function of the surface area of leaves within the canopy fixing carbon. Thus, if the rate of dry matter increase is expressed on a per unit area of leaf basis, changes in the effectiveness of a unit of leaf area can be explored in relation to other variables. Tsuno and Fujise[19] found that there was a linear relationship between leaf area and the net assimilation (Figure 27). As the leaf area of the plant increases to a point where shading of the lower leaves within the canopy is significant, the overall net assimilation rate decreases. When plants were grown on soil low in fertility, and as a consequence had lower leaf area indexes, the net assimilation rate was substantially higher than those planted on fertile soils. Maximum net assimilation rates were between 60 and 70 g m^{-2} leaf area week^{-1}. The net assimilation rate also changes markedly during the season, the degree of which is a function of factors that alter the leaf area of the plant. Early in the growing season, the rate of change is high and the net assimilation rate increases with time. This later tends to peak or plateau and subsequently decline.

IX. ENVIRONMENTAL FACTORS AFFECTING YIELD

A. Radiation

Radiant energy from the sun provides the energy required by plants to fix carbon from carbon dioxide and, as a consequence, is a critical factor affecting yield. There are three general components of light that are of importance to sweet potato production: intensity, duration, and quality.

Individual leaves of the sweet potato when exposed to increasing light intensities exhibit typical light response kinetics. Photosynthesis of individual leaves of the sweet potato saturates at approximately 30 klux, about one-third of full sunlight.[25] Mutual shading of the lower leaves in the canopy, however, decreases the light intensity. While the upper leaves of the canopy may be light saturated, the entire plant seldom is.

There have been several reports of the effect of photoperiod on the energy acquisition by and allocation within the sweet potato.[103,104] Kim[104] found that long days stimulate storage root formation. This appears to be largely an effect on net carbon fixed, rather than a photoperiodic effect *per se*. This position is supported by the decreased length of the growing season as the same cultivar as grown progressively further from the equator. Thus, the cropping season tends to be longer in the tropics where the daylength is shorter than in the temperature zone.

Quality of the radiant energy has not been studied in the sweet potato. This is due largely to the fact that it is a parameter over which we can exert little influence. It is known that light quality changes as the radiation penetrates down into the canopy. This has been studied predominantly on a light quantity rather than quality basis.

B. Temperature

The growth and yield of sweet potatoes is highly temperature dependent.[105-110] Low temperatures in the field (i.e., less than 20°C) decrease yield[107] while higher temperatures favor storage root formation.[106,111] However, soil temperatures above around 30°C substantially decrease storage root yield.[112] Thermoperiodism appears to be critical in modulating the allocation of carbon within the plant.[112,113] Under conditions of high day temperatures (29°C) and low night temperatures (21°C) storage root development was facilitated; with continual high temperatures (day and night) vine growth was dominant with the virtual inhibition of allocation of carbon into storage roots. High soil temperatures give a similar effect.[113]

C. Length of Growing Season

In that yield increases with time, the longer the growing season, the higher the final gross yield. As the area of production moves north or south from the equator, the growing season becomes progressively shorter. In the equatorial tropics the sweet potato may be grown year round, while in the more northern and southern areas of the temperate zones the season may be as short as 120 days. Because of this, a much more intensive selection pressure has been utilized in temperate zone breeding programs for earliness. Some tropical cultivars require up to 350 days[100] while in the temperate areas 120 to 150 days are common. Early maturing cultivars grown in the tropics typically require longer growing periods than they do in the temperate zone. This is due to the progressively shorter photoperiod during the growing season as one approaches the equator. Thus, the total light energy per day is less.

D. Rainfall

A major portion of the sweet potatoes grown in the world are produced on nonirrigated land. Because of this, the natural rainfall pattern and amount are important environmental factors affecting yield. Under conditions of low rainfall, vine growth is repressed, decreasing the photosynthetic area of the plant which in turn decreases root yields. In areas of high rainfall, the amount of rain is negatively correlated with final root yield.[114] Under these conditions vine growth is stimulated at the expense of the storage roots. Thus, when there are distinct annual cycles in moisture availability (i.e., wet and dry periods) it is often advantageous to have the wet period coincide with the early part of the growth cycle of the crop.[115] During this period, vine growth is rapid and facilitated by substantial rainfall. Drier, sunny weather during the storage root development period, especially during the latter part of the season, enhances final yields.

X. PRODUCTION FACTORS AFFECTING YIELD

Unlike environmental factors over which we exert little influence, there are a number of production considerations that we can to a large extent control. These include soil factors (nutrition, pH, type, oxygen concentration, ridging), irrigation, plant population, length of growing period, cultivar, pest control, and growth regulators. Each have been shown to have a significant influence on yield.

A. Soils
1. Soil Fertility

The relationship between soil fertility and subsequent yield has received perhaps the greatest attention of all the production factors (for a detailed review see Tsuno[102]). Soil and plant concentration of both potassium and nitrogen appear to be critical. Potassium has a strong influence on storage root growth;[116-126] generally as the plant potassium level increases so do yields. Under low potassium conditions long slender storage roots are formed.[119] Increasing potassium does not appear to change the leaf area index of the plant, however, it does increase dry matter production.[120] Yield, therefore, appears to be highly correlated with potassium concentration. Ho et al.[121] observed a correlation coefficient of 0.88 between yield index and leaf blade potassium content 40 days after planting and 0.98 at 100 days after planting. By the end of the season, the storage roots contain the major portion of the total amount of potassium in the plant.[122] The effect of potassium is thought to be, in part, due its influence on storage root enlargement,[123] i.e., sink strength. While increases in potassium result in a decrease in the percent dry matter of the storage roots,[124] the substantially higher yield of the plants gives a net effect of more dry matter per hectare. Potassium also influences the rate of photosynthesis of the leaves. As the potassium concentration of the leaf increases so does the photosynthetic rate ($r = 0.82$).[29]

The direct effect of potassium on the enzyme starch synthetase (adenosine-diphosphate glucose- and uridine-diphosphate glucose transglucosylase) and subsequently on the rate of starch synthesis could account for these effects of potassium on the plant.[125] The addition of 0.1 M of potassium, roughly equivalent to the cellular concentration of potassium in storage root, increased the enzyme activity by sevenfold. Thus, high potassium would increase the concentration gradient between source and sink with secondary effects on the rates of photosynthesis and transport.

Nitrogen also appears to exert a dominant influence on dry matter production and distribution. High levels of soil nitrogen stimulate vine growth at the expense of the storage roots.[118,126] Togari[2] suggests that this is due to an increase in lignification of the steles of the young adventitious roots (see Figure 2) resulting in the formation of fibrous roots in lieu of storage roots. Such conditions result in higher than optimum leaf area indexes[19] and greater leaf shedding in the canopy. When nitrogen is deficient, vine growth is minimized, decreasing the photosynthetic surface area of the plant which, in turn, suppresses yields. Leaf nitrogen concentration appears to be closely correlated with photosynthetic rate. When the concentration of nitrogen is below 2.2%, photosynthesis declined dramatically.[19] Thus, nitrogen has a pronounced effect not only on the acquisition of carbon, but also upon its allocation within the plant.

A number of other nutrients have been shown to modulate both net productivity and dry matter distribution in the plant under controlled conditions. For example, B, Ca, S, Mg, Mn, P, K, N, and Fe deficiencies all have been shown to decrease yields.[127-133] Mn, P, K, N, and Fe appear to have the greatest effect on the top to storage root ratio.[128] The leaf blade nutrient concentration at which deficiency symptoms occur are N - 1.5%, S - 0.8%, P - 0.1%, Mg - 0.05%, K - 0.5%, and Ca - 0.2%.[134]

2. Soil pH

The sweet potato tolerates a wide range in soil pH,[135] however, optimum pH is generally thought to be in the 6.1 to 7.7 range.[136] In the tropical soils of Puerto Rico, highest yields were found in soils with a pH above 5 and an aluminum saturation of less than 20% with aluminum exchangeable bases of less than 2.[137] These conditions varied with soil type.

3. Soil Type

Soil type is also known to influence the distribution of dry matter within the plant. Heavy clay soils tend to result in more stems and leaves in relation to storage roots than do more friable soils.[138] This appears to be due, in part, to the higher moisture level and lower oxygen concentration often found in the tighter soils. In addition, soil type can display a significant effect on storage roots shape and grade.[139,140]

4. Soil Oxygen Concentration

Sweet potatoes grown under waterlogged conditions or in solution culture,[141] typically fail to produce storage roots. In addition, with field production of sweet potatoes on many soils, yields are markedly improved with ridging.[142] This effect appears to be largely due to the influence of these treatments on the oxygen concentration within the root zone.[3,142] Storage root induction and/or development is almost totally repressed by a root zone environment of 2.5% oxygen[3] while top growth was substantially increased. Development (enlargement) of existing storage roots is also severely inhibited by conditions resulting in a low oxygen concentration within the root.[144-146] Thus wet conditions decrease the root dry weight per unit area of leaf.[147] Removal of the low oxygen conditions results in a resumption of the normal growth pattern.[3]

5. Ridging or Mounding

Use of ridges or mounds has long been a practice in sweet potato culture.[132,137,138] Yield increases with the use of ridges varies with soil type and the amount of rainfall within the growing season. The advantage appears to be due to better soil oxygen status, lower mechanical resistance and in the temperate zones, earlier soil warming in the spring.

B. Irrigation

Irrigation requirements of the sweet potato vary widely depending upon the amount of

Table 3
RELATIONSHIP BETWEEN PLANT POPULATION, YIELD AND PLANTS REQUIRED PER HECTARE[157]

In row Spacing (cm)	Yield[a] (T/ha)	Difference (T/ha)	Plants needed (No./ha)	Difference (No./ha)
20	15.2		46,128	
		1.1		15,374
30	14.1		30,754	
		1.8		23,064
		0.7		7,690
40	13.4		23,064	

[a] Yield of U.S. #1 and #2 roots at 25 kg/bu.

natural rainfall, the level of evapotranspiration, the water-holding capacity of the soil root zone and the stage of development of the plant. Both insufficient and excess irrigation have been shown to depress yields. Under insufficient moisture conditions, vine growth is inhibited[150] while excess soil moisture results in an overabundance of vine growth.[144] Dry conditions have been shown to also result in the formation of more pencil roots[2] and storage roots at greater depths in the soil.[8] The response to water deficits varies with cultivar.[151] Those that do well appear to develop an extensive root system early in their growth cycle.[152] Thus, irrigation can significantly influence both total carbon acquisition and its allocation within the plant. Optimum soil moisture level for storage root production is approximately 45% of field capacity,[153-155] with higher and lower levels decreasing yields.[64] Higher soil moisture levels have been shown to alter the allocation of [^{14}C] from the leaves to sites other than the storage roots.[156]

C. Plant Population

Generally, as the in-row spacing increases (20 to 81 cm) the total yield decreases slightly while the number of oversize roots increases.[157] Often the additional yields obtained at the closer plant spacings do not compensate for by the relatively higher production costs due to the additional transplants or root pieces required per hectare (Table 3).

D. Length of Growing Period

It is apparent from studies on the crop and yield growth rates of the sweet potato that the plant has a relatively plastic sink and the longer it remains in the field, under suitable environmental conditions, the greater the yield. In more temperate areas, a decline in the rate of growth of the storage roots is commonly seen in the late fall. Much of this appears to be due to concurrent environmental changes which decrease the net assimilation rate of the plant. There is some indication that sink strength may decline when the storage roots become quite large.

While storage root gross yields increase with time, the yield of marketable roots for the fresh market peaks and then declines. This point in the development of the roots depends to a large extent upon their end use and the quality criteria utilized. When harvest comes before the end of the growing season, gross yields are often substantially below their potential. Thus, much of the yield data for fresh market sweet potatoes from production areas that have rigid size requirements, significantly underestimates the true yield potential of the crop.

Cultivar has a significant effect upon the number of growing days required to produce a marketable yield. Late developing types may require as much as 350 days.[100] Some cultivars selected in areas with short growing season, however, can be harvested as early as 90 days after planting.

E. Insects, Diseases, and Nematodes

The sweet potato is subject to yield decreases due to insects, diseases, and nematodes. The effect on yield ranges from slight to total crop loss depending upon the organism and severity of infestations. Since this subject is covered in detail in Chapters 2 and 3 of this text, it will not be discussed here.

F. Plant Growth Regulators

A number of studies have focused upon the endogenous hormones and factors modulating their concentration within the sweet potato[158-164] and the use of synthetic compounds to improve growth of the plant.[165-176] While gibberellic acid, auxin, ethylene, cytokinin, and abscisic acid have been isolated from the plant, their specific role has not been elucidated. Several attempts have been made to develop a role for specific endogenous hormones within the sweet potato to explain some physiological event,[158,177] however, these have not been substantiated. Progress in this area of research on the sweet potato awaits significant advances in the general understanding of the molecular mode of action of these hormones in plants.

Production uses of plant growth regulators have met with little success. Ethylene, whether applied as a gas or via an ethylene releasing or inducing compound has been shown to enhance sprout production from storage roots. This, along with the other uses of agricultural growth regulators to modify the growth of the plant have not found their way into commercial use.

XI. CONCLUSIONS

While a number of important contributions to our knowledge of the physiology of yield in the sweet potato have been made over the past 20 years, our understanding of the plant factors limiting yield remains insufficient. Physiologists, as a consequence, have the potential to play a key role in the development of new high yielding cultivars. This, however, requires being able to both identify major rate limiting processes and integrating this information into ongoing breeding programs. Two key areas for improvement appear to be leaf geometry and the photosynthate transport and storage system.

REFERENCES

1. **Weaver, J. E. and Bruner, W. E.,** *Root Development of Vegetable Crops,* McGraw-Hill, New York, 1927, 351.
2. **Togari, Y.,** A study in the tuberous root formation of sweet potatoes, *Nat. Agric. Exp. Stn. Bull. (Tokyo),* 68, 1950.
3. **Chua, L. K. and Kays, S. J.,** Effects of soil oxygen concentration on sweet potato storage root induction and/or development, *HortScience,* 16, 71, 1981.
4. **Artschwager, E.,** On the anatomy of the sweet potato root, with notes on internal breakdown, *J. Agric. Res.,* 23, 157, 1924.
5. **Wilson, L. A. and Lowe, S. B.,** The anatomy of the root system in West Indian sweet potato *Ipomoea batatas* (L.) Lam. cultivars, *Ann. Bot.,* 37, 633, 1973.
6. **Esau, K.,** *Anatomy of Seed Plants,* John Wiley & Sons, New York, 1960.
7. **Chua, L. K. and Kays, S. J.,** Assimilation patterns of [^{14}C]-photosynthate in developing sweet potato storage roots, *J. Am. Soc. Hortic. Sci.,* 107, 866, 1982.
8. **Garcia, F.,** Sweet potato culture, *N.M. Agric. Exp. Stn. Bull.,* 70, 1909.
9. **Bouwkamp, J. C. and Scott, L. E.,** Production of sweet potatoes from root pieces, *HortScience,* 7, 271, 1972.

10. **Kobayaski, M.**, Studies on breeding and vegetative propagation of sweet potato varieties adapted to direct planting, *Bull. Chugohu Agric. Exp. Stn.*, 16, 245, 1968.
11. **Kobayaski, M. and Akita, S.**, Studies on breeding of sweet potato varieties adapted to direct planting. 3. Comparison of plant with different root types, *Jpn. J. Breed.*, 19, 144, 1969.
12. **Kays, S. J. and Stuttle, G. W.**, Proximal dominance and sprout formation in sweet potato *(Ipomoea batatas* (L.) Lam.) root pieces, *5th Int. Symp. Trop. Root and Tuber Crops*, Manila, Philippines, 1979, 41.
13. **Watanabe, K. and Nakayama, K.**, Studies on the effects of soil physical conditions on the growth and yield of crop plants. X. Effects of nontuberous roots on the distribution of dry matter in sweet potatoes, *J. Crop Sci. Soc. Jpn.*, 39, 446, 1970.
14. **Yoshida, T., Hozyo, Y., and Murata, T.**, Studies on the development of tuberous roots in sweet potato *(Ipomoea batatas*, Lam. var. *edulis*, Mak.). The effect of deep placement of mineral nutrients on the tuber yield of sweet potato, *Proc. Crop Sci. Soc. Jpn.*, 39, 105, 1970.
15. **Chapman, T. and Cowling, D. J.**, A preliminary investigation into the effects of leaf distribution on the yields of sweet potato *(Ipomoea batatas)*, *Trop. Agric. (Trinidad)*, 42, 199, 1965.
16. **Austin, M. E. and Aung, L. H.**, Patterns of dry matter distribution during development of sweet potato *(Ipomoea batatas)*, *J. Hortic. Sci.*, 48, 11, 1973.
17. **Johnson, W. A. and Ware, L. M.**, Effects of rates of nitrogen on the relative yields of sweet potato vines and roots, *Proc. Am. Soc. Hortic. Sci.*, 52, 313, 1948.
18. **Kotama, S., Chuman, K., and Tanoue, M.**, On the growth differentials of sweet potato to different soil fertility, *Bull. Kyushu Agric. Exp. Stn.*, 15, 493, 1970.
19. **Tsuno, Y. and Fujise, K.**, Studies on the dry matter production of sweet potato, *Bull. Nat. Agric. Sci. (Jpn.)*, D13, 1, 1965.
20. **Yen, D. E.**, *The Sweet Potato and Oceania*, Bishop Museum Press, Honolulu, Hawaii, 1974, 389.
21. **Somda, Z. C. and Kays, S. J.**, manuscript in preparation, 1985.
22. **Tsunoda, S.**, A developmental analysis of yielding ability in varieties of field crops. II. The assimilation-system of plants as affected by the form, direction and arrangement of single leaves, *Jpn. J. Breed.*, 9, 237, 1959.
23. **Inada, K.**, Action spectra for photosynthesis in higher plants, *Plant Cell Physiol.*, 17, 355, 1976.
24. **Vines, H. M., Tu, Z.-P., Armitage, A. M., Chen, S.-S., and Black, C. C., Jr.**, Environmental responses of the post-lower illumination CO_2 burst as related to leaf photorespiration, *Plant Physiol.*, 73, 25, 1983.
25. **Fujise, K. and Tsuno, Y.**, Study on the dry matter production of sweet potato. I. Photosynthesis in the sweet potato with special reference to measuring of the intact leaves under natural conditions, *Proc. Crop Sci. Soc. Jpn.*, 31, 145, 1962.
26. **Magnuson, C. E., Fares, Y., Goeschl, J. D., Nelson, C. E., Strain, B. R., Jaeger, C. H., and Bilpuch, E. G.**, An integrated tracer kinetics system for studying carbon uptake and allocation in plants using continuously produced $^{11}CO_2$, *Radiat. Environ. Biophys.*, 21, 51, 1982.
27. **Kays, S. J., Chua, L. K., Goeschl, J. D., Magnuson, C. E., and Fares, Y.**, Assimilation patterns of carbon in developing sweet potatoes using ^{11}C and ^{14}C, in *Sweet Potato Proc. 1st Int. Symp.*, Villareal, R. L. and Griggs, T. D., Eds., Asian Vegetable Research and Development Center, Shanhua, Tainan, Taiwan, 1982, 95.
28. **Pallas, J. E., Jr. and Kays, S. J.**, Inhibition of photosynthesis by ethylene — a stomatal effect, *Plant Physiol.*, 70, 598, 1982.
29. **Tsuno, Y. and Fujise, K.**, Studies on the dry matter production of the sweet potato. VIII. The internal factors influence on photosynthetic activity of sweet potato leaf, *Proc. Crop Sci. Soc. Jpn.*, 33, 230, 1965.
30. **Bhagsari, A. S. and Harmon, S. A.**, Photosynthesis and photosynthate partitioning in sweet potato genotypes, *J. Am. Soc. Hortic. Sci.*, 107, 506, 1982.
31. **Kato, S., Hozyo, Y., and Shimotsuko, K.**, Translocation of ^{14}C-photosynthates from the leaves at different stages of development in *Ipomoea* grafts, *Jpn. J. Crop. Sci.*, 48, 254, 1979.
32. **Katayama, Y. and Shida, S.**, Studies on the variation in leaf pigments by means of paper chromatography. III. Leaf pigments and carbon assimilation in some strains of sweet potato, *Mem. Fac. Agric. Univ. Miyasaki*, 3, 11, 1961.
33. **Tsuno, Y.**, The influence of transpiration upon the photosynthesis in several crop plants, *Proc. Crop Sci. Soc. Jpn.*, 44, 44, 1975.
34. **Kuraishi, S. and Nito, N.**, The maximum leaf surface temperatures of the higher plants observed in the Island Sea area, *Bot. Mag.*, 93, 209, 1982.
35. **Kato, S. and Hozyo, Y.**, The interrelationship between translocation of ^{14}C-photosynthate and $^{14}CO_2$ exposed leaf position on the grafts of *Ipomea*, *Proc. Crop Sci. Soc. Jpn.*, 45, 351, 1976.
36. **Tsuno, Y.**, Dry mattter production of sweet potatoes and yield increasing techniques, *Fertilite*, 38, 3, 1970.

37. **Naka, J. and Tamaki, K.,** Studies on the physiological nature of sweet potato plants. VII. On the relations between the variations of carbohydrates in the shoots and roots during the growing season, *Tech. Bull. Fac. Agric. Kagawa Univ.*, 9, 47, 1957.
38. **Kays, S. J., Goeschl, J. D., Magnuson, C. E., and Fares, Y.,** manuscript in preparation, 1985.
39. **Goeschl, J. D., Magnuson, C. E., and Fares, Y.,** unpublished data, 1984.
40. **Agata, T. and Takeda, T.,** Studies on dry matter production in sweet potato plants. 2. Changes of gross and net photosynthesis, dark respiration and solar energy utilization with growth under field conditions, *J. Fac. Agric. Kyushu Univ.*, 27, 75, 1982.
41. **Haseba, T. and Ito, D.,** Leaf temperature in relation to meteorological factors. 1. Leaf temperature variations with wind speed and solar radiation, *Mem. Coll. Agric. Ehime Unvi.*, 25, 29, 1980.
42. **Haseba, T. and Ito, D.,** Studies of transpiration in relation to the environment. 7. Variation of transpiration rate and leaf temperature with meteorological elements, *J. Agric. Meteorol.*, 30, 173, 1975.
43. **Sung, F. J. M.,** Effect of leaf water status on stomatal activity, transpiration and nitrate reductase of sweet potato, *Agric. Water Manage.*, 4, 465, 1981.
44. **Hozyo, Y.,** Photosynthetic activity and carbon dioxide diffusion resistance as factors in plant production in sweet potato plants, in *Sweet Potato Proc. 1st Int. Symp.*, Villareal, R. L. and Griggs, T. D., Eds., Asian Vegetable Research and Development Center, Shanhua, Tainan, Taiwan, 1982, 129.
45. **Hozyo, Y. and Kato, K.,** The effects of carbon dioxide concentration on photosynthetic activity of leaf blade in *Ipomoea* grafts, *Proc. Crop Sci. Soc. Jpn.*, 46 (extra issue), 107, 1976.
46. **Harazono, Y. and Yabuki, K.,** Studies on the effect of leaf boundry layer resistance on the production of a crop. 1. Effects of wind direction to leaf and angle of attack to leaf on the boundry layer resistance of sweet potato leaf, *J. Agric. Meteorol.*, 37, 103, 1981.
47. **Münch, E.,** *Die Stoffbewegungen in der Pflanze*, Fisher, Jena, 1930.
48. **Giaquinta, R. T.,** Phloem loading of sucrose. Involvement of membrane ATPase and proton transport, *Plant Physiol.*, 63, 744, 1979.
49. **Kato, S. and Hozyo, Y.,** The speed and coefficient of ^{14}C-photosynthates translocation in the stem of grafts between improved variety and wild type plant in *Ipomoea*, *Bull. Nat. Inst. Agric. Sci. Ser. D*, 29, 113, 1978.
50. **De Calderon, C., Acock, M., and Garner, J. O., Jr.,** Phloem development in sweet potato cultivars, *HortScience*, 18, 335, 1983.
51. **Hahn, S. K.,** A quantitative approach to source potentials and sink capacities amoung reciprocal grafts of sweet potato varieties, *Crop Sci.*, 17, 559, 1977.
52. **Hozyo, Y.,** Growth and development of tuberous root in sweet potato, *Proc. 2nd Intern. Symp. Trop. Root and Tuber Crops Hawaii*, 1, 24, 1970.
53. **Hozyo, Y.,** The influence of source and sink on plant production of *Ipomoea* grafts, *Jpn. Agric. Res. Quart.*, 11, 77, 1977.
54. **Wilson, L. A.,** The use of rooted leaves and grafted plants for the study of carbohydrate metabolism in sweet potato, *Proc. Intern. Symp. Trop. Root Crops Trinidad*, 1(II), 46, 1967.
55. **Neales, T. J. and Incoll, L. D.,** The control of leaf photosynthesis rate by the level of assimilate concentration in the leaf: a review of the hypothesis, *Bot. Rev.*, 34, 107, 1968.
56. **Hozyo, Y. and Kato, S.,** The interrelationship between source and sink of the grafts of wild type and improved variety of *Ipomoea*, *Proc. Crop Sci. Soc. Jpn.*, 45, 117, 1976.
57. **Kato, S. and Hozyo, Y.,** Translocation of ^{14}C-photosynthates in grafts between the wild type and improved variety in *Ipomoea*, *Proc. Crop Sci. Soc. Jpn.*, 41, 496, 1972.
58. **Kato, S. and Hozyo, Y.,** Translocation of ^{14}C-photosynthates in several growth stages of the grafts between improved variety and wild type plants in *Ipomoea*, *Bull. Nat. Inst. Agric. Sci. Ser. D*, 25, 31, 1974.
59. **Jenner, C. F. and Rathjen, A. J.,** Factors limiting the supply of sucrose to the developing wheat grain, *Ann. Bot. (London)*, 36, 729, 1972.
60. **Jenner, C. F. and Rathjen, A. J.,** Limitations to the accumulation of starch in the developing wheat grain, *Ann. Bot. (London)*, 36, 743, 1972.
61. **Jenner, C. F.,** Factors in the grain regulating the accumulation of starch, in *Mechanism of Regulation in Plant Growth*, Bieleski, R. L., Ferguson, A. R., and Cresswell, M. M., Eds., *Royal Soc. N.Z. Bull.*, 12, 901, 1974.
62. **Jenner, C. F. and Rathjen, A. J.,** Factors regulating the accumulation of starch in ripening wheat grain, *Aust. J. Plant Physiol.*, 2, 311, 1975.
63. **ap Rees, T.,** Integration of pathways of synthesis and degradation of hexose phosphates, in *The Biochemistry of Plants*, Vol. 3, Stumpf, P. K. and Conn, E. E., Eds., Academic Press, New York, 1980, 1.
64. **Hirai, G.,** A physiological study on variations of dry matter percent of tuber in sweet potato, *The Educt. Univ. Osaka*, 17 (Part 11, No. 2), 33, 1968.
65. **Goeschl, J. D., Magnuson, C. E., DeMichele, D. W., and Sharpe, P. J. H.,** Concentration-dependent unloading as a necessary assumption for a closed form mathematical model of osmotically driven pressure flow in phloem, *Plant Physiol.*, 58, 556, 1976.

66. **Magnuson, C. E., Goeschl, J. D., Sharpe, P. H., and DeMichele, D. W.**, Consequences of insufficient equation in models of the Münch hypothesis of phloem transport, *Plant, Cell Environ.*, 2, 181, 1979.
67. **Hozyo, Y. and Park, C. Y.**, Plant production in grafted plants between wild type and improved variety of *Ipomoea*, *Bull. Nat. Inst. Agric. Sci. Ser. D*, 22, 145, 1971.
68. **Boussingault, J. B.**, *Agronomie, Chimie Agricole et Physiologie*, 5 Vols., 2nd ed., Mallet Bachelier, Paris, 1868.
69. **Hozyo, Y. and Kato, S.**, Thickening growth inhibition and re-thickening growth of tuberous roots of sweet potato plants *(Ipomoea batatas*, Poiret). *Proc. Crop Sci. Soc. Jpn.*, 45, 131, 1976.
70. **Hozyo, Y., Murata, T., and Yoshida, T.**, The development of tuberous roots in grafting sweet potato plants, *Ipomoea batatas* Lam. *Bull. Nat. Agric. Sci. Ser. D*, 22, 165, 1971.
71. **Chua, L. K.**, Studies on the Developmental Physiology of the Sweet Potato Storage Root, M.S. Thesis, University of Georgia, Athens, 1980, 75.
72. **Agata, T. and Takeda, T.**, Studies on matter production in sweet potato plants. 1. The characteristics of dry matter and yield production under field conditions, *J. Fac. Agric. Kyushu Univ.*, 27, 65, 1982.
73. **Scott, L. E. and Bouwkamp, J. C.**, Seasonal mineral accumulation by the sweet potato, *HortScience*, 9, 233, 1974.
74. **Theologis, A. and Laties, G. G.**, Cyanide-resistant respiration in fresh and aged sweet potato slices, *Plant Physiol.*, 62, 243, 1978.
75. **Okamoto, S.**, The growth and the respiration in roots of sweet potato plants under a moderate potassium deficiency, *Soil Sci. Plant Nutr.*, 15, 175, 1969.
76. **Okamoto, S., Ikeda, K., and Yoshikura, J.**, Effect of potassium nutrition on contents of chemical constituents in sweet potato plants, *Sci. Rep. Fac. Agric. Kobe Univ.*, 10, 103, 1971.
77. **Collins, W. W. and Pharr, D. M.**, Flooding damage in sweet potatoes, in *Breeding New Sweet Potatoes for the Tropics*, Martin, F. W., Ed., Amer. Soc. Hortic. Sci., Tropical Region, 27(B), 107, 1983.
78. **Ton, C. S. and Hernandez, T. P.**, Wet soil stress effects on sweet potato, *J. Am. Soc. Hortic. Sci.*, 103, 600, 1978.
79. **Kushman, L. J. and Deonier, M. T.**, Relation of internal gas content and respiration to keeping quality of Porto Rico sweetpotatoes, *Proc. Am. Soc. Hortic. Sci.*, 74, 622, 1959.
80. **Kushman, L. J. and Deonier, M. T.**, Effects of weather, date of harvest, and curing treatments on keeping qualities of Puerto Rico sweet potatoes, *Proc. Am. Soc. Hortic. Sci.*, 71, 369, 1958.
81. **Ahn, J. K., Collins, W. W., and Pharr, D. M.**, Influence of preharvest temperature and flooding on sweet potato roots in storage, *HortScience*, 15, 261, 1980.
82. **Ahn, J. K., Collins, W. W., and Pharr, D. M.**, Gas atmosphere in submerged sweet potato roots, *HortScience*, 15, 795, 1980.
83. **Chang, L. A., Hammett, L. K., and Pharr, D. M.**, Ethanol, alcohol dehydrogenase, and pyruvate decarboxylase in storage roots of four sweet potato cultivars during simulated flood damage and storage, *J. Am. Soc. Hortic. Sci.*, 107, 674, 1982.
84. **Corey, K. A. and Collins, W. W.**, Effects of vine removal prior to flooding on ethanol concentration and storage quality of sweet potato roots, *HortScience*, 17, 631, 1982.
85. **Corey, K. A., Collins, W. W., and Pharr, D. M.**, Effect of duration of soil saturation on ethanol concentration and storage loss of sweet potato roots, *J. Am. Soc. Hortic. Sci.*, 107, 195, 1982.
86. **Martin, F. W.**, Variation of sweet potatoes with respect to effects of waterlogging, *Trop. Agric. (Trinidad)*, 60, 117, 1983.
87. **Chang, L. A. and Kays, S. J.**, Effect of low oxygen storage on sweet potatoes, *J. Am. Soc. Hortic. Sci.*, 106, 481, 1981.
88. **Chang, L. A., Hammett, L. K., and Pharr, D. M.**, Carbon dioxide effects of ethanol production, pyruvate decarboxylase, and alcohol dehydrogenase activities in anaerobic sweet potato roots, *Plant Physiol.*, 71, 59, 1983.
89. **Lu, S. Y., Zheng, X. R., Li, W. J., and Wu, C. G.**, Morphology during the process of tuber formation and investigation of a method for early appraisal in high-starch breeding of sweet potato, *Acta Agric. Univ. Pekinensis*, 7(3), 13, 1981.
90. **Aimi, R. and Nishio, T.**, Cell-physiological studies on the synthesis and accumulation of starch in sweet potato, *Proc. Crop Sci. Soc. Jpn.*, 24, 201, 1955.
91. **Yoshida, S. and Parao, F. T.**, Climactic influence on yield and yield components of lowland rice in the tropics, in *Climate and Rice*, International Rice Research Institute, Los Baños, Laguna, Philippines, 1976.
92. **Bartholomew, D. P. and Kadzimin, S. B.**, Pineapple, in *Ecophysiology of Tropical Crops*, Alvim, P. de T. and Kozlowski, T. T., Eds., Academic Press, New York, 1977, 113.
93. **Tsuno, Y. and Fujise, K.**, Studies on the dry matter production of sweet potato. II. Aspect of dry matter production in the field, *Proc. Crop Sci. Soc. Jpn.*, 26, 285, 1963.
94. **Haynes, P. H., Spence, J. A., and Walter, C. J.**, The use of physiological studies in the agronomy of root crops, *Proc. Int. Symp. Trop. Root Crops*, Trinidad, 1967, III, 1.

95. **Hahn, S. K. and Hozyo, Y.,** Sweet potato, and yams, in *Symposium on Potential Productivity of Field Crops Under Different Environments,* International Rice Research Institute, Los Baños, Laguna, Philippines, 1980.
96. **Enyi, B. A. C.,** Analysis of growth and tuber yield in sweet potato *(Ipomoea batatas)* cultivars, *J. Agric. Sci.,* 89, 421, 1977.
97. **Lowe, S. B. and Wilson, L. A.,** Comparative analysis of tuber development in six sweet potato *(Ipomoea batatas* (L.) Lam.) cultivars. 2. Interrelationships between tuber shape and yield, *Ann. Bot.,* 38, 319, 1974.
98. **Denis, R. J.,** Yield and Shape of Sweet Potatoes as Affected by Variety, Date of Planting, Plant Population and Petroleum Mulch, M.S. Thesis, Rutgers University, New Brunswick, N.J., 1973.
99. **Anderson, W. S. and Randolph, J. W.,** Sweet potatoes, *Miss. St. Agric. Exp. Stn. Bull.,* 378, 1943, 24.
100. **Huett, D. O.,** Evaluation of yield, variablity and quality of sweet potato cultivars in sub-tropical Australia, *Exp. Agric.,* 12, 9, 1976.
101. International Institute of Tropical Agriculture, *Annu. Rep. of Root and Tuber Improvement Program for 1974,* Ibadan, Nigeria, 1974.
102. **Tsuno, Y.,** *Sweet Potato: Nutrient Physiology and Cultivation,* International Potash Institute, Berne, Switzerland, n.d., 73.
103. **Edmond, J. B. and Ammerman, G. R.,** *Sweet Potatoes: Production, Processing and Marketing,* AVI Publishing, Westport, Conn., 1971, 334.
104. **Kim, Y. C.,** Studies on the photoperiodical control for tuber formation in sweet potato, *Korean J. Bot.,* 2, 35, 1957.
105. **Togari, Y. and Akimine, H.,** Studies on the tuber formation of sweet potato, *Jpn. Agric. Hortic.,* 20, 95, 1945.
106. **Ipo, H. and Toi, S.,** Studies on the tuber formation of sweet potato, *Jpn. J. Hortic.,* 16, 1, 1947.
107. **Isigugo, D. and Matsuhara, S.,** The relation between climate and tuber growth of sweet potato, *Jpn. Agric. Hortic.,* 12, 571, 1937.
108. **Kim. Y. C.,** Studies on the tuber formation of *Ipomoea batatas, Bull. Korean Agric. Soc.,* 1, 1, 1954.
109. **Kim, Y. C.,** Effects of thermoperiodism on tuber formation in *Ipomoea batatas* under controlled conditions, *Plant Physiol.,* 36, 680, 1961.
110. **Sakr, S. M.,** Effect of temperature on yield of the sweet potato, *Proc. Am. Soc. Hortic. Sci.,* 42, 517, 1943.
111. **Sekoika, H.,** Effect of temperature on the translocation of surcose-C^{14} in the sweet potato plant, *Proc. Crop Sci. Soc. Jpn.,* 30, 27, 1961.
112. **Hasegawa, H. and Yahiro, T.,** Effects of high soil temperatures on the growth of the sweet potato plant, *Proc. Crop Sci. Soc. Jpn.,* 26, 37, 1957.
113. **Sekioka, H.,** The effect of temperature on the translocation and accumulation of carbohydrates in sweet potato, *Proc. 2nd Int. Symp. on Trop. Root and Tuber Crops,* Hawaii, 2, 37, 1970.
114. **Gollifer, D. E.,** A time of planting trial with sweet potatoes, *Trop. Agric. (Trinidad),* 57, 363, 1980.
115. **Kennard, G. B.,** Sweet potato variety experiments at I. C. T. A. 1927-43, *Trop. Agric.,* 24, 69, 1944.
116. **Fujise, K. and Tsuno, Y.,** Effect of potassium on the dry matter production of sweet potato, *Proc. Intern. Symp. Trop. Root Crops,* Trinidad, 1(II), 20, 1967.
117. **Robbins, W. R.,** How potassium affects sweet potatoes, *N.J. Agric.,* 11(6), 10, 1929.
118. **Stino, K. R. and Lashin, M. E.,** Effect of fertilizers on the yield and vegetative growth of sweet potatoes, *Proc. Am. Soc. Hortic. Sci.,* 61, 367, 1953.
119. **Schermerhorn, L. G.,** Influence of fertilizers on the yield and form of the sweet potato, *Proc. Am. Soc. Hortic. Sci.,* 20, 162, 1923.
120. **Tsuno, Y. and Fujise, K.,** Studies on the dry matter production of sweet potato. III. The relation between the dry matter production and the absorption of mineral nutrients, *Proc. Crop Sci. Soc. Jpn.,* 32, 297, 1964.
121. **Ho, C. T., Su, N. R., Tang, C. N., and Sheng, C. Y.,** Studies on correlation of soil and plant potassium with response of sweet potato to added potash in Chiayi prefecture, *Soil and Fertilizers in Taiwan,* 14, 32, 1967.
122. **Togari, Y. and Shiraswa, Y.,** Changes of principal components in the sweet potato plant in the growing period, *Proc. Crop Sci. Soc. Jpn.,* 24, 99, 1955.
123. **Sugawara, T.,** Comparative studies on the water culture in sweet potatoes with special reference to the concentration of potash salts, *Nippon Dojyo-Hiryogaku Zasshi,* 12, 154, 1938.
124. **Constantin, R. J., Jones, L. G., and Hernandez, T. P.,** Effects of potassium and phosphorous fertilization on quality of sweet potatoes, *J. Am. Soc. Hortic. Sci.,* 102, 779, 1977.
125. **Murata, T. and Akazawa, T.,** Enzymatic mechanism of starch synthesis in sweet potato roots. 1. Requirement of potassium ions for starch synthatase, *Arch. Biochem. Biophys.,* 126, 873, 1968.

126. **Schermerhorn, L. G.**, Sweet potato studies in New Jersey, *N.J. Agric. Exp. Stn. Bull.*, 398, 1924.
127. **Badillo-Feliciano, J. and Lugo-López, M. A.**, Effect of four levels of N, P, K, and micronutrients on sweet potato yields in an oxisol, *J. Agric. Univ. P.R.*, 60, 597, 1976.
128. **Cibes, H. and Samuels, G.**, Mineral-deficiency symptoms displayed by sweet potato plants grown under controlled conditions, *Univ. P. R. Agric. Exp. Stn. Tech. Paper*, 20, 1, 1957.
129. **Leonard, O. A., Anderson, W. S., and Gieger, M.**, Effect of nutrient level on the growth and chemical composition of sweet potatoes in sand culture, *Plant Physiol.*, 23, 223, 1948.
130. **Li, L.**, Studies on the response surfaces and economic optima in fertilizer experiments of sweet potato, *J. Agric. Assoc. China*, 66, 30, 1969.
131. **Spence, J. A. and Ahmad, N.**, Plant nutrient deficiencies and related tissue composition of the sweet potato, *Agron. J.*, 59, 59, 1967.
132. **Tsuno, Y. and Fujise, K.**, Studies on the dry matter production of sweet potato. IV. The relationship between the concentration of mineral nutrients in plant distribution ratio over dry matter production, *Proc. Crop Sci. Soc. Jpn.*, 27, 301, 1964.
133. **Tsuno, Y. and Fujise, K.**, Studies on the dry matter production of sweet potato. X. An explanation of the determination of dry matter percentage in tuber with viewpoint of nutrient conditions and dry matter production, *Proc. Crop Sci. Soc. Jpn.*, 37, 12, 1968.
134. **Bolle-Jones, E. W. and Ismunadji, M.**, Mineral-deficiency symptoms of sweet potato, *Emp. J. Expt. Agric.*, 31, 60, 1963.
135. **Steinbauer, C. E. and Beattie, J. H.**, Influence of lime and calcium chloride applications on growth and yield of sweet potatoes, *Proc. Am. Soc. Hortic. Sci.*, 36, 526, 1939.
136. **Kotama, T.**, Culture of sweet potato, in *Encyclopedia of Crops*, Root Crop Section, Yokendo Inc., Tokyo, 1962.
137. **Abruña, F., Vicente-Chandler, J., Rodriguez, J., Badillo, J., and Silva, S.**, Crop response to soil acidity factors in ultisols and oxisols in Puerto Rico. V. Sweet potato, *J. Agric. Univ. P.R.*, 63, 250, 1978.
138. **Wang, S.-T. and Yu, Z.-Q.**, A preliminary study of high-yield in sweet potato in different soil types, *Scientia Agric. Sinica*, 1, 49, 1981.
139. **Hines, W.**, Effect of Location on the Yield, Grade and Quality of Unit 1 Porto Rico Sweet Potato, M.S. Thesis, Louisiana State University, Baton Rouge, 1949.
140. **Peterson, M. J., Aull, G. H., Knobel, E. W., and Downing, J. C.**, Sweet potato production possibilities in South Carolina (with special reference to soil suitability), *S.C. Agric. Exp. Stn. Bull.*, 364, 1946.
141. **Wilson, L. A.**, The process of tuberization in sweet potato *(Ipomoea batatas* (L.) Lam.), *Proc. 2nd Int. Symp. Trop. Root and Tuber Crops*, Hawaii, 1, 24, 1970.
142. **Edmond, J. B., Garrison, O. B., Wright, R. E., Woodard, O., Steinbauer, C. E., and Deonier, M. T.**, Cooperative studies of the effects of height of ridge, nitrogen supply, and time of harvest on yield and flesh color of the Porto Rico sweet potato, *U.S. Dept. Agric. Circ.*, 832, 40, 1950.
143. **Watanabe, K., Ozaki, K., and Yashiki, T.**, Studies on the effects of soil physical conditions on the growth and yield of crop plants. VII. Effects of soil air composition and soil bulk density and their interaction on the growth of the sweet potato, *Proc. Crop Sci. Soc. Jpn.*, 37, 65, 1968.
144. **Inden, T.**, The influence of soil moisture on root tuber formation of sweet potatoes and anatomical studies of root, *J. Hortic. Assoc. Jpn.*, 19, 49, 1952.
145. **Naka, J.**, Physiological studies on the growing process of sweet potato plant, *Mem. Fac. Agric. Kagawa Univ.*, 9, 1, 1962.
146. **Watanabe, E.**, Agronomic studies on the mechanism of excessive vegetative growth in sweet potato *(Ipomoea batatas* Lam.), *J. Cent. Agric. Exp. Stn.*, 29, 1, 1979.
147. **Spence, J. A. and Humphries, E. C.**, Effect of moisture supply, root temperature and growth regulators on photosynthesis of isolated root leaves in sweet potato *(Ipomoea batatas)*, *Ann. Bot.*, 36, 115, 1972.
148. **Bourke, R. M.**, Sweet potato in Papua, New Guinea, in *Sweet Potato Proc. 1st Int. Symp.*, Villareal, R. L. and Griggs, T. D., Eds., Asian Vegetable Research and Development Center, Shanhua, Tainan, Taiwan, 1982, 45.
149. **Woodard, O.**, Sweet potato culture in the Coastal Plain of Georgia, *Ga. Coastal Plain Exp. Stn. Bull.*, 17, 1932.
150. **Miller, J. C.**, Sweet potato breeding and yield studies for 1958, *La. Agric. Exp. Stn. Hortic. Res. Circ.*, 41, 1958.
151. **Villareal, R. L., Lin, S. K., and Lai, S. H.**, Variations in the yielding ability of sweet-potato under drought stress and minimum input conditions, *HortScience*, 14, 31, 1979.
152. **Yen, C.-T., Chu, C.-V., and Sheng, C.-L.**, Studies on the drought resistance of sweet potato varieties, *Crop Sci. (Peking)*, 3, 183, 1964.
153. **Bowers, J. L., Benedit, R. H., and Watts, V. M.**, Supplemental irrigation of sweetpotatoes, *Ark. Agric. Exp. Stn. Bull.*, 578, 1956.

154. **Lambeth, V. N.,** Studies in moisture relationships and irrigation of vegetables, *Miss. Agric. Exp. Stn. Res. Bull.*, 605, 1956.
155. **Lana, E. P. and Peterson, L. E.,** The effect of fertilizer-irrigation combinations on sweet potatoes in Buckner coarse sand, *Proc. Am. Soc. Hortic. Sci.*, 68, 400, 1956.
156. **Ehara, K. and Sekioka, H.,** Effect of atmospheric humidity and soil moisture on the translocation of sucrose-C^{14} in sweet potato plant, *Proc. Crop Sci. Soc. Jpn.*, 31, 41, 1962.
157. **Anderson, W. S., Cockran, H. L., Edmond, J. B., Garrison, O. B., Wright, R. E., and Boswell, V. R.,** Regional studies of time of planting and hill spacing of sweet potatoes, *U.S. Dept. Agric. Circ.*, 725, 1945.
158. **Akita, S., Yamamoto, F., Ono, M., Kushara, M., and Ikemoto, S.,** Studies on the small tuber set method in sweet potato cultivation, *Bull. Chugoku Agric. Exp. Stn.*, 8, 75, 1962.
159. **Imbert, M. P. and Wilson, L. A.,** Simulatory and inhibitory effects of scopoletin on IAA oxidase preparation from sweet potato, *Phytochemistry*, 9, 1787, 1970.
160. **Imbert, M. P. and Wilson, L. A.,** Effects of chlorogenic and caffeic acids on IAA oxidase prepartations from sweet potato roots, *Phytochemistry*, 11, 2671, 1972.
161. **Murakami, Y.,** Gibberellin-like substances in roots of *Oryza sativa, Pharbitis nil,* and *Ipomoea batatas,* and the site of their synthesis in the plant, *Bot. Mag. (Tokyo),* 81, 334, 1968.
162. **Oritani, T., Oritani, T., and Yoshida, R.,** Growth inhibitors in tuberous roots of *Ipomoea batatas, Proc. Crop Sci. Soc. Jpn.*, 41, 166, 1972.
163. **Oritani, T. and Yoshida, R.,** Studies on nitrogen metabolism in crop plants. XI. The changes of abscisic acid and cytokinin-like activity accompanying with growth and senescence in the crop plants, *Proc. Crop Sci. Soc. Jpn.,* 40, 325, 1971.
164. **Sirju, G. and Wilson, L. A.,** IAA oxidase preparations from fresh and aged *Ipomoea batatas* tuber discs, *Phytochemistry,* 13, 111, 1974.
165. **Alagappan, R.,** Growth responses in sweet potato to indole-3-acetic acid and naphthalene acetic acid, *Annamalai Univ. Agric. Res. Annu.*, 2, 60, 1970.
166. **Alvarez, M. N., Whatley, B. T., Anthony, R. M., and Henderson, J. H.,** The effect of polaris, n-n bis phosphonomethyl glycine, on yield and quality of sweet potato roots, *HortScience,* 11, 228, 1976.
167. **Andrews, C. P., Love, J. E., Fontenot, J. F., and Williams, B. R.,** Physiological responses of Centennial sweet potato plants to ethephon, *HortScience,* 11, 229, 1976.
168. **Austin, M. E. and Aung, L. H.,** Influence of growth regulators on the development of *Ipomoea batatas, J. Hortic. Sci.,* 48, 271, 1973.
169. **Bouwkamp, J. C. and McArdle, R. N.,** Effects of triacontanol on sweet potatoes, *Ipomoea batatas, HortScience,* 15, 69, 1980.
170. **Kabi, T. and Sarma, R.,** An analysis of IBA, GA_3 and CCC interactions in the growth of leaves of *Ipomoea batatas* L., *Indian J. Plant Physiol.,* 16, 140, 1973.
171. **Muthukrishnan, C. R., Thambura, J. S., Shanmugam, A., and Shanmugavelu, K. G.,** Effect of certain growth regulators on tapioca *(Manihot esculenta* Crantz.) and sweet potato *(Ipomoea batatas* (L.) Lam.), *J. Root Crops,* 2, 52, 1976.
172. **Paterson, D. R., Fuqua, M. C., and Earhart, D. R.,** Influence of depth of watering, silver ion and 2-chloroethyl phosphonic acid on growth and storage root development in *Ipomoea batatas, HortScience,* 12, 236, 1972.
173. **Samantari, B. and Mohanty, C. R.,** Role of coumarin on the process of tuber formation from the isolated hormone treated leaves, *Indian Sci. Cong. Assoc. Proc.,* 57, 310, 1970.
174. **Shanmugam, A. and Srinivasan, C.,** Influence of ethephon on the growth and yield of sweet potato *(Ipomoea batatas* Lam.), *Hortic. Res.,* 13, 143, 1974.
175. **Stino, K. R., Hegazy, A. T., Gaafar, A. K., and El-Gharbawi, A. A.,** Studies on the effect of gibberellic acid on growth, flowering and chemical composition of sweet potatoes, *J. Bot. U. A. R.,* 12, 123, 1969.
176. **Tompkins, D. R. and Horton, R. D.,** Plant production by sweet potato roots as influenced by ethephon, *HortScience,* 8, 415, 1973.
177. **Wilson, L. A.,** Tuberization in sweet potato *(Ipomoea batatas* (L.) Lam.), in *Sweet Potato Proc. 1st Int. Symp.,* Villareal, R. L. and Griggs, T. D., Eds., Asian Vegetable Research and Development Center, Shanhua, Tainan, Taiwan, 79, 1982.
178. **Schlimme, D. V.,** Anatomical and physiological aspects of sprout initiation and development in sweet potato storage roots, *Diss. Abstr.,* 26(9), 4947, 1966.

CONCLUSIONS — PART I

Although a reading of the chapters in this section will reveal that much is presently known with regard to production of sweet potatoes, it may also be noted that there are several serious gaps or deficiencies in our knowledge. Much of the research reported has been conducted in temperate areas. This does not necessarily invalidate the results, but it does suggest that it may be appropriate to confirm some results before they can be applied to tropical conditions. It is hoped that as more trained scientists become available to national programs in the tropics and as agricultural development policies become more broadly based and less tightly focused on cereal grain and legume production, that the research to confirm and extend the results obtained in temperate areas will be accomplished.

Far too little is known regarding the physiological determinants of yield and the effects of environmental and edaphic constraints on these processes. It is particularly important in the tropics, where yields are generally lower than in temperate zones, to determine the relationship of the effects of environment and genotypes and to establish breeding programs to overcome these constraints.

The control of the sweet potato weevil is perhaps the single most important problem to be resolved. Much of the potential importance of sweet potato production in development strategy will depend on controlling the depredation of the weevil. Whether controls are based on chemicals, biological control or host plant resistance, they must be safe, efficient, and adaptable to the technology available throughout the tropics. The use of chemical insecticides in order to control the sweet potato weevil would seem to carry the potential of long term damage to various ecosystems. Although this threat cannot be dismissed, the recent report to Talekar et al.[1] suggests that persistence of some chemicals in tropical environments may be less than in temperate environments.

The exchange of vegetative germplasm has posed problems on many occasions. In order to prevent the entry of potentially destructive viruses, many countries have imposed strict quarantine procedures. This procedure greatly impedes the exchange of cultivars for cooperative testing and evaluation since it is slow, expensive, and requires a fairly high level of technological expertise. Those countries lacking funds and/or trained personnel for an effective quarantine program must either prohibit entry or risk the entry of potentially destructive viruses. One solution is to establish an international clearing house for disinfecting, maintaining, and distributing virus-free germplasm. Such a plan is presently under discussion. Thus the possibilities for the resolution of this problem seem to be encouraging.

In many cases, production problems may be of a local nature. The highlighting of those problems considered to be most important generally, should not be construed as to discourage efforts to solve other problems. It should be presumed that the individual scientists are best equipped to determine their research priorities.

REFERENCES

1. **Talekar, N. S., Chen, J. S., and Kao, H. T.**, Long term persistance of selected insecticides in subtropical soil: their absorption by crop plants, *J. Econ. Entomol.*, 76, 207, 1983.

Part II
Utilization

INTRODUCTION — PART II

The realization of the production potentials of sweet potatoes as calorie and nutrient sources will depend, to a considerable extent, on efficient and appropriate utilization of the product. The semiperishable nature of the crop would seem to present some difficulties but sweet potatoes are no more perishable than other high-moisture staples such as cassava, potatoes, and yams and less perishable than many vegetables commonly marketed in the tropics. Notwithstanding, methods to stabilize quality may be useful to encourage the utilization of the crop.

The various chapters on utilization of sweet potatoes and sweet potato products should not be considered as competing strategies or methods. It is hoped that the various methods will highlight the potential versatility in the utilization of this crop and serve to stimulate investigations of these and other utilization methods. It may be noted that many staple crops are rarely consumed directly in the form in which they are produced. The importance of corn, wheat, and soybeans in the agricultural economics of temperate zone countries is related in part to the versatility in their utilization. Sweet potatoes, among other typical crops, may be expected to play similar roles in tropical countries as production and utilization techniques are developed.

The value of sweet potatoes as a natural resource for the tropics will depend on how the resource is utilized. The versatility of utilization would suggest that some combination of utilization methods may be optimum. It is likely that the methods of utilization have not been exhausted and that new ways to utilize the resource may add substantially to the value of the resource. It is hoped the reader will consider the chapters in this part with the viewpoint that, what is presently known represents but a fraction of the information needed to effectively and efficiently utilize this resource.

Chapter 5

SWEET POTATO PRODUCTION AND UTILIZATION IN ASIA AND THE PACIFIC*

Steve S. M. Lin, Creighton C. Peet, Der-Min Chen, and Hsiao-Feng Lo

TABLE OF CONTENTS

I.	Introduction	140
II.	Current Status and Production and Utilization Trends	140
	A. Production	140
	B. Utilization	140
	1. Sweet Potato Roots	140
	2. Sweet Potato Greens (Tips)	141
III.	Growing Environment	141
IV.	Desired Varietal Characteristics	144
	A. Skin and Flesh Color	145
	B. Eating Texture and Sweetness	145
	C. Sweet Potato Greens	145
V.	Major Biological and Environmental Constraints	145
	A. Disease and Insect Pests	145
	B. Environmental Constraints	145
VI.	Current Status of the Variety Improvement Programs	146
VII.	Conclusions	147
	A. For Human Consumption	147
	B. As Animal Feed or Industrial Materials	147
References		148

---

* Due to the death of Mr. Lin, correspondence concerning this material should be sent to Dr. Bouwkamp. Mr. Peet's present address is Katmandu (ID), Department of State, Washington, D.C. 20520.

I. INTRODUCTION

Historically, sweet potato (*Ipomoea batatas* (L.) Lam) has played an indispensable role as a source of food in Asia and the Pacific Islands. Today, more than 94% of the world's sweet potato is produced in these regions, with annual production estimated at 98.6 million metric tons. In 1980, the crop was ranked number four in Asia in terms of total food production following rice, sugar cane, and wheat. In China, the world's most populated state, sweet potato is second only to rice, and accounts for more than 80% of the world's total production![1]

Regardless of the importance of sweet potato in Asia, the production area actually decreased steadily in most countries in these areas during the past 10 years (Figure 1; Table 1). More importantly, average yield have remained virtually the same (8 t/ha) in Asia for the past 2 decades (Figure 2). In fact, the low average yield was not the major reason responsible for the lack of development of this traditional crop in some areas. For example, Japan, Taiwan, and Korea have the highest average yields (17 to 20 t/ha); they, however, decreased in sweet potato growing area most drastically in recent years (Figure 3). It is apparent that the lack of demand and change of utilization pattern of sweet potato were the major cause for the reduction of this crop in these places.

This paper was prepared to discuss the current production and utilization methods of sweet potato in Asia and the Pacific. Most of the data presented here were based on the survey recently conducted by the Asian Vegetable Research and Development Center.[2] Some results were also presented in the workshop on "Breeding New Sweet Potatoes for the Tropics" held in Mayaguez, Puerto Rico, 1983.

II. CURRENT STATUS AND PRODUCTION AND UTILIZATION TRENDS

A. Production

According to the result of the survey, Indonesia, the Philippines, and India have the largest amount of land under sweet potato production (excluding Mainland China). Among the countries surveyed, only Bangladesh, India, the Philippines, Papua New Guinea (PNG) and most of the Pacific Islands increased the amount of land devoted to sweet potato during the past 10 years (Table 1). The major reasons cited for the increases were the development of new land and the replacement of traditional food crops of lower social status, e.g., taro and banana in the case of Papua New Guinea. In Indonesia and Malaysia, sweet potato was replaced by more "prestigious" crops such as corn, legumes, and vegetables.

Average yields in these areas are generally very low, however. For example, the yields in Papua New Guinea are only 4.6 t/ha and 11.4 t/ha in the Pacific Islands. Japan has the highest average yield, 20 t/ha, followed by 19.6 t/ha in Korea and 16.6 t/ha in Taiwan (Table 2).

In terms of importance, sweet potato accounted for 65% of the total agricultural land in Papua New Guinea. In Tonga and the Solomon Islands, sweet potato was ranked as the most important staple food crop with per capita consumption of the fresh roots calculated at 532.5 kg per year; number one in the world (Table 1).

B. Utilization

1. Sweet Potato Roots

Except for Japan, Korea, and Taiwan, sweet potato roots were reported to be used mainly for human consumption (70 to 100%) in most of the countries surveyed. Only small portions of the sweet potatoes produced were used for animal feed; negligible amounts were used for industrial purposes (Table 2). Only 11 and 38% of all roots are produced for human consumption in Taiwan and Japan, respectively. In Taiwan, dried sweet potato chips used

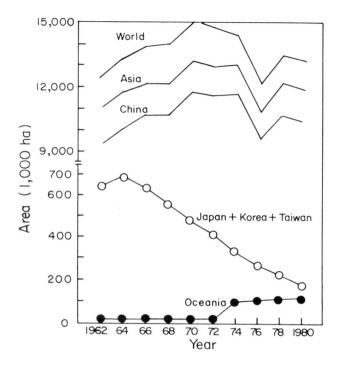

FIGURE 1. Harvested area of sweet potato in Asia and the Pacific. (From *FAO Production Year Book 1962-1980*, Food and Agriculture Organization of the United Nations, Rome.)

for hog feed were cited as an important use of the crop in rural areas. In Japan and Korea about 35 to 36% of all sweet potatoes produced are used for industrial purposes, mainly for starch extraction and fermentation. In Oceania, sweet potato is mainly consumed as a staple food; about 9 to 15% is used for feed (mainly for hogs) (Table 2).

2. Sweet Potato Greens (Tips)

Sweet potato greens are consumed as a vegetable in most of the surveyed countries and territories except Sri Lanka and a few of the Pacific Islands. Consumption differs according to area, however. For example, the Japanese and Koreans consume only petioles and the Taiwanese mainly use the tender leaves. The terminal tips, usually 10 to 15 cm long (Table 2) are consumed in most of the other surveyed areas. In Taiwan, sweet potato greens are often the only vegetable available in the market after a typhoon. Low yield and the lack of tender cultivars were listed as major production constraints.

III. GROWING ENVIRONMENT

The mean maximum and minimum temperatures in Asia and the Pacific were listed at 29.6°C and 18.5°C during the major sweet potato growing seasons. Most surveyed places have minimum temperatures about 10°C (the chilling threshold for sweet potato). Frost damage and growth retardation due to low temperatures were reported in the highlands of Papua New Guinea and in the early spring in Japan and Korea. In Japan, clear plastic mulches are used to raise the soil temperatures.

Most production areas in Asia and the Pacific were characterized by both high day and night temperatures. Differences between day and night temperatures are usually small,

Table 1
PRODUCTION AREA, AVERAGE YIELD, AND PER CAPITA CONSUMPTION OF SWEET POTATO IN ASIA AND THE PACIFIC

Country/territory	SP area (1000 ha)	SP area/ Agr area (%)	Average yield (t/ha)	Consumption (kg/person/year)
Japan	70 (−)	2.1	20.0	12.1
Korea	70 (−)	6.2	19.8	37.2
Taiwan	74 (−)	8.1	16.6	70.1
Bangladesh	73 (+)	5.0	10.9	9.2
India	225 (+)	0.6	6.9	2.3
Indonesia	309 (−)	5.8	7.6	15.8
Malaysia	4	1.2	9.7	2.7
Philippines	228 (+)	20.5	4.6	21.0
Sri Lanka	21	3.9	6.2	8.9
Thailand	36 (−)	1.4	9.7	7.6
Papua New Guinea	96 (+)	64.7	4.6	145.2
Pacific Islands	13[a] (+)	23.8[b]	11.4	532.3[b]
World Total	13,638	—	8.4	—
Asia Total	12,330	—	8.5	—
China	10,860	—	8.5	—

Note: (+) = increased in the past 10 years and (−) = decrease in the past 10 years.

Sources: FAO, Production Year Book, 1980 and the AVRDC Survey, 1982.

[a] Fiji, New Caledonia, Tonga, and Solomon Is.
[b] Tonga and Solomon Is.

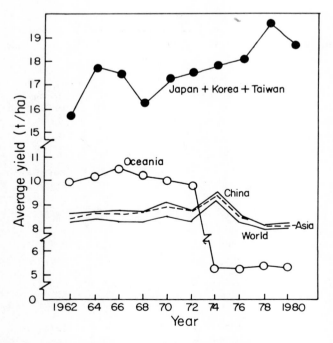

FIGURE 2. Average yield of sweet potato in Asia and the Pacific. (From *FAO Production Year Book 1962-1980*, Food and Agriculture Organization of the United Nations, Rome.)

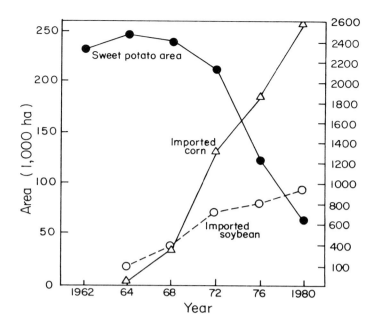

FIGURE 3. Sweet potato planting area and volume of imported corn and soybean in Taiwan, 1962-1980. (From Taiwan Agricultural Year Book; Upland Food Crops and Animal Husbandry 1981, Taiwan, R.O.C.)

Table 2
UTILIZATION METHODS OF SWEET POTATO ROOTS AND GREENS IN ASIA AND THE PACIFIC[a]

Country/territory	Roots				Greens
	Food	Feed	Indus. (%)	(Starch)	
Japan	38	18	35	(29)	Few (petiole)
Korea	56	5	36	(9)	Yes (petiole)
Taiwan	11	73	16	(16)	Few (leaves)
Bangladesh	100				Yes (tips)
India	90	10			Few (tips)
Indonesia	90	10			Yes (leaves and tips)
Malaysia	70	30			Few (tips)
Philippines	80	10	10	(10)	Yes (tips)
Sri Lanka	100				No
Thailand	80	15	5	(5)	Yes (tips)
Papua New Guinea	85	15			Few
Pacific Islands[b]	91	9			Some (tips and leaves)

[a] AVRDC Survey on sweet potato production and utilization, 1982.
[b] Niue, Palau, Cook Is., Fiji, Tahiti, Vanuatu, Tonga, Guam, and Ponape.

particularly in Oceania, and this condition was often cited as favorable to excessive vine growth and detrimental to storage root development.

Rainfall was reported to vary greatly according to location and planting season. Mean

Table 3
MOST IMPORTANT CHARACTERISTICS THAT NEED IMPROVEMENT ON STAPLE- AND SNACK-TYPE SWEET POTATOES

Characteristics	Frequency[a] Staple	Snack
Eating qualities[b]	17 (1)	17 (1)
Nutritional composition[c]	16 (2)	11 (2)
Insect resistance	13 (3)	6 (5)
High yield	12 (4)	6 (5)
Early maturity	11 (5)	8 (4)
Appearance and uniformity	10 (6)	10 (3)
Disease resistance	7 (7)	3 (6)
High dry matter	5 (8)	3 (6)
High starch	4 (9)	2 (7)
Keeping quality	3 (10)	3 (6)
Plant type	3 (11)	0
Adaptability	1 (12)	1 (8)
Others	3	1

[a] Numbers in parenthesis are ranks of importance.
[b] Flavor, taste, texture, and sweetness.
[c] Mainly referred to higher protein and β-carotene content.

Source: AVRDC Survey, 1982.

monthly rainfall during the sweet potato growing season ranged from 35 to 235 mm per month. Most sweet potatoes are grown without irrigation, and drought was ranked as the most serious environmental stress. In many areas, flooding or excessive moisture was also cited as an important constraint.

Sweet potatoes are reported most often grown in sandy and clay loam soils, although they are also produced in volcanic soil in Japan and coral sands on some of the Pacific Islands.

IV. DESIRED VARIETAL CHARACTERISTICS

Variety improvement was ranked as the number one priority in all countries and territories surveyed, and about 60% of all respondents felt that the lack of improved cultivars is a major constraint in their particular regions. The most important variety characteristics listed for improvement are: eating quality and nutritional content (ranked first for both staple- and snack-type sweet potatoes) resistance, high yield (staple type) and appearance, uniformity and early maturity for the snack types (Table 3).

Interestingly, eating quality and nutrition were ranked as more important than high-yield in the survey. Apparently, sweet potato is considered more as a vegetable or supplemental commodity than a major food crop. Texture (dry or moist), flavor, sweetness, and fiber were frequently mentioned. Among these, dryness was the most important; 60 to 87% of the respondents preferred dry or moderately dry textures. Moist texture was only preferred in the developed countries. Protein and vitamin A (beta-carotene) content were the major nutritional concerns.

A. Skin and Flesh Color

Red is the preferred skin color for staple sweet potatoes followed by white and purple. For flesh color, more than 50% of the questionnaires noted a preference for white or yellow, compared with about 30% for orange. Deep orange was the least favored flesh color; only 8% of the persons surveyed preferred this type in Asia and the Pacific. Light orange seemed to be more acceptable in these areas; 18% responded favorably to this type.

B. Eating Texture and Sweetness

In Asia and the Pacific, 77% selected dry and moderately dry textures as desirable for the staple-type sweet potato; a moist texture was only preferred in the U.S. Seventy-six percent of the respondents felt that sweet and moderately sweet roots were preferable to the nonsweet types (except in South and Central America).

C. Sweet Potato Greens

Tender terminal tips are the major consumption portion of sweet potatoes when used for its vegetable greens. The consumable length of the terminal tips ranged from 10 to 15 cm, which indicates a lack of tender cultivars for this purpose. In areas where leaves and petioles are used, cultivars with large leaves and petioles are preferred. Green or purplish-green were reported to be preferred stem and leaf colors. Yellow or yellow-green colored cultivars were not considered acceptable, even though they are usually more tender than the green types.

Leaf shape proved to be a major concern; the number of respondents who preferred deep lobing and nonlobing leaves were similar. However, preparation method for cooking sweet potato greens was reported to be boiling (47%); browning problems after boiling should be considered by breeders.

V. MAJOR BIOLOGICAL AND ENVIRONMENTAL CONSTRAINTS

A. Disease and Insect Pests

In Asia and the Pacific, disease and insect pests were ranked as the most important production constraints for sweet potato, followed by storage, processing and marketing problems, and the lack of improved cultivars. In terms of disease problems, the sweet potato scab (*Elsinoe batatas*) was the most prevalent pathogen. This disease was important in almost every country throughout the tropical regions of Asia and the Pacific, except Taiwan (Table 4). Actual yield losses are not known, and no resistant cultivars have yet been developed. Fusarium wilt (*Fusarium oxyperum*) and witches' broom or the so called "little leaf" disease occurs in India, Indonesia, Philippines, Tonga, Vanuatu, Papua New Guinea, and Penghu Islands near Taiwan.

Sweet potato weevil (*Cylas fomicarius*) was listed as the most destructive insect pest and was rated far more important than any other insects in this survey. It is a major predator almost everywhere except Japan, Korea, and New Zealand. The stem borer (*Omphisa anastomosalis*) was rated as the second most important insect. It was reported to be a serious problem in Fiji, Indonesia, Malaysia, and Niue and Palau Islands. The distribution of these pests by country/territory are listed in Table 4.

B. Environmental Constraints

Of the important environmental constraints, drought and excessive rainfall were listed as the two most severe problems. In Asia and the Pacific, damage or growth retardation due to frost or low temperatures was ranked third; it happens in Japan, Korea, and the highlands of Papua New Guinea and Indonesia.

Seventy-seven percent of the respondents indicated that sweet potato is frequently grown under dry or drought conditions and 79% felt strongly that there was a need to develop drought-tolerant cultivars in Asia and the Pacific.

Table 4
DISTRIBUTION OF MAJOR SWEET POTATO DISEASES AND INSECTS

Country/territory	Witches broom	Fusarium wilt	Scab	Nematodes	Viruses	Stem borer	Weevil
Bangladesh	+	−	+	(−)	+	(−)	+++
Cook Is.	(−)	−	+	(−)	(−)	−	+
Fiji	−	+++	+++	+	+	+++	+++
Guam	−	−	−	(−)	−	(−)	+++
India	+	(−)	+	(−)	(−)	(−)	+++
Indonesia	+	(+)	+++	(−)	+	+++	+++
Japan	−	+++	(−)	+	+	−	−
Korea	−	+	(−)	(−)	+	−	(−)
Malaysia	−	+	+	(−)	+	+++	+++
New Zealand	(−)	+		+	(−)	−	−
Niue	−	−	−	−	(−)	+++	+++
Palau	(−)	+++	+++	(−)	(−)	+++	+++
Papua New Guinea	+	−	+++	(+++)	(+)	(+)	(+++)
Philippines	+	(+)	+++	+	+	+	+++
Ponape	−	−	+++	−	−	(−)	−
Puerto Rico	−	−	(−)	+	+	+	+++
Sri Lanka	−	−	−	(−)	−	+	+++
Tahiti	−	−	−	−	−	−	+
Taiwan	+	+	+	(−)	+	(+++)	+++
Thailand	(−)	(−)	(+)	(+)	(−)	(+++)	+++
Tonga	+	+	+++	+	+	+	+
USA	(−)	(+++)	(−)	+	+	(+)	+++
Vanuatu	+++			+			+

Note: − = no; (−) = no or don't know
+ = yes, not serious; (+) = yes or don't know
+++ = very serious; (+++) = yes or very serious.

Source: AVRDC Survey, 1982.

Excessive soil moisture or flooding was also reported to be a serious problem for sweet potato grown in Asia and the Pacific. Twenty-nine percent responded that sweet potatoes were grown under frequently flooded conditions, and 44% felt strongly that there was a need to develop cultivars that can be grown under high rainfall conditions.

VI. CURRENT STATUS OF THE VARIETY IMPROVEMENT PROGRAMS

Although variety improvement was recognized by the survey participants as the most important research priority for sweet potatoes in Asia and the Pacific, very few variety improvement programs are in fact being conducted in the region's tropical areas. According to the survey data, about 300,000 true sweet potato seeds are being screened each year in sweet potato breeding programs in Asia and the Pacific. Less than 10,000 seeds were screened per year for areas other than Taiwan and the Philippines (tropical regions). This figure may help to explain the consistently low yields of sweet potato in tropical Asia. In addition, many countries and territories have rigorous quarantine laws that restrict the introduction of foreign plant materials.

Except for international centers, very few research programs are aimed at developing cultivars or cultural practices suited to tropical environments. For example, excess soil moisture and high temperatures are major problems for sweet potato production in the tropics, yet very few cultivars are suited to the environments where these constraints occur. In Taiwan, average yield for sweet potato planted in the hot-wet season is only one third of that of the cool-dry season.[3]

VII. CONCLUSIONS

There is ample room for the improvement of sweet potato in the tropics. By comparing the farmer's yields and those of the experimental station, Villareal[4] indicated that sweet potato has the largest yield improvement potential of any field crop in Asia. As a traditional crop already adapted to the tropical environments and familiar to farmers, sweet potato has the potential to play an important role as a nutrition and energy source. Due to the current status of this crop, however, this will require a systematic approach involving close international cooperation.

Based on the current sweet potato production and utilization in the various countries surveyed, two different utilization types should be developed.

A. For Human Consumption

These types should be developed for use as a snack or supplementary food aimed primarily at improving eating and marketing qualities as well as nutritional content. For some areas in Asia, these varieties could be bred for maximum yield and response to higher management inputs. The snack type of sweet potatoes used in Japan and Taiwan may also prove important, particularly in urban areas. The new varieties should have the following characteristics: (1) good storage and marketing qualities, (2) good eating quality and nutritional content, (3) early maturity and high yield, and (4) tolerance to major insects, disease pests, and environment stresses.

B. As Animal Feed or Industrial Materials

Varieties must also be bred for animal feed and industrial uses. Their principal characteristics should be high yield and tolerance to major environmental and biological stresses. As a subsistence crop well-adapted to Asia and Oceania, sweet potato has a better chance than grain crops to be developed as a major energy source. Although the yield potential of sweet potatoes in tropical Asia and the Pacific has never been clearly demonstrated, cultivars capable of producing 46 metric tons (13 tons of dry-matter) under favorable conditions, and 20 tons under extremely hot and wet conditions (over 1000 mm of precipitation and 27°C mean temperature during the 4-month growing period) have been developed at Asian Vegetable Research and Development Center (AVRDC). These cultivars were reported to have double or triple the yield of local cultivars used by cooperators in tropical Asia and the Pacific (AVRDC 1983). The development of these types of sweet potato will have advantages in utilizing the region's vast areas of nonirrigated lands and in hastening rural development. The high value products (e.g., meat, starch, etc.) produced directly and indirectly from sweet potato roots and vines can either be marketed locally or shipped to a distant urban market. Most important, this type of sweet potato can be easily utilized for human consumption during times of famine or war.

REFERENCES

1. FAO, *FAO Production Yearbook (1980)*, Food and Agriculture Organization of the United Nations, Rome, 1981.
2. AVRDC, *Progress Report for 1982*, The Asian Vegetable Research and Development Center, Shanhua, Tainan, Taiwan, 1983.
3. **Calkins, P. H., Huang, S. Y., and Hong, J. F.,** *Farmer's Viewpoint of Sweet Potato Production in Taiwan*, Tech. Bull. No. 4, Asian Vegetable Research and Development Center, Shanhua, Tainan, Taiwan, 1977.
4. **Villareal, R. L.,** Sweet potato in the tropics—progress and problems, in *Sweet Potato, Proc. of the First Int. Symp.*, Villareal, R. L. and Griggs, T. D. Eds., The Asian Vegetable Research and Development Center, Shanhua, Tainan, Taiwan, 1982, 3.

Chapter 6

RELATIVE NUTRIENT COSTS AND NUTRITIONAL VALUE

S. C. S. Tsou

Basically there are three major ways to utilize sweet potato: as food, feed, or starch production. The utilization pattern of sweet potato varies from country to country. In Taiwan, about 60% of the sweet potato root is used as feed, 20% as food, and 20% for starch production.[1] In most developing countries, the primary use of sweet potato is for food.[2] The nutritional value of sweet potato thus is of fundamental concern.

The nutritional value of a given commodity is often presented in food composition tables or expressed as per cent of Recommended Daily Allowance. This type of information is useful when consumers have some basic knowledge of nutrition and know which nutrient is lacking in their diet. Since most consumers in the developing countries are lacking of this knowledge, the utility of food composition tables is limited to nutritionists and dietitians. In addition, household income and nutritional status are often related: in general, higher income groups have better nutrition. The ratio between food expenditure and household expenditure can be used as a measure of the consumer's freedom in selecting food and is termed the Engel coefficient. Most developing countries have a high Engel coefficient, which indicates that the price is an important influence on purchasers.

An index system, which can express the nutritional value of any given commodity in dollars has been developed at Asian Vegetable Research and Development Center (AVRDC[3]). This makes it possible to take advantage of the consumer's economic concerns to promote nutrition education and better food habits.

The nutrition index system developed is based on relative nutrient cost, which is defined as cost per unit intake of a given nutrient. The relative nutrient cost (RNC_j) for nutrient j can be expressed as:

$$RNC_j = \frac{\text{Food expenditure}}{\text{intake of nutrient j}} \quad (1)$$

The nutrition value per unit weight of a given commodity then, can be calculated as:

$$\text{Commodity nutrition value} = \frac{\sum_{j=1}^{n} RNC_j \times [C]_j}{n} \quad (2)$$

where [C] is the content of nutrient j per unit weight of the commodity. This index system provides a means to express the nutrition value of a commodity by a single value that may include as many nutrients as desired. The nutritional value can be readily compared to the market price.

Both the relative nutrient cost and commodity nutrition value are dependent on food prices, food pattern, and intake of the respective nutrient, and are thus time and location specific. A food pattern with low intake of nutrient j has a high relative cost for j, and foods which are high in j have a high nutrition value. As an example of the use of these indexes the relative nutrient cost of food energy and eight major nutrients, calculated from 1980 Taiwan

Table 1
RELATIVE NUTRIENT COST AND RELATIVE COST TO MEET RDA OF MAJOR NUTRIENTS IN TAIWAN DIET[1,4,5]

Nutrient	Relative nutrient cost	% Intake to RDA[a]	Relative cost to meet RDA
Energy	17.3 NT$/$10^3$ cal	96.96	26.10
Protein	0.62 NT$/g	120.31	20.61
Calcium	0.1 NT$/mg	95.63	25.95
Iron	3.58 NT$/mg	113.58	21.84
Vitamin A	7.68 NT$/$10^3$ IU	126.89	19.55
Thiamine	29.01 NT$/mg	112	22.20
Riboflavin	51.30 NT$/mg	59.38	41.60
Ascorbic acid	0.37 NT$/mg	218.67	11.34

[a] Recommended daily allowance for male adult of Taiwan.

diet, are presented in Table 1. The percent intake of recommended daily allowance (RDA) (male, adult) and relative cost to meet RDA for each nutrients are listed in the same table. Nutrients deficient in Taiwan diet have a higher cost to meet the RDA. Food commodities rich in those nutrients are high in their nutritional value calculated from Equation 2. Nutrition value and market price of selected commodities of Taiwan (1980)[4] are listed in Table 2. Orange flesh sweet potato appears to be a beneficial commodity in terms of its nutrition value and market price. Rice may not necessarily be the most economical staple food under Taiwan conditions. Guava, papaya, and orange are fruits of choice for consumers from nutrition and price point of view. The nutrition value of vegetables varied from 5.1 of wax gourd to 109 of spinach. Sweet potato vines have the highest nutrition value per unit cost. Commodities of animal origin are generally expensive commodities. Small dry fish and eggs are low cost nutrient sources that are good for low income families.

The percent contribution of major nutrients to nutrition value of sweet potato are compared in Table 3. Vitamin A content of orange flesh sweet potato root and sweet potato tips contribute 70% and 50% to their nutrition value, respectively. Sweet potato vine tips are also rich in riboflavin, calcium, and iron.

There are two major constraints to the use of a simple value for expressing the overall nutrition value of a given commodity: (1) each nutrient has its own biological function, and is expressed in different units, so nutrient content is not additive, and (2) the importance of a given nutrient is dependent on food pattern. Both of these constraints are overcome in this index system. It may, however, be possible to further refine the method to better reflect the importance of given nutrients in the diet.

It is assumed in this system that all listed nutrients are all equally valuable. The listed nutrients are all given equal weight in the calculation of commodity nutrition values. This assumption may not always be reasonable. One way to modify the formula to reflect the relative importance of nutrients is to include a coefficient, aj, for each nutrient:

$$\text{Commodity nutrition value} = \frac{\sum_{j=1}^{n} a_j \times RNC_j \times [C]_j}{n} \qquad (3)$$

The coefficient, aj, can be assigned subjectively by the user. It can be zero when one does not intend to include certain nutrient in the calculation or it can be percent intake of RDC to further emphasize those nutrients deficient in the diet. This system can be applied for other purposes. One may use it to calculate the nutrition value production per ha as:

Table 2
NUTRITION VALUE AND MARKET PRICE OF
SELECTED COMMODITIES IN TAIWAN 1980[1,4,5]

Food group	Commodity	Nutrition value NT$/kg	Market price NT$/kg
Cereal	Rice	23.61	24.1
	Wheat flour	37.32	18.6
Starchy root	Sweet potato, orange	97.97	8(e)
	Sweet potato, white	23.59	8(e)
	Potato	15.75	16.7
Pulses and nuts	Soybean	131.69	21.4
	Peanut	94.83	46.7
	Mungbean	95.47	32.9
Vegetables	Spinach	163.61	38.3
	Sweet potato tips	133.47	10(e)
	Carrot	144.44	16.1
	Tomato	23.38	21.3
	Chinese cabbage	23.35	19.4
	Cucumber	6.03	22.8
	Wax gourd	10.66	14.6
Fruit	Guava	115.97	18.5
	Papaya	57.41	19.6
	Orange	54.67	28.0
	Banana	17.56	15.6
Meat, fish, milk, and eggs	Small dried fish	297.24	138
	Pork lean	58.30	125
	Chicken egg	66.21	45.4
	Beef, medium	36.26	145—246
	Chicken	41.77	66.9

Note: 1 US$ is approximately equal to NT$40. e = estimated.

Table 3
CONTRIBUTION OF MAJOR NUTRIENTS TO NUTRITION VALUE OF
SELECTED COMMODITIES[1,4,5]

Commodity	Nutrition value (NT$/kg)	% Contribution							
		Energy	Protein	Calcium	Iron	Vit. A	Vit. B_1	Vit. B_2	Vit. C
Rice	2.38	32.16	21.17	7.88	11.28	0	16.76	10.78	0
Sweet potato (white)	2.36	10.17	5.91	8.47	17.07	0	15.37	13.59	29.40
Sweet potato (orange)	9.80	2.49	1.82	5.87	4.57	69.55	2.96	3.27	9.44
Sweet potato tip	13.35	0.34	1.74	14.33	12.07	50.34	3.80	10.09	7.28
Soybean	13.17	5.34	21.66	20.50	25.14	0.15	12.12	15.10	0
Common cabbage	3.72	0.99	3.96	16.47	6.01	12.90	4.87	5.17	49.73
Papaya	5.74	1.43	0.68	4.80	2.34	26.09	2.53	3.35	58.82

Nutrition value production = commodity nutrition value × yield/ha or to calculate the production cost of nutrition value as:

Production cost of nutrition value = production cost/nutrition value production.

The nutrition value production and production cost of nutrition value of selected vegetables under AVRDC home garden system are listed in Table 4. Such information can be used for designing cropping systems of backyard garden.

The use of nutrition cost and nutritional value indexes may be effectively used to develop

Table 4
NUTRITIONAL VALUE AND NUTRITIONAL VALUE PRODUCTIVITY OF SELECTED VEGETABLES FROM AVRDC GARDEN (JUNE 1981 TO MAY 1982)[1,4,5]

Vegetable	Nutrition value (NT$/100 g)	Nutritional value productivity (Unit yield/m^2)	Production cost (NT$/100 Unit)
Carrot	117.35	274.7	2.55
Chinese cabbage	29.92	26.9	49.39
Chinses leek	73.51	503.4	2.28
Garland chrysanthemum	83.96	311.7	1.60
Cowpea	40.21	15.4	373.4
Kale	145.63	128.6	9.02
Radish	12.16	16.9	172.5

research priorities and strategies. Since these indexes depend on cost and eating patterns, progress in achieving planning goals can be charted and newly arising conditions can be noted.

REFERENCES

1. Council of Agricultural Planning and Development, Executive Yuan, Taiwan Food Balance Sheet, 1983, Taipei, Taiwan, 1983.
2. **Lin, S. S. M., Peet, C. C., Chen, D. M., and Lo, H. F.,** Breeding goals for sweet potato in Asia and the Pacific: a survey on sweet potato production and utilization, *Proc. Am. Soc. Hortic. Sci., Trop. Region,* 27B, 42, 1983.
3. **Tsou, S. C. S.,** The role of vegetables in Asian diets, in *Tropical Foods,* Inglett, G. E. and Charalambous, G., Eds., Academic Press, New York, 1979, 401.
4. Bureau of Budget, Accounting and Statistics, Taipei City Government, Taipei City Household Expenditure Survey in 1980, Vol. 37, 1981.
5. Council of Agriculture Planning and Development, Executive Yuan, Taiwan Food Balance Sheet 1980, Taipei, Taiwan, 1981.
6. **Tung, D. C., Huang, P. C., Lee, H. C., and Chen, H. L.,** Composition tables of foods used in Taiwan, *Taiwan Med. J.,* 60, 973, 1961.

Chapter 7

FRESH ROOTS FOR HUMAN CONSUMPTION*

Wanda W. Collins and W. M. Walter, Jr.

TABLE OF CONTENTS

I. Introduction 154

II. Nutritional Components of Sweet Potatoes 154
    A. Carbohydrates 154
    B. Dietary Fiber 154
    C. Protein 154
    D. Vitamins 155
    E. Lipids 158
    F. Minerals 158
    G. Antinutritional Factors 159

III. Harvest and Postharvest Handling Methods and their Effect on Quality and Nutrition of Fresh Roots 159
    A. Harvest Procedures 160
    B. Postharvest Handling 160
        1. Curing 160
        2. Storage 162

IV. Preparation Systems and their Effect on Nutritional Components 165

V. Summary 169

References 169

---

* Paper No. 9138 of the journal series of the North Carolina Agricultural Research Service, Raleigh, N.C. Mention of a trademark or proprietary product does not constitute a guarantee or warranty of the product by the U.S. Department of Agriculture or North Carolina Agricultural Research Service, nor does it imply approval to the exclusion of other products that may be suitable. This chapter was prepared in part by a U.S. Government employee as a part of his official duties and legally cannot be copyrighted.

I. INTRODUCTION

Human consumption is one of the most important uses of sweet potatoes in both temperate and tropical growing areas and they are used either as fresh or processed products. In at least one area of the tropics, Papua New Guinea, it is the main staple crop of the population.[1,2] In addition to marketing for local use, tropical growers are increasingly interested in the exportation of tropical vegetable crops including sweet potatoes.[3] Various researchers agree that the major constraints to export are the failure to maintain a high quality product after harvest and inappropriate handling techniques resulting in high postharvest losses.[3,4]

This chapter will present (1) a short review of nutritional components of sweet potatoes, and (2) the effect of harvest and postharvest handling and preparation techniques on the quality and nutritional components of fresh roots for human consumption.

II. NUTRITIONAL COMPONENTS OF SWEET POTATOES

A. Carbohydrates

The sweet potato at harvest contains between 16 and 40% dry matter.[5,6] Of this dry matter, 75 to 90% is carbohydrate. This carbohydrate contains starch, sugar, cellulose, pectins, and hemicellulose (Table 1). Data on carbohydrate composition must be regarded as approximate because of high variability among cultivars due to genetic, environmental, storage, and sample preparation factors. The starch is composed of 60 to 70% amylopectin, and the remainder is amylose.[7,8]

Sucrose is the most abundant sugar in raw sweet potatoes with smaller amounts of glucose and fructose (Table 2). Maltose was not detected and this is in agreement with McDonald[9] who reported that the maltose found by others was an artifact produced by treatment of raw sweet potatoes with hot solvent. Because of their relative abundance, the carbohydrates make up a large part of the caloric value of sweet potatoes, which is reported to be 4.1 kcal/g dry weight.[10]

B. Dietary Fiber

Much less has been reported about the nonamyloid carbohydrates than the amyloids. These materials, including pectins, cellulose, and hemicellulose, together with lignin are loosely classified as dietary fiber. There has been a recent increase in interest in dietary fiber due to studies which have implied that increased dietary fiber may reduce the incidence of such diseases as colonic cancer, diabetes, heart disease, and certain digestive diseases.[11,12] Complete fiber analyses for sweet potato are limited. The data in Table 3 illustrate two examples. Cellulose was high in both reports,[13,14] whereas hemicellulose content for the variety 'Garnet'[13] was an order of magnitude greater than hemicellulose in the variety 'Porto Rico'.[14] Lignin[14] and pectin[13] were present in similar amounts. Other reports[15] are available for total fiber, but due to the recent changes in methodology, the results are not directly comparable.

Pectins have been studied more extensively than any of the other fibrous materials due to their role in the rheological properties of cooked sweet potatoes.[16,17] The mean total pectic content of eight cultivars[17] was 5.1% fresh weight, estimated to be about 20% dry weight at harvest.

C. Protein

The sweet potato has long supplied a significant amount of the caloric requirements in the tropics. However, the protein content and its contribution to overall nutrition have been overlooked until recently. The crude protein content ranges from 1.3 to > 10% (dry basis).[18,19,20] Variability in protein content is due to production practices,[21] environmental conditions,[22,23] and genetic factors.[22,23] It has recently been recognized that the potential

Table 1
**CARBOHYDRATE CONTENT
IN RAW AND BAKED
'GARNET' SWEET POTATOES
(% DRY WEIGHT)**

	Raw	Baked
Starch	46.2	2.6
Sugars	22.4	37.6
Hemicellulose	3.8	1.0
Cellulose	2.7	2.5
Water insoluble pectin	0.47	0.31

Some data from Shen, M. C. and Sterling, C., *Starch*, 33, 261, 1981.

exists to increase the protein content and protein quality through selection for those genetic traits.[22,23]

The protein of sweet potato is quite evenly distributed throughout the root; there are no statistically significant differences in circumferential or radial distribution.[24] Thus, it would not be possible to prepare high protein products by selectively cutting sweet potatoes.

Appreciable amounts of nitrogen are found in the nonprotein nitrogen (NPN) fraction. The NPN fraction is defined as nitrogen not precipitated by 12% trichloroacetic acid and, thus, is of low molecular weight. NPN content ranges from 15 to 35% at harvest.[25] The NPN for 'Jewel' cultivar is comprised of asparagine (61%), aspartic acid (11%), glutamic acid (4%), serine (4%), and threonine (3%).[26] These components accounted for 88.5% of the nitrogen in the fraction. Thus, from a nutritional standpoint, most of the sweet potato NPN is available to satisfy the requirement for total utilizable nitrogen, but provides only small amounts of essential amino acids.

Most of the protein of sweet potato is reported[27] to be globulin, "ipomoein." Upon storage of the root, the ipomoein is partially converted into a polypeptide that is considerably different from the parent globulin in its physical and chemical properties.

A limited number of reports are available concerning the nutritional quality of isolated sweet potato protein. Amino acid analyses that are available indicate that protein from some sweet potato cultivars may be deficient in total sulfur and lysine (Table 4).[18,28,29,30] For 'Jewel' (Table 4), Walter and Catignani[28] reported both total sulfur and lysine to be limiting, while Purcell et al.[18] reported only total sulfur to be limiting for 'Jewel'. Nagase[29] reported no limiting amino acids for a Japanese cultivar. The data (Table 4) indicate that there is some amino acid variability both between cultivars and within the same cultivar. In addition, the data of Purcell et al.[18] for five other cultivars showed total sulfur to be limiting in all cases and that there was considerable between-cultivar variability in content of several amino acids.

There are limited data available for the amino acid content of the whole sweet potato. Among the essential amino acids, only the aromatic amino acids are nonlimiting.[31] This difference in essential amino acid patterns between isolated protein and whole sweet potato is due to the presence of significant amounts of nonprotein nitrogen[26] in the latter. This material effectively dilutes the essential amino acids, thereby lowering the concentration.

As the amino acid analyses indicate, protein nutritional quality is high. Walter and Catignani[28] reported that the protein efficiency ration (PER) of isolates and concentrates is equal to that of casein. Whole sweet potato flour was reported[32] to have a PER of 2.2 and 1.8 (relative to 2.5 for casein) for 'Centennial' and 'Jewel', respectively. The PER was found to be highly dependent on the severity of the heat treatment used in the manufacture of the flour.

Table 2
STARCH, FRUCTOSE, GLUCOSE, AND SUCROSE CONTENT[a] OF RAW SWEET POTATOES[b,c,d]

Time after harvest	Starch		Fructose		Glucose		Sucrose	
	'Jewel'	'Centennial'	'Jewel'	'Centennial'	'Jewel'	'Centennial'	'Jewel'	'Centennial'
0 time	15.1A	15.9A	0.08D	0.08D	0.05D	0.07C	2.41C	2.58D
1 week	17.7B	10.8B	0.49C	0.25C	0.55C	0.14C	3.48B	3.40C
2 months	11.3B	10.8B	1.39A,B	0.35B	1.31B	0.35B	3.23B	5.16B
4 months	8.96C	6.19C	1.17A	0.61A	1.25B	0.66A	4.84A	5.44A
6 months	7.76C	6.24C	1.32A,B	0.56A	1.52A	0.68A	5.18A	5.69A
LSD 0.05	1.53	0.71	0.41	0.10	0.09	0.09	0.54	0.28

[a] Grams in 100 g fresh weight.
[b] Means within columns followed by the same letter are not significantly different at $p < 0.05$.
[c] Maltose not detected.
[d] Walter and Hoover.[94]

Table 3
FIBER[a] IN RAW SWEET POTATOES

	Lund and Smoot[14,b]	Shen and Sterling[13,c]
Cellulose	3.76	3.26
Hemicellulose	0.46	4.95
Insoluble pectin	NR	0.50
Lignin	0.44	NR

[a] Percent of dry matter.
[b] Cultivar not identified.
[c] 'Jersey' cultivar.

Table 4
AMINO ACID COMPOSITION OF PROTEIN ISOLATES (g OF AMINO ACID PER 100 g PROTEIN)

	Walter and Catignani[28][a]	Purcell et al.[18][a]	Nagase[6,29]	FAO[30]
Essential				
Threonine	6.4	5.5	4.6	4.0
Valine	7.9	6.8	7.9	5.0
Methionine	2.0	2.6	2.5	—
Total sulfur	3.1	3.0	4.1	3.5
Isoleucine	5.6	5.3	5.3	4.0
Leucine	7.4	7.8	8.7	7.0
Tyrosine	6.9	5.2	3.6	6.0
Phenylalanine	8.2	6.7	6.0	
Lysine	5.2	6.8	6.5	5.5
Tryptophan	1.2[c]	1.1[c]	1.8[c]	1.0
Chemical score				
Total sulfur	88.0	86.0	100.0	
Lysine	95.0	100.0	100.0	
Nonessential				
Aspartic acid	18.9	14.4	13.1	
Serine	6.6	5.1	5.5	
Glutamic acid	9.6	8.6	11.8	
Proline	4.2	5.4	4.3	
Glycine	5.3	0.3	2.6	
Alanine	5.4	4.6	6.1	
Histidine	2.7	2.4	4.2	
NH_3	1.6	—[d]	—[d]	
Arginine	5.9	6.0	6.4	

[a] 'Jewel' cultivar.
[b] Cultivar unknown.
[c] Tryptophan content measured colorimetrically on enzyme hydrolyzed material.
[d] NH_3 not reported.

From Walter, W. M., Jr. and Catignani, G. L., *J. Agric. Food Chem.*, 29, 797, 1981. With permission.

Horigome et al.[33] reported that protein isolated from a starch production facility had a PER of 1.9, which was increased to 2.5 by the addition of lysine and methionine, indicating that these amino acids are either deficient or are destroyed by the process.

Table 5
VITAMINS IN SWEET POTATOES[a,b,c]

	Raw	Baked	Boiled
Thiamin	0.09	0.07	0.07
Riboflavin	0.05	0.05	0.05
Niacin	0.53	0.55	0.51

[a] Provitamin A, vitamin C not included.
[b] Watt and Merrill.[43] Cultivars not identified.
[c] Mg in 100 g.

It has been reported[34] that when the sweet potato was the only source of nitrogen in the human diet, the nitrogen balance was maintained. In addition, there are areas in New Guinea in which the population obtains a significant portion of its protein requirement from sweet potatoes.[1] Huang[35] reported that sweet potato alone is not sufficiently rich in protein to satisfy the requirements of growing age children but that the crop is a source of high quality protein that should not be overlooked. In fact, a 13% equicaloric replacement of sweet potato for rice was shown[36] to enhance human nitrogen balance. The evidence indicates that the sweet potato not only is a source of calories, but also can improve protein nutrition when added to diets consisting mainly of cereals and grains.

D. Vitamins

The most abundant vitamins from a human nutrition standpoint are beta-carotene (provitamin A) and ascorbic acid (vitamin C). Beta-carotene is the major pigment of the orange flesh in those cultivars that have been studied.[37-40] For 'Goldrush', 90% of the carotenoids had vitamin A activity,[38] and for 'Centennial', 88% of the carotene had vitamin A activity. Genetic selection of cultivars is the most important factor in determining the carotenoid content, but variations in carotene content with location have been observed.[41]

The level of carotenes in the orange-fleshed varieties preferred in the United States are sufficiently high to provide several days' supply of vitamin A per serving. However, the preferred type in much of the tropics is either white-fleshed or cream-colored. This type of sweet potato is obviously a poor source of pro-vitamin A. Since high dry matter and light flesh color appear to be genetically linked, it may be difficult to select for an orange, dry-fleshed variety. However, this should be a goal of plant breeders because of the high incidence of vitamin A deficiency noted in some parts of the tropics.

Vitamin C is fairly abundant in sweet potatoes ranging from 20 to 50 mg per 100 g of fresh weight.[42,43] Thiamin, riboflavin, and niacin contents are shown in Table 5. There is one report that the vitamin E content of raw sweet potatoes is 4 mg/100 gram.[44]

E. Lipids

Lipids are a minor component of the sweet potato ranging from 0.29 to 2.7% (dry basis).[45,46,47] Linolenic acid is the major fatty acid followed by palmitic, linolenic, and stearic acids.[46,47] Boggess et al.[45] separated the lipids into three fractions, nonphospholipids (85.1% of the total), cephaline (9.6% of the total), and lecithin (5.3% of the total). Walter et al.[47] found 42.1% neutral lipid, 30.85% glycolipid, and 27.1% phospholipids.

F. Minerals

The mineral content of sweet potatoes has been measured by several researchers. The values (Table 6) show the variability present, as is the case in most of the components of

Table 6
MINERAL CONTENT OF RAW SWEET POTATOES

	Reported values (mg/100 g)		
	Elkins[48,a]	Lopez et al.[49,b]	RDA (mg)[c]
Calcium	32.7	17.4	870
Phosphorus	48.9	39.2	800
Magnesium	22.2	18.3	352
Sodium	8.0	30.3	1000
Potassium	228.0	360.0	2000
Iron	0.85	9.59	100
Copper	0.25	0.13	2.0
Magnesium	0.59	0.24	4.0
Zinc	0.26	0.27	15.0

[a] Mean from analysis of 3 lots of 'Centennial' and 1 lot of 'Nemagold'.
[b] 'Jewel' cultivar.
[c] Recommended dietary allowance (50).

the sweet potato. The amount of the human requirement furnished in 100 g is small for all of the minerals, with the possible exception of potassium, which furnishes about 11.4%[48] or 18%[49] of the recommended dietary allowance (RDA)[50] depending upon the study. Sodium is low and, thus, would provide no problem for those on a restricted sodium intake diet.

G. Antinutritional Factors

The only antinutritional factor reported for sweet potato is a trypsin inhibitor.[51,52] The presence of this material in a food causes an adverse nutritional effect by inhibition of the proteolytic action of trypsin during the digestion process.

III. HARVEST AND POSTHARVEST HANDLING METHODS AND THEIR EFFECT ON QUALITY AND NUTRITION OF FRESH ROOTS

Harvest and postharvest handling procedures differ greatly between temperate and tropical growing areas. Mechanical harvesting and handling is practiced extensively in temperate areas. While this mechanical harvesting may result in more injury to roots than the more labor intensive harvest methods used in the tropics, the economic loss due to injury is minimized by careful postharvest treatment.[53] Roots grown in temperate areas are normally cured for 4 to 7 days in specially designed facilities where the temperature is maintained around 29°C and relative humidity is kept at 85 to 90%. This promotes wound periderm (cork) formation on injured surfaces of the root which in turn prevents excessive water loss and prevents pathological organisms from invading the injured roots. Once the curing procedure is complete, roots are stored at temperatures of 13 to 16°C with 85 to 95% relative humidity. Under these carefully controlled conditions roots may be kept for 12 months or longer depending on cultivar. Therefore, high quality roots may be marketed year-round in temperate areas.

In the humid tropics, sweet potato can be grown during most of the year. Although it is a perennial plant, it is grown as an annual with a growing season of 3 to 8 months and two crops a year. In some areas where rainfall (too much or too little) is a problem only one crop can be grown. The availability of fresh sweet potato roots on a continuous basis for marketing has resulted in a minimum of curing and storage procedures for roots in the tropics. However, postharvest losses are extremely high and much of the crop may be lost before it is sold.[3,4] The potential for export of tropical vegetable crops including sweet potatoes from tropical countries is often limited by the failure to maintain quality and by the lack of proper handling techniques to reduce losses after harvest.[3]

A. Harvest Procedures

In general, harvesting in the tropics is manual with a variety of relatively simple digging implements including sticks, spades, 4 to 6 pronged potato hoes, and occasionally knives.[54,55] Mechanical harvesting is usually practiced only on large-scale production areas where the terrain is suitable for machinery. Mechanization may range from a variety of plows,[56,57] either tractor drawn or animal drawn, to machines which remove the roots from the soil and deposit them in a collection container.[58] Harvest methods have a direct and dramatic effect on the quality of fresh roots for market. At time of harvest, roots are extremely susceptible to skinning and bruising. Heavy losses may occur unless digging is done very carefully.[59] Roots should not be thrown into piles for later packing but should be carefully placed.

Time of harvest is dictated by weather, disease and insect problems, and the marketability of the roots. Roots must be harvested before flooding occurs in areas with a definite rainy season. If sweet potato weevil is a problem, then roots are harvested before the damage becomes severe.[4] If roots are harvested too early, yields will be reduced; if they are left too long, they may rot or become very fibrous and finally inedible.[60] Because they do grow as perennials, sweet potatoes may be harvested according to how much a grower can sell at any particular time. Progressive harvest of individual plots or individual plants for this purpose or for home use occurs in many areas.[1,2,55,61,62] Also, progressive harvest of individual plots insures a steady supply of fresh roots to the market and avoids the need for long-term storage of surplus roots. Progressive harvest of individual plants is practiced extensively in Papua New Guinea. Roots of a desirable size are removed from each plant without disturbing smaller roots on the plant. Bourke[1] reports that individual plants can be harvested up to four times in a year and that progressive harvest of the same plants can be carried on for a period of months or even years. However, a single harvest was shown to result in higher numbers of marketable size roots than progressive harvest although progressive harvesting resulted in a higher overall yield.[63] In Uganda the largest roots are removed with digging sticks; 2 to 3 months later the remainder of the crop is removed.[64]

Once the sweet potato roots have been harvested, they may be left in the sun to dry for a short length of time in some areas but usually not overnight.[57,65,66] More often they are transported to the market place and sold with a minimum of storage time, usually less than a week, and with no curing other than that which occurs naturally. However, without proper curing and storage procedures even on a short-term basis, losses can be severe. Most losses result from bruising and injury to the roots during harvesting, packing, and transport. These losses are mainly due to water loss and the invasion of fungi and bacteria. The effect on quality is dramatic and many roots are of unmarketable quality by the time they reach the marketplace.

B. Post Harvest Handling
1. Curing

Curing is rarely practiced in the tropics except that which occurs naturally. If sweet potato weevil is not a problem in a particular area, then the tops of sweet potato plants may be removed a week before harvest.[4,66] This allows the roots to achieve some degree of curing in the ground. However, when they are removed from the ground the new injuries again render them susceptible to disease organisms and water loss.

Proper curing occurs ideally at 29°C, and 85 to 90% relative humidity for 4 to 7 days.[53] The normal temperature and relative humidity in the humid tropics are very similar to those

Table 7
EFFECT OF CURING TREATMENT OF SWEET POTATOES STORED 90 DAYS (24 TO 29°C) AVERAGE OF FOUR REPLICATIONS, THREE CULTIVARS[a], APRIL TO JUNE 1974[b]

Treatment	Weight loss (%)	Marketable (%)
Standard A	11.6	82
Modified B	9.3	85
Control C	18.2**	48**

[a] Cultivars Centennial, Georgia Red, Goldrush.
[b] Means followed by ** are significantly different at the 1% level.

From Gull, D. D. and Duarte, O., *Proc. Trop. Region Am. Soc. Hortic. Sci.*, 1974, 168, 170. With permission.

Table 8
THE EFFECT OF CURING DURATION ON MARKETABILITY OF CENTENNIAL, GEORGIA RED, AND GOLDRUSH SWEET POTATOES STORED 90 DAYS AFTER CURING AVERAGE OF FOUR REPLICATIONS[a]

	Days cured			
Cultivar	0	3	6	9
	(% marketable)			
Centennial	26*	54*	79	83
Georgia Red	66	78	85	82
Gold Rush	35*	38	82	83

[a] Cultivar means followed by * are significantly different at the 5% level.

From Gull, D. D. and Duarte, O., *Proc. Trop. Region, Am. Soc. Hortic. Sci.*, 1974, 168, 170. With permission.

required for curing and some natural curing does occur provided roots are given adequate ventilation. Roots are occasionally cured by placing under a shade for 8 to 10 days at ambient temperatures.[57] Gull and Duarte[67] conducted tests using several sweet potato cultivars to determine if a simple, economical method of curing under tropical conditions could be established to prolong the storage life of roots. They used three cultivars and three methods of curing: standard method (A) which is the method most commonly used in temperate areas (30°C, 85 to 90% RH); modified method (B) consisting of crates lined with plastic at 27 to 32°C, 40 to 60% RH; and control method (C) consisting of regular crates, 27 to 32°C, 40 to 60% RH. After curing roots were placed at different storage temperatures (15°C, 25°C, and ambient). The ambient temperature ranged from 24 to 29°C in a well-constructed masonry building. Their results are shown in Tables 7 and 8.

The results in Table 7 indicate that adequate curing would result if plastic box liners were used to maintain high relative humidity (modified method B). This modified procedure resulted in lower weight loss and a higher percentage of marketable roots than the control technique C and it was comparable to the more expensive standard method A. Results shown in Table 8 indicate that no benefit was derived from curing past 6 days and that there was a differential varietal response to curing. Thus, effective breeding procedures could be used to develop varieties more suitable to curing under tropical conditions. The authors concluded that curing temperatures of 28 to 31°C in tropical areas would be no major problem and that the use of plastic box liners to maintain high relative humidity would result in adequate, economical curing of sweet potato roots.

The effect of curing on the quality of freshly harvested roots has been investigated extensively in temperate growing areas. Curing sweet potatoes reduced decay[68] and resulted in less weight loss during the storage period after curing.[69] Although temperature and relative humidity may be near ideal for curing in tropical areas, roots which are immediately packed for market in sacks or crates may not have received adequate ventilation for the healing process to occur. Ventilation is necessary to remove free moisture and prevent CO_2 accumulation and O_2 depletion.[69] Curing apparently has no dramatic effect on nutritional components of sweet potato although slight changes have been recorded in carbohydrates, carotene and ascorbic acid.[70] Reports of these changes are often contradictory and may depend on such factors as cultivar, growing environment, and exact conditions during curing.

1. Storage

The continuous year-round supply of fresh roots to the market in almost all tropical countries limits the necessity for storage. In Taiwan, for example, 75% of growers in one survey sold their crop immediately after harvest with the remainder of the growers chipping the roots before sale.[71] In addition many growers use methods to avoid storage as much as possible.[66,72] To avoid storage, growers practice ground storage (progressive harvest)[4,55,73] or, in some areas, they use a succession of early and late cultivars including different ones for wet and dry seasons.[66] When ground storage is practiced roots are harvested only in amounts that can be successfully marketed; the remainder of the crop is left until needed. This is the most economical method of storage, but there are a number of problems associated with it. If sweet potato weevils or other insect pests such as beetles or termites are present, ground storage can be particularly hazardous.[64,73,74] In addition, roots stored for too long can become very fibrous and can crack making them unsuitable for marketing although some varieties are better suited to ground storage than other varieties.[4] A study of cultural practices such as crop rotation along with development of insect-resistant varieties has been suggested to develop increased and more effective ground storage.[4]

Once roots are removed from the ground, they are susceptible to immediate quality decline if not properly cured. The roots are living organs and remain metabolically active after harvest. Respiration continues and water losses occur. Roots that are bruised or injured during harvest frequently are more metabolically active than uninjured roots and are more likely to show a decline in quality. In addition, the sites of injuries are ideal invasion points for pathogenic and secondary-invasion organisms. These factors limit storage potential of the roots for any length of time.

Several methods of short-term storage (up to one week) are practiced in tropical growing areas. Typically, roots are packed soon after harvest into sacks, boxes or crates for transport to market (Figure 1).[65,75] The roots may remain in these containers for up to a week; however, after a few days quality begins to deteriorate rapidly. Much of this loss is due to rapid water loss resulting from harvest damage with no healing process of the injuries. In addition, toxic stress metabolites may be formed in the edible root as a result of infection by certain microorganisms.[76] These toxins have been isolated from roots purchased from local markets with only minor blemishes.[76]

Long-term storage is rarely used. The major constraints to long-term storage in the tropics are (1) moisture and dry matter losses (Olurunda reported losses up to 95%),[4,73] (2) rots due to injury during harvest and handling,[75] (3) sweet potato weevil damage,[4,72] and (4) sprouting.[4,65,77] All of these problems lead to a product which is of poor quality and unmarketable or inedible. Attempts have been made to utilize several types of structures to overcome these constraints and lengthen the storage period and have met with some success. Among these structures are ground pits, mounds, sheds, caves, and houses.

Ground pits have been used extensively. Best[78] describes two types of pits used by early New Zealand Maoris: (1) semi-subterranean pits consisting of excavations into sloping ground or on terraces and lined with dried plant material with a roof of timber covered with earth so that the pit could be sealed; and (2) subterranean pits which were dug into the ground, filled and sealed. Little or no ventilation was provided in either case and heavy losses resulted due to decay and possible rodent damage. In Barbados, sweet potatoes have been stored for up to 4 months in pits,[59] but, in general, storage life is extended only to one or two months

FIGURE 1. Freshly harvested sweet potatoes prepared for export in St. Vincent, West Indies. (Photograph by W. W. Collins.)

because of spoilage and sprouting.[60] In the Philippines, roots were stored in a trench 50 cm deep covered with sand and sheltered by a roof; however 30% of the roots decayed and 45% sprouted.[57] In Trinidad, West Indies, immediate storage in pits when the temperature was about 24.5°C and the relative humidity about 82.5% resulted in less weight loss than if the roots were cured for 2 to 6 days in a house or in the sun.[57]

Mounds (also called clamps) are also used for storage. Keleny[79] suggests a well-designed aboveground mound structure with ventilation provided by trenches beneath a bed of perforated boards. A small flue structure through the center of the pile provides additional ventilation. The pile base is then covered with dried plant material with sweet potato roots placed in a conical arrangement around the flue. The entire pile is covered with 20 cm of additional dried plant material and a 15 cm layer of soil. The completed mound can be protected from weather and rodents. In some mounds, the trench may contain a hurricane lantern to provide heat for some degree of curing.

The Department of Science and Agriculture of the Federation of the West Indies developed a successful method of clamp storage for Barbados.[80] Clamps are located on well-drained sites which are dug out to a depth of 3 to 4 feet over a rectangular area 3 feet wide and as long as necessary. The floor is lined with dried grass and other dried vegetation. Roots are stacked to a depth of 3 feet; the heap is then covered with dried vegetation, soil at least 1 foot deep, and more dried vegetation to prevent washing. Sweet potato roots lost as much as 20% of their weight but were still palatable after 4 months storage. Only roots free of diseases and insects should be stored in this way.

Other methods of storage include stacking in sheds,[55] in well-ventilated storehouses,[65] on raised platforms,[60] in heaps on floors of barns or houses,[73] and in baskets or in roof spaces.[66]

It is generally recognized that these storage areas must be dry and well-ventilated. Roots must be free of pests. Often fires are built to provide some additional heat and roots may become slightly smoked;[72] in other areas roots may be covered with ashes.[73] Roots have been successfully stored in these types of structures for up to 4 weeks.[66]

Research results comparing clamps with house and pit storage in Papua New Guinea showed that sweet potatoes can be successfully stored for 30 to 50 days depending on location and exact structure.[72] Clamp storage is suggested only for the highland tropical areas where temperatures are lower. Lowland storage in Papua New Guinea was found to be just as efficient and less expensive in specially constructed houses utilizing a center fire for curing and hanging wet bags for maintaining high relative humidity. Storage in these houses was successful for 2 to 3 weeks.

It is apparent that storage presents a major problem to the availability of fresh high quality roots in the tropical marketplace. Without cool storage temperatures roots cannot be stored for more than 3 to 4 months under the most ideal available conditions and usually not for more than a few weeks. Sweet potato roots consistently decrease in quality even under ideal conditions. Under the less than ideal conditions available to most tropical growers, the changes are even more drastic.

The effects of storage on quality of fresh roots are increased rotting, shriveling, and pithiness due to weight and water loss, sprouting, and insect and rodent damage. Rotting, shriveling, and pithiness can be minimized by placing in storage only roots free of disease and injury.

Several methods have been used to decrease sprouting in storage. Sprouting was reduced in roots stored for 4 to 8 weeks by spraying the crop two weeks before harvest with a maleic hydrazide solution or putting freshly harvested roots in containers with confetti that had been treated with a methyl ester of naphthalene acetic acid in acetone.[57] No effect on weight loss was apparent. CIPC[81] and thiourea[82] have also been used for this purpose with variable success.

One of the most serious problems in storage of sweet potatoes is the sweet potato weevil. Hahn and Anota[83] report a weevil mortality rate of 89.5% if the storage temperature of sweet potato roots can be reduced to 20°C.[83] This would control reproduction and spread of the weevils in storage. In addition, they studied the effect of immersing sweet potato weevil-infested roots in water to prevent subsequent weevil damage in storage. Adult weevils died within 10 min of immersion in water at 52 or 62°C and within 30 min in water at 42°C. All larvae died within 10 min in hot water and within 12 hr in tap water. In the same study, they showed that weevils will survive in underground storage (buried) but length of survival depends on depth of storage. Quick death occurs within 5 cm of the soil surface (90% within 5 days). All weevils died within 3 days if kept at the soil surface. These mortality rates are probably due to high daytime temperatures occurring at those depths. At lower levels, insects survived longer but most were dead after eight days. In addition to the devasting effect on storage life and quality of roots, weevil infestation also results in the production of certain terpene compounds which make damaged roots unsuitable for human consumption.[84]

The effect of storage on nutrition has been studied extensively under temperate storage conditions. During storage starch is lost through metabolism, while the levels of the sugars increase. After 6 months' storage under ideal conditions, sucrose levels in two cultivars exceeded 5%, while in 'Jewel' fructose and glucose levels are greater than 1.3% (Table 2). The data in Table 2 illustrate changes during storage and variability between cultivars.

Some nitrogen is lost during storage, but the rate of loss is less than the rate at which carbohydrates are lost. Thus, the relative concentration of protein increases during storage.[25] The limit to the degree of concentration in storage is not known. In our laboratory, we have measured stored roots with 16% protein, which we estimate contained about 6% protein at harvest. These roots were pithy, and microscopic examination showed a greatly decreased number of starch grains, all of which were very small. Nonprotein nitrogen decreases during the early part of storage and then increases.[25]

Storage is reported to cause an increase in carotene[42] probably because the dry matter content of the roots decreases during storage. For orange-fleshed cultivars, this carotene content ranges from 5 to 20 mg of vitamin A per 100 g sweet potato.[42] For vitamin C in the cultivars Triumph and Nancy Hall, declines of 28 and 48%, respectively, were measured during storage.[42] Others have reported a loss of vitamin C during storage with values falling from 46 to 28 mg/100 g in 4 months.[85] Very little is known about the effect of storage on other vitamins.

Pectins decrease during storage from 5.1% fresh weight at harvest to 3.5% fresh weight after 6 months of storage.[17] Most of the decrease was due to changes in the hydrochloric acid-soluble fraction, while the ammonium oxalate-soluble and water-soluble fractions did not change significantly. The degree of esterification decreased during storage.[17]

IV. PREPARATION SYSTEMS AND THEIR EFFECT ON NUTRITIONAL COMPONENTS

The sweet potato is consumed throughout the tropics mainly as a home-prepared dish. Baking, boiling, steaming, and frying are the forms of heat processing used. Preparation practices vary according to the location. For example, in New Guinea, boiling and baking are common. In East Africa, roots are boiled unpeeled or roasted unpeeled in the ashes of a fire before being eaten, or less commonly, the sweet potato is boiled or fried with other vegetable or root crops. In Taiwan, most sweet potatoes are eaten boiled or boiled and mixed with white rice. Thus, the heat treatments used in the tropics are relatively mild when compared to canning or dehydration on heated drums. Consequently, nutrient retention should be excellent.

The preferred type of sweet potato in most of the world including the tropics is a root which when cooked has "dry" mouthfeel similar to the white potato (*Solanum* spp.), a white to light yellow flesh color, and a moderately sweet taste. Villareal[86] and co-workers at the Asian Vegetable Research and Development Center (AVRDC) established several categories of sweet potato, depending upon the utilization goal. For human consumption, as a staple part of the diet, the preferred type would be a white-fleshed, low sugar, high starch type, while a dessert type root would be an orange, moist-fleshed sweet type. The logic behind this classification is to increase consumption and take advantage of the potential for high per-unit productivity possible with this crop. The classification and goals seem to have a sound basis. The orange color and sweet taste relegate sweet potatoes to a special use category and limit its consumption. If cultivars are developed embodying the concept of high starch, white color, and low sugar, the sweet potato may indeed become accepted in the tropics as a staple item and, thus, serve to alleviate malnutrition. This strategy is based on the concept that a food, no matter how nutritious, is of no value unless it is eaten.

Textural properties of cooked sweet potato have been the subject of many investigations. Gore[87,88] reported that a diastase in sweet potato converted large amounts of starch into maltose during slow cooking. Later, workers[89] showed that sweet potato also contains an α-amylase, Walter et al.[90] reported that the degree of moist (or dry) texture is due to the amount of starch left after baking, the amounts and sizes of dextrins and the amount of sugar present. All of these parameters are influenced by the activity of the amylolytic enzymes. Shen and Sterling[13] reported that amylolytic enzyme activity was less affected by heat in the moist type than in the dry type and that hemicellulosic fibrils of the cell wall break down more rapidly in the moist types. Thus, the degree of moist mouthfeel appears to be a function of the starch, dextrin, and sugar content of the cooked roots. These components, in turn, are dependent upon the activity of the amylolytic enzymes during cooking. There also appears to be some difference in the cell wall structure of moist and dry types.

The quality factor of color (or lack of color) has been extensively studied by plant breeders. Orange flesh has been shown to be negatively associated with high dry matter.[91,92] Thus, the goal of producing a white, high dry matter low sugar-type sweet potato is genetically favored. Conversely, development of orange-fleshed (high vitamin A) cultivars with high dry matter will be more difficult. A recent study[93] showed that acceptability varies according to nationality of the consumer, but that flavor (sweetness) and color are good quality characteristics for predicting general acceptability for steamed sweet potatoes.

Significant changes in the carbohydrate fraction occur during cooking. If the roots are cut into strips and rapidly cooked, significant amounts of starch remain after cooking (Table 9), whereas, if whole roots are baked, starch is more completely converted into dextrins and sugars.[13,94,95] In baked roots, the degree of starch conversion is dependent upon the cultivar. The "dry" types do not convert nearly as much starch during cooking as do the "moist" types.[13,90,94,95] It should be noted (Table 9) that maltose is the main sugar produced during cooking. The quantities of the other sugars decrease during cooking probably due to reaction between the reducing groups of the sugars and nitrogenous material. Sucrose, of course, would have to first undergo heat-mediated hydrolysis, thereby freeing the reducing groups for reaction. A paper which appeared in 1931[96] reported very dissimilar results. These workers examined sweet potatoes cooked by boiling, steaming, and baking. They found that sugars and dextrins decreased during cooking and starch content increased. We have no explanation for these results except that the methods used did not give an accurate measure of the various carbohydrates.

The changes in amylolytic carbohydrate levels are attributed to the action of α- and β-amylases which are naturally present in the roots.[89,97] These enzymes probably are involved in mobilizing carbohydrates for respiration during storage, but they evidently do not become fully active until starch is gelatinized. Both enzymes seem to have appreciable tolerance to high temperature and remain active for several minutes at temperatures which disrupt the starch granules. The amount of enzyme and, consequently, the magnitude of carbohydrate conversion during cooking varies according to cultivar and postharvest treatment as well as conditions of cooking.[90] There appear to be no direct changes in nutritional value due to carbohydrate conversion. Baking and processing decrease the amount of pectins and the degree of esterification. However, no direct relationship was found that correlated rheological and sensory changes of baked sweet potatoes with changes in pectin content or molecular size of the pectins. Cooking decreases the amount of insoluble pectin and thereby decreases its role as a component of fiber. Cellulose is decreased slightly by cooking, and hemicellulose in 'Garnet' variety is significantly decreased (Table 4). In a dry-fleshed variety, 'Jersey', both components decreased slightly during baking.[13] There is no report of the effect of cooking on lignin content. The storage history of the root also affects fiber content. Hemicellulose of cooked sweet potatoes is about 40% less in roots stored 7 months before cooking than in roots cooked at harvest.[95]

Trypsin inhibitors are affected by cooking. Presently, the preponderance of evidence[51,98] indicates that any cooking process which causes the root to reach 90°C or more for several minutes will effectively inactivate the inhibitor. However, it has been found that a short-time, high-temperature treatment used in the preparation of animal feed is not an effective way to destroy the inhibitor.[99] The trypsin inhibitor has been implicated as a factor in the disease *Enteritis necroticans*.[100] However, unless the roots are consumed raw over long periods, it does not appear that the inhibitor from sweet potatoes could be a factor in the occurrence of this disease.

Since heat treatments used in the tropics are relatively mild, little damage to the protein nutritional quality is expected. Purcell and Walter[101] reported that baking caused less nutritional damage than did either canning or drum drying. The major nutritional change due to heat processing is the loss of lysine, probably via reaction with reducing groups of sugars.

Table 9
STARCH, FRUCTOSE, GLUCOSE, SUCROSE, AND TOTAL PECTIN CONTENT OF COOKED[a] SWEET POTATO STRIPS[b,c]

Time after harvest	Starch		Fructose		Glucose		Sucrose		Maltose	
	'Jewel'	'Centennial'	'Jewel'	'Centennial'	'Jewel'	'Centennial'	'Jewel'	'Centennial'	'Jewel'	'Centennial'
0 time	9.24A	9.83A	0.07D	0.05E	0.04C	0.03E	2.11C	2.58C	4.43A	5.46A
1 week	7.30B	6.91B	0.44C	0.17D	0.48B	0.11D	2.82B,C	3.51C	3.72A,B	4.61A,B
2 months	7.08B	7.28B	1.15A	1.27C	1.22A	0.30C	3.13B	4.40B	3.75A,B	4.35B
4 months	6.22B,C	3.98C	1.09A,B	0.56A	1.19A	0.60A	4.72A	5.20A	2.89B	2.10C
6 months	5.77C	3.75C	1.14A	0.46B	1.29A	0.55B	4.36A	5.32A	1.83C	2.48C
LSD 0.05	1.17	1.45	0.25	0.07	0.17	0.05	0.88	0.66	0.73	0.97

[a] Grams in 100 g fresh weight. Roots cut into strips and steam 5 min.
[b] Means within columns followed by the same letter are not significantly different at $p < 0.05$.
[c] Walter and Hoover.[94]

It is likely that boiling caused some loss of the free amino acids through leaching, as has been reported[102] for roots canned in liquid.

Processing, including baking or boiling, of sweet potatoes may cause minor changes in carotenoid content (vitamin A value) due to heat-mediated isomerization.[103] Although provitamin A has been shown in many processing studies to be remarkably heat-stable and to be only slightly affected by cooking or processing, there are reports of 20 to 25% losses during baking.[104] Consequently, losses of vitamin A value for sweet potatoes during processing are minimal. Occasionally, processing appears to increase the amount of carotenoids. Most of the increase can be rationalized as loss of water content or leaching of water-soluble dry matter.

Vitamin C is lost during cooking. McNair reported[104] that from 12 to 20% of the vitamin C was lost during baking and boiling; neither cooking method appeared to be more destructive than the other. The lowest amounts of vitamin C have been shown[105] to be located in the outer 4 mm of tissue. Thus, peeling would not remove large amounts of the vitamin. It appears that the food preparation practices in the tropics would not cause serious loss of vitamin C.

Thiamin is the least heat-stable of the other three water-soluble vitamins with about 22% destroyed by either baking or boiling. Riboflavin and niacin appear to be stable to heat processing (Table 5). Very little is known about the levels of the other water-soluble vitamins, and about variability and effect of storage on all water-soluble vitamins except vitamin C. Except for beta-carotene, the fat-soluble vitamins have received little attention. One report cites the vitamin E content at 4 mg per 100 g.[106]

The sweet potato produces a series of stress metabolites in response to injury, physiological stimuli, and infectious agents. The most important, from a toxicological standpoint, are a family of furanoterpenoids[107] which have been shown to contain pulmonary toxins[108] and hepatotoxins.[109] These toxins are found in sweet potatoes infected by *Ceratocystis fimbriata* (black rot fungus) and by several *Fusarium* species (rot fungi). The toxins are also produced by cut injury or exposure of wounded tissue to mercuric chloride.[109] Since we are concerned in this chapter with the food supply, we shall discuss only toxin production via microbial infection and mechanical injury.

The most abundant of the toxins isolated from infected tissue, ipomeamarone, is a hepatotoxin. It has been shown[110] that baking destroyed more than 90% of the ipomeamarone and significant amounts of the lung edema toxin, 4-ipomeanol. This is in conflict with the report of Wilson et al.[76] who reported that cooking did not destroy significant amounts of ipomeamarone. From a human health perspective, the research results of Catalano et al.[111] are important. These workers reported peeling and trimming sweet potatoes from 3 to 10 mm beyond the infected area effectively reduces the toxin content to barely detectable or not detectable even when the infected tissue contained more than 1000 ppm. More importantly, normal commercial processing conditions of lye peeling, followed by trimming, also removed the toxin. In fact, no ipomeamarone has been detected in canned sweet potatoes randomly selected from the normal marketing chains.[110] The series of papers[76,108,109] regarding the presence of toxins in sweet potatoes apparently made the problem seem much worse than is actually the case. For example, Boyd and Wilson[112] reported that selected sweet potatoes from food markets in Nashville, Tenn., and Lexington, Ky., had high levels of ipomeamarone. However, no mention was made of the extent of decay and of whether or not trimming away diseased areas had any effect on toxin levels. It is highly unlikely that a consumer would eat a partially rotted or seriously blemished sweet potato without first trimming away the damaged area. The main danger then is that the infected roots might be fed to animals. In fact, in the U.S. the only documented cases of poisoning outside the laboratory have been for farm animals fed moldy sweet potatoes.[76]

V. SUMMARY

Commercial potential of the sweet potato in the tropics has been limited by low yields, failure under tropical conditions to maintain quality of the fresh roots, and failure to reduce total losses due to the perishability of this crop. High costs of cooling and/or heating and ventilation make it impractical to use handling methods which have been shown to maximize storage life and minimize storage losses. Therefore, efforts should most likely be directed toward developing harvest and handling systems using the resources available in the tropics. Variety development should also proceed to identify types that are more suited to those resources. Proper handling and packaging (i.e., cardboard or wooden ventilated boxes instead of sacks), cultural practices to minimize diseases and even curing under ambient conditions can lead to a commercial product with improved quality and storage life for human consumption.[78,113]

REFERENCES

1. **Bourke, R. M.,** Sweet potato in Papua, New Guinea, in *Sweet Potato: Proc. First Int. Symp.*, Villareal, R. L. and Griggs, T. D., Eds., Asian Vegetable Research and Development Center, Shanhua, Tainan, Taiwan, 45, 1982.
2. **Kimber, A. J.,** The sweet potato in subsistence agriculture, *Papua New Guinea Agric. J.*, 23, 80, 1972.
3. **Pantastico, E. B. and Bautista, O. K.,** Post-harvest handling of tropical vegetable crops, *HortScience*, 11, 122, 1976.
4. **Coursy, D. G. and Jackson, G. V. H.,** Working group report: handling and storage, in *Small Scale Processing and Storage of Tropical Root Crops*, Plucknett, D. L., Ed., Westview Press, Boulder, Colo., 1979, 15.
5. **Martin, W. S. and Griffith, G.,** Annual Report of the Chemical Section, Uganda Dept. of Agric. Rep., Kampala, Part 2, 1938, 50—56.
6. **Purcell, A. E., Pope, D. T., and Walter, W. M., Jr.,** Effect of length of growing season on protein content of sweet potato cultivars, *HortScience*, 11, 31, 1976.
7. **Madamba, L. S. P., Bustrillos, A. R., and San Pedro, E. L.,** Sweet potato starch: physiochemical properties of the whole starch, *Philipp. Agric.*, 54, 350, 1973.
8. **Bertoniere, N. R., McLemore, T. A., and Hasling, V. C.,** The effect of environmental variables on the processing of sweet potatoes into flakes and on some properties of their isolated starches, *J. Food Sci.*, 31, 574, 1966.
9. **McDonald, R. E. and Newson, D. W.,** Extraction and gas-liquid chromatography of sweet potato sugars and inositol, *J. Am. Soc. Hortic. Sci.*, 95, 299, 1970.
10. **Jeffers, H. F. and Haynes, P. H.,** A preliminary study of the nutritive value of some tropical root crops, *Proc. Intern. Symp. Trop. Root and Tuber Crops*, Vol. I, University of the West Indies, St. Augustine, Trinidad, 1967, 72.
11. **Kimura, K. K.,** High fiber diet — who needs it, *Cer. Foods World*, 22, 16, 1977.
12. Select Committee on Nutrition and Human Needs, United States Senate, Dietary Goals for the United States, U.S. Government Printing Office, Washington, D.C., 1978.
13. **Shen, M. C. and Sterling, C.,** Changes in starch and other carbohydrates in baking *Ipomoea batatas*, *Starch*, 33, 261, 1981.
14. **Lund, E. D. and Smoot, J. M.,** Dietary fiber content of some tropical fruits and vegetables, *J. Agric. Food Chem.*, 30, 1123, 1982.
15. **Juritz, C. I.,** Sweet potato. II. Chemical and comparative analysis of tubers, *J. Dept. Agric. S. Afr.*, 2, 340, 1921.
16. **Heinze, P. H. and Appleman, C. O.,** A biochemical study of curing processes in sweet potatoes, *Plant Physiol.*, 18, 548, 1943.
17. **Ahmed, E. M. and Scott, L. E.,** Pectic constituents of the fresh roots of the sweet potato, *Proc. Am. Soc. Hortic. Sci.*, 71, 376, 1957.
18. **Purcell, A. E., Swaisgood, H. E., and Pope, D. H.,** Protein and amino acid content of sweet potato cultivars, *J. Am. Soc. Hortic. Sci.*, 97, 30, 1972.

19. **Li, L.**, Variation in protein content and its relation to other characters in sweet potatoes, *J. Agric. Assoc. China*, 88, 17, 1974.
20. **Splittstoesser, W. E.**, Protein quality and quantity of tropical roots and tubers, *HortScience*, 12, 294, 1977.
21. **Constantin, R. J., Hernandez, T. P., and Jones, L. G.**, Effects of irrigation and nitrogen fertilization on quality of sweet potatoes, *J. Am. Soc. Hortic. Sci.*, 99, 308, 1974.
22. **Collins, W. W. and Walter, W. M., Jr.**, Potential for increasing the nutritional value of sweet potatoes, in *Sweet Potato: Proc. First Int. Symp.*, Villareal, R. L. and Griggs, T. D., Eds., Asian Vegetable Research and Development Center, Shanhua, Tainan, Taiwan, 1982, 355.
23. **Li, L.**, Breeding for increased protein content in sweet potatoes, in *Sweet Potato: Proc. First Int. Symp.*, Villareal, R. L. and Griggs, T. D., Eds., Asian Vegetable Research and Development Center, Shanhua, Tainan, Taiwan, 1982, 345.
24. **Purcell, A. E., Walter, W. M., Jr., and Giesbrecht, F. G.**, Distribution of protein within sweet potato roots, *J. Agric. Food Chem.*, 24, 64, 1976.
25. **Purcell, A. E., Walter, W. M., Jr., and Giesbrecht, F. G.**, Changes in dry matter, protein and nonprotein nitrogen during storage of sweet potatoes, *J. Am. Soc. Hortic. Sci.*, 103, 190, 1978.
26. **Purcell, A. E. and Walter, W. M., Jr.**, Changes in composition of the nonprotein-nitrogen fraction of 'Jewel' sweet potatoes during storage, *J. Agric. Food Chem.*, 28, 842, 1980.
27. **Jones, D. B. and Gersdorff, C. E. F.**, Ipomein, a globulin from sweet potatoes, *Ipomoea batatas*, *J. Biol. Chem.*, 93, 119, 1931.
28. **Walter, W. M., Jr. and Catignani, G. L.**, Biological quality and composition of sweet potato fractions, *J. Agric. Food Chem.*, 29, 797, 1981.
29. **Nagase, T.**, Japanese foods. XXXIV. Amino acid content of the potato protein and the rates of digestibility and absorption of Japanese foods, *Fukuoka-Igaku - Zasshi*, 48, 1828, 1957.
30. **FAO**, Amino Acid Content of Foods and Biological Data on Protein, FAO Nutritional Studies, Food and Agricultural Organization of the United Nations, Rome, 1973.
31. **Walter, W. M., Jr., Collins, W. W., and Purcell, A. E.**, Sweet potato protein: a review, *J. Agric. Food Chem.*, 32, 695, 1984.
32. **Walter, W. M., Jr., Catignani, G. L., Yow, L. L., and Proter, D. H.**, Protein nutritional value of sweet potato flour, *J. Agric. Food Chem.*, 31, 847, 1983.
33. **Horigome, T., Nakayama, N., and Ikeda, M.**, Nutritive value of sweet potato protein produced from the residual products of the sweet potato industry, *Nippon Chikusan Gakkahi-Ho*, 43, 432, 1972.
34. **Adolph, W. H. and Liu, H. C.**, The value of sweet potato in human nutrition, *Chin. Med. J.*, 55, 337, 1939.
35. **Huang, P. C., Lee, N. Y., and Chen, S. H.**, Evidences suggestive of no intestinal nitrogen fixation for improving nutrition status in sweet potato eaters, *Am. J. Clin. Nutr.*, 32, 1741, 1979.
36. **Yang, T. H. and Blackwell, R. Q.**, Nutritional evaluation of diets containing varying proportions of rice and sweet potato, in *Proc. Eleventh Pacific Science Congress*, Tokyo, Japan, 1966.
37. **Miller, J. C. and Covington, H. M.**, Some factors affecting the carotene content of sweet potatoes, *Proc. Am. Soc. Hortic. Sci.*, 40, 519, 1942.
38. **Matlach, M. B.**, The carotenoid pigments of the sweet potato, *J. Wash. Acad. Sci.*, 27, 493, 1937.
39. **Purcell, A. E.**, Carotenoids of 'Goldrush' sweet potato flakes, *Food Technol. (Chicago)*, 16, 99, 1962.
40. **Purcell, A. E. and Walter, W. M., Jr.**, Carotenoids of 'Centennial' variety sweet potato, *J. Agric. Food Chem.*, 16, 769, 1968.
41. **Edmond, J. B.**, Cooperative studies of the effect of height of ridge, nitrogen supply and time of harvest on the yield and flesh color of 'Porto Rico' sweet potato, *U.S. Dept. Agric. Circ.*, 832, 1950.
42. **Ezell, B. D. and Wilcox, M. S.**, Influence of storage temperature on carotene, total carotenoids, and ascorbic acid content of sweet potatoes, *Plant Physiol.*, 27, 81, 1952.
43. **Watt, B. K. and Merrill, A. L.**, *Composition of Foods*, Handbook No. 8, U.S. Department of Agriculture, Washington, D.C., 1963.
44. **Emmerie, A. and Engel, C.**, Colorimetric determination of alpha-tocopherol (Vitamin E), *Rec. Trav. Chem.*, 57, 1351, 1938.
45. **Boggess, T. S., Marion, J. E., Woodroof, J. G., and Dempsey, A. H.**, Changes in lipid composition as affected by controlled storage, *J. Food Sci.*, 32, 554, 1967.
46. **Boggess, T. S., Marion, J. E., and Dempsey, A. H.**, Lipid and other compositional changes in nine varieties of sweet potatoes during storage, *J. Food Sci.*, 35, 306, 1970.
47. **Walter, W. M., Jr., Hansen, A. P., and Purcell, A. E.**, Lipids of cured 'Centennial' sweet potatoes, *J. Food Sci.*, 36, 795, 1971.
48. **Elkins, E. R.**, Nutrient content of raw and canned green beans, peaches, and sweet potatoes, *Food Technol. (Chicago)*, 2, 66, 1979.

49. **Lopez, A., Williams, H. L., and Cooler, F. W.,** Essential elements in fresh and canned sweet potatoes, *J. Food Sci.,* 45, 675, 1980.
50. **Anonymous,** Food and Nutrition Board "Recommended Daily Dietary Allowances." National Academy of Sciences, National Research Council, Washington, D.C., 1974.
51. **Sohonnie, K. and Bhandarker, A. P.,** Trypsin inhibitors in Indian foodstuffs. I. Inhibitors in vegetables, *J. Sci. Ind. Res.,* 138, 500, 1954.
52. **Dickey, L. F. and Collins, W. W.,** Cultivar differences in trypsin inhibitors of sweet potato roots, *J. Am. Soc. Hortic. Sci.,* 109, 750, 1984.
53. **Wilson, L. G., Averre, C. W., Baird, J. V., Estes, E. A., Sorensen, K. A., Beasley, E. O., and Skroch, W. A.,** Growing and marketing quality sweet potatoes, *N.C. Agric. Ext. Service Bull.,* AG-09, 1979.
54. **Jana, R. K.,** Sweet potatoes in East Africa, in *Sweet Potato: Proc. First Int. Symp.,* Villareal, R. L. and Griggs, T. D., Eds., Asian Vegetable Research and Development Center, Shanhua, Tainan, Taiwan, 1982, 63.
55. **Villanueva, M. R.,** Processing and storage of sweet potato and aroids in the Philippines, in *Small Scale Processing and Storage of Tropical Root Crops,* Plucknett, D. L., Ed., Westview Press, Boulder, Colo., 1979, 83.
56. **Cross, L.,** Methods for the Production of Root Crops in Trinidad and Tobago: Yams and Sweet Potatoes, Farmers Bull. No. 4, Ministry of Agriculture, Lands, and Fisheries, Trinidad and Tobago, W.I., 1968.
57. **Knott, J. E. and Deamon, J. R., Jr.,** *Vegetable Production in Southeast Asia,* University of the Philippines, College of Agriculture, Los Baños, Laguna, Philippines, 1967.
58. **Tanaka, J. S. and Sekioka, T. T.,** Sweet potato production in Hawaii, in *Proc. 4th Int. Symp. on Trop. Root and Tuber Crops,* Cook, J., MacIntyre, R., and Graham, M., Eds., CIAT, Cali, Colombia, 1976, 150.
59. **Kay, D. E.,** Sweet potato, in *Crop and Product Digest 2: Root Crops,* Tropical Products Institute, London, 1973, 144.
60. **Onweumi, I. C.,** Sweet potato, in *Tropical Tuber Crops,* John Wiley & Sons, New York, 1978, 167.
61. **Tavioni, Michael,** Kumara (sweet potato) in the Cook Islands, in *Proc. 5th Int. Symp. on Trop. Root and Tuber Crops,* Belen, E. H. and Villanueva, M., Eds., Philippine Council on Agriculture and Resources Research, Los Baños, Laguna, Philippines, 1979, 67.
62. **MacDonald, A. S.,** Sweet potatoes with particular reference to the tropics, *Field Crops Abstr.,* 220, 225, 1963.
63. **Rose, C. J.,** Comparison of single and progressive harvesting of sweet potato, *Papua New Guinea Agric. J.,* 30, 61, 1979.
64. **Aldrich, D. T. A.,** The sweet potato crop in Uganda, *East Afr. Agric. For. J.,* 29, 41, 1963.
65. **Winarno, F. G.,** Sweet potato processing and by-product utilization in the tropics, in *Sweet Potato: Proc. First Int. Symp.,* Villareal, R. L. and Griggs, T. D., Eds., Asian Vegetable Research and Development Center, Shanhua, Tainan, Taiwan, 1982, 373.
66. **Irvine, F. R.,** *West African Crops,* Oxford University Press, London, 1969, chap. 13.
67. **Gull, D. and Duarte, O.,** Curing sweet potatoes (*Ipomoea batatas*) under tropical conditions, in *Proc. Trop. Region, Am. Soc. Hortic. Sci.,* 1974, 168, 1970.
68. **Lutz, J. M.,** Influence of temperature and length of curing period on keeping quality of Porto Rico sweet potatoes, *Proc. Am. Soc. Hortic. Sci.,* 59, 421, 1952.
69. **Kushman, L. J. and Wright, F. S.,** Overhead Ventilation of Sweet Potato Storage Rooms, Tech. Bull. No. 166, N.C. Agric. Exp. Stn. and U.S. Dep. of Agric., Raleigh, N.C., 1965.
70. **S-101 Technical Committee,** Sweet Potato Quality, Southern Coop. Series Bull. No. 249, U.S. Dep. Agric., Athens, Ga., 1980, 6.
71. **Calkins, P. H. and Wang, H. M.,** Improving the Marketing of Perishable Commodities: A Case Study of Selected Vegetables in Taiwan, AVRDC Tech. Bull. No. 9(78-86), Asian Vegetable Research and Development Center, Shanhua, Tainan, Taiwan, 1978.
72. **Siki, B. F.,** Processing and storage of root crops in Papua New Guinea, in *Small Scale Processing and Storage of Tropical Root Crops,* Plucknett, D. L., Ed., Westview Press, Boulder, Colo., 1979, 64.
73. **Olurunda, A.,** Storage and processing of some Nigerian root crops, in *Small Scale Processing and Storage of Tropical Root Crops,* Plucknett, D. L., Ed., Westview Press, Boulder, Colo., 1979, 90.
74. **Anonymous,** Sweet potato, *Malayan Agric. J.,* 28, 221, 1940.
75. **Hrishi, N. and Balagopal, C.,** Storage problems in aroids and sweet potatoes in India, in *Small Scale Processing and Storage of Tropical Root Crops,* Plucknett, D. L., Ed., Westview Press, Boulder, Colo., 1979, 127.
76. **Wilson, B. J., Yang, D. T. C., and Byrd, M. R.,** Toxicity of mould-damaged sweet potatoes, *Nature (London),* 227, 521, 1970.

77. **Booth, R. H.,** Post-harvest deterioration of tropical root crops: losses and their control, *Trop. Sci.,* 16, 49, 1974.
78. **Best, E.,** Maori agriculture, *N.Z. Dominion Museum Bull.,* No. 9, 1925.
79. **Keleny, G. P.,** Sweet potato storage, *Papua New Guinea Agric. J.,* 17, 102, 1965.
80. **Anonymous,** Storing sweet potatoes, *World Crops,* 2, 73, 1960.
81. **Kushman, L. J.,** Inhibition of sprouting in sweet potatoes by treatment with CIPC, *HortScience,* 4, 61, 1969.
82. **Schlimme, D.,** Anatomical and physiological aspects of sprout initiation and development in sweet potato storage roots, *Diss. Abstr.,* 26, 4947, 1966.
83. **Hahn, S. K. and Anota, T.,** Int. Inst. Trop. Agric. Annu. Rep. for 1981, Ibadan, Nigeria, 1982, 73.
84. **Uritani, I., Saito, T., Honda, H., and Kim, W. K.,** Induction of furano-terpenoids in sweet potato roots by the larvae components of the sweet potato weevils, *Agric. Biol. Chem.,* 37, 1875, 1975.
85. **Hollinger, M. E.,** Ascorbic acid of the sweet potato as affected by variety, storage, and cooking, *Food Res.,* 9, 76, 1944.
86. **Villareal, R. L.,** Sweet potato in the tropics — progress and problems, in *Sweet Potato: Proc. First Int. Symp.,* Villareal, R. L. and Griggs, T. D., Eds., Asian Vegetable Research and Development Center, Shanhua, Tainan, Taiwan, 1982, 8.
87. **Gore, H. C.,** Occurrence of diastase in sweet potato in relation to the preparation of sweet potato syrup, *J. Biol. Chem.,* 44, 19, 1920.
88. **Gore, H. C.,** Formation of maltose in sweet potatoes on cooking, *Ind. Eng. Chem.,* 15, 938, 1923.
89. **Ikemiya, M. and Deobald, H. J.,** New characteristic alpha-amylase in sweet potatoes, *J. Agric. Food Chem.,* 14, 237, 1966.
90. **Walter, W. M., Jr., Purcell, A. E., and Nelson, A. M.,** Effects of amylolytic enzymes on "moistness" and carbohydrate changes of baked sweet potato cultivars, *J. Food Sci.,* 40, 793, 1975.
91. **Jones, A., Steinbauer, C. E., and Pope, D. J.,** Quantitative inheritance of ten root traits in sweet potatoes, *J. Am. Soc. Hortic. Sci.,* 94, 271, 1969.
92. **Jones, A.,** Heritabilities of seven sweet potato root traits, *J. Am. Soc. Hortic. Sci.,* 102, 440, 1977.
93. **Villareal, R. L., Tsou, S. C., Lai, S. H., and Chiu, S. I.,** Selection criteria for eating criteria in steamed sweet potato roots, *J. Am. Soc. Hortic. Sci.,* 104, 31, 1979.
94. **Walter, W. M., Jr. and Hoover, M. W.,** unpublished data, 1983.
95. **Reddy, N. N. and Sistrunk, W. A.,** Effect of cultivar, size, storage, and cooking method on carbohydrates and some nutrients of sweet potatoes, *J. Food Sci.,* 45, 682, 1980.
96. **Sinoda, K. O., Kodera, A., and Oya, C.,** The chemical changes of carbohydrates in the sweet potato according to various methods of cooking, *Biochem. J.,* 25, 1973, 1931.
97. **Balls, A. K., Walden, M. K., and Thompson, R. R.,** A crystalline beta-amylase from sweet potatoes, *J. Biol. Chem.,* 173, 9, 1948.
98. **Hseu, C. T., Huang, C. J., Tsai, Y. C., and Yang, T. H.,** Studies on the effect of preheating of sweet potato raw material for the elimination of antitryptic factor and the improvement of feed efficiency, *Mem. Coll. Agric. Nat. Taiwan Univ.,* 19, 2, 1979.
99. **Yeh, T. P.,** Utilization of sweet potatoes for animal feed and industrial use: potential and problems, in *Sweet Potato: Proc. First Int. Symp.,* Villareal, R. L. and Griggs, T. D., Eds., Asian Vegetable Research and Development Center, Shanhua, Tainan, Taiwan, 1982, 385.
100. **Lawrence, G. and Walker, P. D.,** Pathogenesis of *Enteritis necroticans* in Papua, New Guinea, *The Lancet,* 2, 125, 1976.
101. **Purcell, A. E. and Walter, W. M., Jr.,** Stability of amino acids during cooking and processing of sweet potatoes, *J. Agric. Food Chem.,* 30, 443, 1982.
102. **Meredith, F. I. and Dull, G.,** Amino acid levels in canned sweet potatoes and snap beans, *Food Technol.,* 2, 55, 1979.
103. **Sweeney, J. P. and March, A. C.,** Effect of processing on pro vitamin A in vegetables, *J. Am. Diet. Assoc.,* 59, 238, 1971.
104. **McNair, V.,** Effect of storage and cooking on carotene and ascorbic acid of some sweet potatoes grown in Arkansas, *Arkansas Agric. Exp. Stn. Bull.,* 574, 1956.
105. **Jenkins, W. F. and Moore, E. L.,** The distribution of ascorbic acid and latex vessels in three regions of the sweet potato, *Proc. Am. Soc. Hortic. Sci.,* 63, 389, 1954.
106. **Crosby, D. G.,** Organic constituents of food. III. Sweet potato, *J. Food Sci.,* 29, 287, 1964.
107. **Uritani, I.,** Abnormal substance produced in fungus-contaminated foodstuffs, *J. Assoc. Offic. Anal. Chem.,* 50, 105, 1967.
108. **Boyd, M. R., Burka, L. T., Harris, T. M., and Wilson, B. J.,** Lung toxic furano terpenoids produced by sweet potatoes, *Biochem. Biophys. Acta,* 337, 184, 1973.
109. **Burka, L. T. and Wilson, B. J.,** Toxic furanosesquiterpenoids from mold-damaged sweet potatoes, in *Mycotoxins and Other Fungal Related Food Problems,* Rodricks, J., Ed., Adv. Chem. Ser. 149, American Chemical Society, Washington, D.C., 387, 1976.

110. **Cody, M. and Haard, N. F.**, Influence of cooking on toxic stress metabolites in sweet potato root, *J. Food Sci.*, 41, 469, 1976.
111. **Catalano, E. A., Hasling, V. C., Dupuy, H. P., and Constantin, R. J.**, Ipomeamarone in blemished and diseased sweet potatoes, *J. Agric. Food Chem.*, 25, 94, 1977.
112. **Boyd, M. R. and Wilson, B. W.**, Preparation and analytical gas chromatography of ipomeamarone, a toxic metabolite of sweet potatoes, *J. Agric. Food Chem.*, 19, 547, 1971.
113. **Luh, C. L. and Moomaw, J. C.**, Present and future outlook for sweet potato in Asia; research and development needs, *Proc. 5th Int. Symp. on Trop. Root and Tuber Crops*, Belen, E. H. and Villanueva, M., Eds., Philippine Council on Agricultural Research, Los Baños, Laguna, the Philippines, 1979, 9.

Chapter 8

SWEET POTATO VINE TIPS AS VEGETABLES

R. L. Villareal, S. C. Tsou, H. F. Lo, and S. C. Chiu

TABLE OF CONTENTS

I.	Introduction	176
II.	Problems	176
III.	Progress So Far	176
	A. Survey of Morphological Traits	177
	B. Evaluation for Yield, Nutritional, and Eating Qualities	177
IV.	Research Priorities	180
	A. Development of an Attractive Variety for Sweet Potato Tips	180
	B. Development of a Variety with a Mild Flavor	180
	C. Evaluation of Toughness in Tips	180
	D. Determination of Chemical Constituents	180
	E. Development of Ways of Maintaining Postharvest Quality	182
	F. Development of Varieties with Erect Plant Habit	182
V.	Conclusions	183
Acknowledgment		183
References		183

I. INTRODUCTION

When the Philippines suffered serious flooding in 1972, only two green leafy vegetables could be regularly found in both urban and rural markets — sweet potato tips and water convolvulus or "Kangkong" *(Ipomoea aquatica)*. All other vegetables were destroyed by about a month's continuous rain. Tips in this context refer to that portion of the top used for human consumption. The term top is used to denote the entire aboveground part of the plant.

During the Second World War, many American and Filipino soldiers in the prison camps in the Philippines subsisted on rice and sweet potato tips.[1] Those fed with tips retained fairly good vision, whereas those who were fed a rice diet alone developed skin diseases and poor eyesight, due to lack of vitamin A. Many Filipinos, for that matter, survived famine and malnutrition during the war because of sweet potatoes. During those days, virtually every Filipino family had a vegetable garden and sweet potato was a crop included in every garden. We know now that if even a very active man eats about 100 g of tips a day he gets his supply of vitamin A for about two days, one-quarter of his vitamin B_2 requirement, and more than half of his vitamin C and Fe needs in addition to some quantities of Ca, ash, and fiber.

Weather in the Asian tropics is hot and humid in summer, and thus vegetable production is low and prices are high. Sweet potato tips, able to survive adverse conditions, could serve as an additional leafy green vegetable during this time of the year if not the whole year round. The tips' high amounts of vitamins A, B_2, C and of some minerals could also help alleviate nutritional problems such as night blindness, scurvy, and anemia among the inhabitants of the poorer areas of the tropics.

The purpose of this paper is to summarize the approaches which we follow at the Asian Vegetable Research and Development Center (AVRDC) in developing sweet potato varieties for tips.

II. PROBLEMS

Although sweet potato tips have been used as vegetables in the Philippines, Indonesia, Thailand, Malaysia,[2] many rural areas of Kwangtung and Fukien Provinces in southern China,[17] New Guinea, Korea,[3] Zaire,[4] Sierra Leone,[5] Tanzania,[6] and Liberia,[7] their usefulness as human food has been neglected in sweet potato research. Consequently, there were many questions that needed answers when AVRDC started its pioneering work on this underrated vegetable. For example, which leaf shapes, and color of stems, veins, or leaves are preferred? Are nonhairy preferred to hairy leaves? How can one evaluate for tenderness, since this trait is always sought after in a leafy vegetable? Which of these quality characteristics contribute to acceptability? Which parts are eaten — the stems, the leaves, or both? Information was also needed on nutritional quality, productivity, and how productivity could be improved. In short, sweet potato vine tips had to be evaluated in the several ways in which conventional leafy vegetables have normally been evaluated.

III. PROGRESS SO FAR

As a starting point the senior author conducted a limited consumer survey in the Philippines in 1974. He observed that acceptable sweet potato tips should be tender, glabrous (nonhairy), and purplish, the preference for purple being to add color to the table.[2] The parts eaten consisted of the apical 10 cm vine tips including both stem and leaves. These are the parts generally eaten in many other countries, too. In Korea the petioles are eaten instead of tips whereas among Haka people living in Taiwan, the mature leaves and petioles are eaten.

Table 1
MORPHOLOGICAL TRAITS OF TIPS OF 369 SWEET POTATO CULTIVARS

	Number of cultivars		
	Stems	Vines	Leaves
Color			
Purple	61	229	43
Green-purple	135	13	169[a]
Green	173	127	152
Others	0	0	5
Hairiness			
Glabrous			139
With light hairs			153
With medium hairs			56
With heavy hairs			21
Leaf shape			230
Entire (one point)			183
Trifoliate (3 points/leaf)			54
Palmate (5 points/leaf)			107
Others			25

[a] Green with purple edge.

From Villareal, R. L., Lin, S. K., Chang, L. S., and Lai, S. H., *Exp. Agric.*, 15, 113, 1979. With permission.

A. Survey of Morphological Traits

Based on the aforementioned survey, four separate screenings were conducted between August and September in 1974, 1976, and 1977 evaluating 177, 136, and 56 accessions, respectively. The aim was to identify accessions with acceptable morphological traits for human consumption. The results of these screenings are summarized in Table 1 and Figure 1.

A wide variety of color and shape is available with or without glabrous leaves (Table 1). Some cultivars with only slightly hairy leaves could be selected if glabrous cultivars would not combine with other desirable traits.

The general preference for 10 cm tips as greens in many countries where tips are eaten as vegetables is logical, since a large proportion of the leaves in the top 10 cm are new and thus tender. It was noted from the study that leaves with more than 1 cm long petioles are older and generally tougher than those with less than 1 cm petioles. We therefore assessed desirability by counting the number of leaves with and without 1 cm petioles on each 10 cm tip on ten random samples per entry. Tips with the largest number of leaves with petioles less than 1 cm long were considered desirable, since they are more tender and good for the table. Results of the study (Figure 1) showed that there were from 2.3 to 8.7 such leaves per tip, and it was possible to select from about 3% of the accessions evaluated so far as this trait was concerned.

Tips are marketed in the Philippines either by weight or by number,[8,9] but only by weight in Malaysia and Taiwan. Therefore the researchers were interested in tips which weighed more, but were tender. Evaluation showed that a 10 cm tip could vary between 0.9 g and 12.5 g (Figure 1).

B. Evaluation for Yield, Nutritional and Eating Qualities

Promising cultivars which combine appropriate color, hairiness and tenderness were then

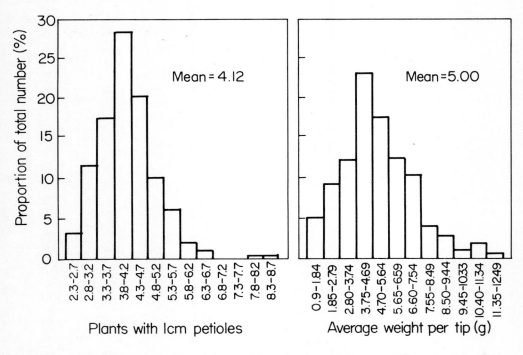

FIGURE 1. Distributions of vine tips with 1 cm petioles and various weights among 369 sweet potato cultivars. (From Villareal, R. L., Lin, S. K., Chang, L. S., and Lai, S. H., *Exp. Agric.*, 15, 113, 1979. With permission.)

evaluated for yield, and for nutritional and eating qualities. We included BNAS White, a popular Philippine cultivar grown for roots and tips, and Kinangkong for tips. Dilaw is also a popular Philippine cultivar for tips because of its attractive yellow leaves.

The results of the yield trial are presented in Table 2. Four entries (Dilaw, Kinangkong, PI 344138, and PI 344120) outyielded BNAS White, the check cultivar in terms of weight/ ha; the other entries were comparable with the check. With the exception of Dilaw the same entries outperformed the check in number of tips/hectare. If a grower needs a dual cultivar (which yields equally well for roots and tips), the choice would be PI 344128, PI 344120, or PI 318856. As evidenced in Table 2, the root yields of other entries were drastically affected by frequent harvesting of tips whereas these three entries gave at least 15 t/ha of marketable roots although they were subjected to the same frequency of harvesting as the others. It is clear from other studies[3,11,12] that loss of sweet potato leaves in general results in reduced root yields.

Another experiment was held to determine whether frequency of harvesting and nitrogen fertilization would influence tip yield and other traits of three cultivars with different leaf types. It was found that frequency of harvesting did not influence yield or other traits, but the effects of nitrogen application and choice of cultivars were significant for some traits (Table 3). The 120 kg N/ha treatment outyielded the control due to production of more and heavier tips. This treatment also produced higher protein contents and lower oxalate levels in the leaves than the control, suggesting that levels of both compounds can be manipulated through nitrogen fertilization. Dry matter and vitamin B_2 were not influenced by additional nitrogen fertilizer. Kinangkong, the fine-leaf type gave the highest number of tips/ha, had the highest protein content, lowest oxalate, and fairly high dry matter, which makes it a very desirable green vegetable. Its vitamin B_2 content is comparable with that of the two other cultivars.

Nutritional evaluation showed variations in dry matter, fiber content, ash, vitamins B_2,

Table 2
TIP AND ROOT YIELD OF TEN SWEET POTATO CULTIVARS[a]

Cultivar name	Tip yield[b]				Root yield[d] (t/ha)
	Weight (t/ha)	% Over control	Number (1000/ha)	T[c]	
Dilaw	16	46	1420	38	2
Kinangkong	14	27	3520	59	1
PI 344138	14	27	1860	43	15
PI 344120	14	27	2080	46	15
Daja 380	13	18	746	27	10
PI 318856	12	9	1250	35	16
Earlyport	11	0	970	31	9
HM 16	11	0	1310	37	6
Rose Centennial	10	−10	610	25	3
BNAS-White (control)	11	—	1230	36	12
Avg.	13		1500	38	9
LSD 0.05	3			5	3

[a] Planted 28 July 1976.
[b] Harvested six times at 14-day intervals from 6 Sept. to 28 Nov.
[c] Transformation using square roots of tips/ha.
[d] Marketable yield; harvested 6 Dec. after 130 days.

From Villareal, R. L., Tsou, S. C. S., Lin, S. K., and Chiu, S. C., *Exp. Agric.*, 15, 117, 1979. With permission.

Table 3
COMPARISON OF TIP YIELD AND OTHER TRAITS OF THREE VARIETIES OF SWEET POTATO[a]

Factors	Number (100/ha)	T[b]	Weight (t/ha)	Avg wt/tip (g)	Protein (%)	Dry matter	Oxalate (%)	Vit. B$_2$[c] (mg/100 g)
			Effects of three levels of nitrogen					
Nitrogen level (kg/ha)								
0	3198	56.6	24	4.35	2.79	10.0	0.46	0.22
60	3039	55.1	22	4.55	2.97	10.0	0.43	0.22
120	3575	59.8	26	4.62	2.98	10.0	0.42	0.23
LSD 0.05		2.4	2	0.14	0.04	NS	0.02	NS
Variety								
Dilaw (broad)	1826	42.7	18	6.72	2.93	9.0	0.49	0.21
AIS 016-1 (medium)	3287	57.3	26	4.10	2.77	10.7	0.48	0.22
Kingankong (fine)	5100	71.4	25	2.67	3.05	10.3	0.34	0.24
LSD 0.05		5.7	5	0.33	0.08	0.3	0.04	NS

[a] Planted 15 June 1977; means of weekly and twice-monthly harvestings (means of two replications).
[b] Transformation using square roots of the number of tips/ha.
[c] Data from weekly harvestings only.

and C and oxalate content.[10] No significant differences were detected among cultivars in Ca, Fe, and vitamin A contents. Sweet potato tips are superior to other leafy vegetables in nutrient contents (Table 4). Potentially their most important contribution to nutrition may be in their high level of B_2, which is generally deficient in most Asian diets. Their lower amounts of oxalate than amaranth and spinach make them appear better in terms of oxalate content than these two vegetables.

One of the most difficult tasks in encouraging greater use of sweet potato tips is to determine the different selection criteria to be used in predicting their general acceptability. The researchers approached this by evaluating the crop's eating qualities such as tenderness, flavor, stem and leaf color, and hairiness.[13] It was found that all these quality attributes are important in predicting general acceptability of blanched tips (Table 5). However, an initial selection could be conducted in terms of stem color and hairiness. In stepwise multiple regression analysis to determine the relative contribution of each of the quality traits to acceptability of sweet potato tips, the coefficient of determination (R^2) indicated that stem color (X_3) accounted for 64% of the variation in acceptability (Table 5). Adding hairiness (X_5) into the equation showed an increase of 22% of accuracy in predicting acceptability.

IV. RESEARCH PRIORITIES

After conducting a series of pioneeering studies on tips for human consumption, AVRDC scientists feel that there is still much work to be done to enhance their utilization. Investigations should be undertaken in the following areas.

A. Development of an Attractive Variety for Sweet Potato Tips

Acceptance of tips is still influenced by psychological factors because they are considered "poor man's salad". Tips with new attractive appearance would help dispel this stigma. Some genotypes with better market appeal are deep purple or deep yellow with fine to medium leaves. So far AVRDC breeders have not succeeded in combining the attractive yellow color of Dilaw with fine or medium leaves. Dilaw is difficult to cross and the gene that controls this leaf color seems to be a deleterious recessive. Of the thousands of seed sown at AVRDC only one seedling appeared yellow, but it was weak and eventually died. Last year, Dilaw produced a lot of open-pollinated seeds. We sowed about 5,000 seeds in May 1981. Four yellow seedlings emerged but the yellow color disappeared in about 15 days from germination. Thus, a month-old seedling of Dilaw which was yellow at germination appeared lighter green on the tips and normal green on the older leaves. There has been little progress at AVRDC on developing a deep purple variety.

B. Development of a Variety with Mild Flavor

The strong flavor of tips make them less easy to cook with other vegetables. There seems to be a limited number of dishes which can be prepared from tips. For instance, cabbage, cauliflower, broccoli, Chinese cabbage, and even water convolvulus when cooked with sweet potato tips tend to lose their natural flavor. One alternative approach might be for home economists to develop tip dishes that exploit this strong flavor.

C. Evaluation of Toughness in Tips

Tips are considered tough compared to other leafy vegetables. The level of fiber that contributes to the toughness of tips should be determined. Does fiber tend to be concentrated in the petiole, stem, or leaves? Are there distinct varietal differences in terms of fiber content of tips? Would heavy nitrogen application and frequent irrigation make tips more tender?

D. Determination of Chemical Constituents

A complete bioassay to determine chemical and other constituents which might be either harmful or beneficial to human health should be carried out. Bitterness in tips, for example,

Table 4
NUTRITIONAL CONSTITUENTS OF SWEET POTATO TIPS AND OF FIVE COMMON LEAFY VEGETABLES

Vegetable	Moisture (%)	Protein (%)	Fiber (%)	Ash (%)	Minerals		Vitamins			Oxalic[a] acid (% DW)
					Ca (mg/100 g)	Fe (mg/100 g)	A (IU/100 g)	B$_2$ (mg/100 g)	C (mg/100 g)	
Sweet potato tips[b]	86.1	2.7	2.0	1.7	74	4	5580	0.32	41	5.1
Water convolvulus[c]	91.8	2.3	0.9	1.0	94	1	4200	0.20	43	4.5
Spinach[c]	92.3	2.3	0.8	1.7	70	2	10500	0.18	60	9.6[d]
Amaranth[c]	87.8	1.8	1.3	2.1	300	6	1800	0.23	17	10.3
Head lettuce[c]	96.3	0.9	0.3	0.2	14	0.2	4300	0.03	6	1.3
Cabbage[a]	92.1	1.7	0.9	0.7	64	0.7	75	0.05	62	0.3

[a] From Food Composition Table, Handbook 1 (4th rev.), Food and Nutrition Research Center, Manila, Philippines, 1968.
[b] Average of ten sweet potato cultivars.[14]
[c] From Reference 15.
[d] From Reference 16.

Table 5
SUMMARY OF STEPWISE MULTIPLE REGRESSION ANALYSIS OF ACCEPTABILITY AND EATING QUALITY CHARACTERISTIC OF SWEET POTATO TIPS

Variable[a]	Accumulative (R^2)
1. Stem color (X_3)	0.64[b]
2. Hairiness (X_5)	0.86[b]
3. Tenderness (X_1)	0.90[b]
4. Leaf color (X_4)	0.92[b]
5. Flavor (X_2)	0.95[b]

[a] Order of variables entered in addition to previous variables.
[b] Highly significant at 1% probability level using F-test.

From Villareal, R. L., Tsou, S. C. S., Chiu, S. C., and Lai, S. H., *Exp. Agric.*, 15, 123, 1979. With permission.

might be due to phenolic compounds. Is excess intake of these compounds poisonous? What screening technique could be used to determine their presence in tips? On the other hand, sweet potato tips are also known to have saluretic properties (increasing the elimination of sodium salt) and diuretic properties (increasing amount of urine) which make them good vegetables for individuals suffering from high blood pressure.[2] Could tips, then, be recommended by doctors as a regular diet for people with high blood pressure? Why is this not done? Are there side effects associated with eating tips?

E. Development of Ways of Maintaining Postharvest Quality

The postharvest appearance of tips deteriorates rapidly. There is a need to identify and develop simple, inexpensive, and effective means of maintaining the quality of tips from producer to consumer and to investigate the postharvest characteristics of different varieties. Do varieties with smaller leaves have longer shelf-life than those with bigger leaves? Do yellow leaves maintain attractiveness longer than either purple or green leaves?

F. Development of Varieties with Erect Plant Habit

The creeping growth habit of sweet potato makes harvesting of tips difficult. This difficulty would be alleviated if the tips were in a standing rather than in a drooping position at time of harvest. Such tips would be easier to harvest and may be amenable to mechanization in the future. This growth habit needs to be transferred to acceptable varieties. Another possibility which should be explored is the raising of tips with the use of bedded roots which normally produce standing sprouts or slips.

Please note that the problems we have identified relate mostly to the needs of the consumers (items A to E). Only item F deals with the problems of the farmer-producer. Production problems of tip growers should also be given more attention.

V. CONCLUSIONS

Undoubtedly the sweet potato tips, like the roots, are under-exploited. But they could have a significant role to play in feeding large numbers of people in the Third World because tips and roots are inexpensive, easy to grow and nutritious. Scientists should help make this vegetable's tips and roots more appealing and acceptable. Plant breeders should develop a desirable sweet potato variety for tips and the food scientists and home economists should prepare more appetizing and appealing dishes out of tips. We hope that other scientists will share their expertise in this worthwhile undertaking.

ACKNOWLEDGMENT

This chapter is reproduced with the kind permission of the Asian Vegetable Research and Development Center, Shanhua, Taiwan from the book "Sweet Potato, Proceedings of the First Symposium."

REFERENCES

1. **Miller, J. C.,** Summary report of world tour, *Hort-Science*, 2, 1967.
2. **Villareal, R. L., Lin, S. K., Chang, L. S., and Lai, S. H.,** Use of sweet potato *(Ipomoea batatas)* leaf tips as vegetable. I. Evaluation of morphological traits, *Exp. Agric.*, 15, 113, 1979.
3. **Dahniya, M. J.,** Use of sweet potato vines and leaves as human food, in First Annual Research Conference, IITA, Ibadan, Nigeria, October 15 to 19, 1979.
4. **Ifefo, B. B.,** Sweet potato in Zaire, in First Annual Research Conference, IITA, Ibadan, Nigeria, October 15 to 19, 1979.
5. **Kamara, S. I., Raymundo, S. A., Jones, R. A. D., and Johnson, S. D.,** Sweet potato improvement in Sierra Leone, in First Annual Research Conference, IITA, Ibadan, Nigeria, October 15 to 19, 1979.
6. **Msabaha, M. A. M.,** Sweet potato in Tanzania, in First Annual Research Conference, IITA, Ibadan, Nigeria, October 15 to 19, 1979.
7. **Saqui, M. A. A.,** Sweet potato improvement in Liberia, in First Annual Research Conference, IITA, Ibadan, Nigeria, October 15 to 19, 1979.
8. AVRDC, Annual Report for 1974, Asian Vegetable Research and Development Center, Shanhua, Tainan, Taiwan, 1975.
9. AVRDC, Annual Report for 1975, Asian Vegetable Research and Development Center, Shanhua, Tainan, Taiwan, 1976.
10. **Villareal, R. L., Tsou, S. C. S., Lin, S. K., and Chiu, S. C.,** Use of sweet potato *(Ipomoea batatas)* leaf tips as vegetables. II. Evaluation of yield and nutritive quality, *Exp. Agric.*, 15, 117, 1979.
11. **Gonzales, F. R., Cadiz, T. G., and Bugawan, M. S.,** Effect of topping and fertilization on the yield and protein content of three varieties of sweet potato, *Philipp. J. Crop Sci.*, 2, 97, 1977.
12. **Yen, F. C.,** The relation between foliage insect pests and yields of sweet potato, peanut, and soybean, *Taiwan Agric. Q.*, 9, 41, 1974.
13. **Villareal, R. L., Tsou, S. C. S., Chiu, S. C., and Lai, S. H.,** Use of sweet potato *(Ipomoea batatas)* leaf tips as vegetables. III. Organoleptic evaluation, *Exp. Agric.*, 15, 123, 1979.
14. **Luh, C. L. and Moomaw, J. C.,** Present role and future outlook for sweet potato in Asia. Research and development needs, in 5th Symp. Int. Soc. Trop. Root Crops, Manila, Philippines, September 17 to 21, 1979.
15. **Tung, T. C., Huang, P. C., Li, H. C., and Chen, H. L.,** Composition of Foods Used in Taiwan, 1975.
16. **Eheart, J. F. and Massey, P. H., Jr.,** Factors affecting the oxalate content of spinach, *Agric. Food Chem.*, 10, 325, 1962.
17. **Luh, C. L.,** Senior Consultant, Taiwan Sugar Corporation, personal communication, 1984.

Chapter 9

PROCESSING OF SWEET POTATOES — CANNING, FREEZING, DEHYDRATING

John C. Bouwkamp

TABLE OF CONTENTS

I.	Introduction	186
II.	General Procedures	186
	A. Receiving and Grading	186
	B. Cleaning	186
	C. Preheating	186
	D. Peeling and Trimming	187
III.	Canning	189
	A. Procedures	189
	1. Sizing and Cutting	189
	2. Blanching, Filling, Syruping	189
	3. Exhausting, Closing, Retorting, Cooling	189
	B. Factors Affecting Firmness	190
	1. Preprocessing Factors	190
	2. Processing Factors	191
	C. Factors Affecting Color	193
	1. Preprocessing Factors	193
	2. Processing Factors	194
	D. Factors Affecting Nutritional Quality	194
	1. Preprocessing Factors	194
	2. Processing Factors	195
IV.	Freezing	196
	A. Procedures	196
	1. Sizing and Cutting	196
	2. Blanching or Cooking	196
	3. Freezing	196
V.	Dehydration	197
	A. Production of Slices, Dices, Strips	197
	B. Production of Dehydrated Flakes	198
VI.	Conclusions	199
References		200

I. INTRODUCTION

The focus of this chapter is on how to produce a processed product rather than on how to decide if processing is an appropriate alternative or adjunct procedure in the utilization of the crop. As such, it is intended as an aid to food scientists rather than processors.

The decision of whether or not to produce a processed product on a commercial basis will depend on economic and social circumstances as well as on the resolution of technological considerations. These and other issues, such as the availability of personnel to manage and maintain equipment, availability of materials and spare parts, marketing infrastructure, etc., although of equal or greater importance are not addressed. We will, instead, describe the procedures commonly used and the effects of certain variables on the quality of the processed products.

II. GENERAL PROCEDURES

A. Receiving and Grading

As the product is received, it is usually inspected to be sure it conforms to the standards set by the processor. Typical categories for which standards are set include freedom from excessive soil or other foreign matter, freedom from rotting and insect damage, and freedom from excessive mechanical damage. Depending on the nature of the processed product standards may be set on size of roots and freedom from growth cracks.

Usually a tolerance is set allowing a certain amount of out-of-grade roots before the price paid to the producer is reduced and a level of defects that will render the product unacceptable. Tolerances and methods of grading should be clearly understood by the suppliers of the raw product, since rejection of a load results in economic and social costs to both the buyer and seller.

To the extent possible, deliveries of sweet potatoes should be scheduled according to daily expected requirements. If necessary, sweet potatoes can be held in small, well-ventilated piles for a few days although some loss to rotting is nearly certain. If the piles are too large or not well-ventilated the oxygen may become depleted in the center of the pile and considerable deterioration and loss of quality could result.

B. Cleaning

As much soil as possible should be removed from the roots before preheating and/or peeling. This is usually accomplished with rod or chain conveyors in combination with rotating drums with slots large enough for soil to fall through. High pressure sprays of water (90 to 100 psi) in the drum wash away the remaining adhering soil. If disposal of the muddy waste water is a problem, dry brushing preceded by sorting to remove soft rotted roots may be effective.

C. Preheating

Preheating involves immersing sweet potato roots in heated water or live steam for a specified time. Several benefits are attributed to preheating. Intercellular gases are driven off, aiding in maintaining good can vacuums and reducing stresses on containers during processing. Problems of enzymatic discoloration are greatly reduced. Peeling times are frequently reduced from both the direct effect of the heat and the reduction of discoloration.

Temperatures and treatment times for preheating have been reported to be 30 min at 70°C when the final product was dehydrated flakes,[1-3] 30 min at 52°C[4] or 8 min at 63°C[5] before lye peeling for canning. Scott[6] found that 40 sec preheating in live steam prior to steam peeling gave good control of enzymatic darkening. Using an experimental steam peeler that permitted the steam pressure to be slowly increased and decreased, preheating was not found to be necessary.[7]

Table 1
PEELING SWEET POTATOES

Material	Time to peel (min)	Temperature (°C)	Ease of removal of peel	Loss in weight on peeling (%)
Boiling water	14	99	Required much rubbing and washing	18—30
Solution sat. salt (NaCl)	10	108	Peel removed easily by washing	20—35
Solution 2-1/2% lye[a]	12	101	Peeling required slight rubbing and much washing	20—35
Solution 5% lye[a]	8	101	Peeling was rapid with fine water spray	20—35
Solution 7% lye[a]	6	102	Peeling was rapid with fine water spray	20—35
Solution 10% lye	6	102	Peeling was rapid with fine water spray	20—35
Oven	20	177	Peel was difficult to remove mechanically	15—20
Mineral oil	2.5	204	Peeling was very easy but oil was difficult to remove	15—30
Gas flame	2.5	538+	Peeling was uneven	15—30
Mechanical	Few	—	Very easy and rapid, losses very heavy	30—60
With knife by hand	—	—	Very slow and tiresome	30—40

[a] Wetting agents added to the lye solution reduced the time of treatment from 20 to 10%. Those used, listed in order of preference, were: Wetsit Single, Aerosol OS, Santomerse, 600 Cleanser, sodium metasilicate, sodium sesquisilicate, trisodium phosphate, Aerosol OT, diglycol laurate, diglycol sterate.

D. Peeling and Trimming

A number of methods have been successfully used to peel sweet potatoes. Woodroof and Atkinson,[8] comparing 11 methods of peeling (Table 1), found that exposing the roots to a 7 to 10% (w:w) concentration of boiling lye for 6 min followed by washing with a high pressure spray produced satisfactory results. Steam peeling is often used for sweet potatoes. Peeling conditions are fairly easily controlled and waste water and solids disposal is much simplified as compared to the disposal of lye. Scott[6] suggested 30 sec at pressures of 70 psi or 20 sec at 90 psi and Woodroof et al.[9] found that 15 to 20 psi for 5 to 6 min followed by sudden release of pressure produced satisfactory results. High pressure steam peeling occasionally produces a phenomenon called "tunneling" described as small holes resembling insect damage. Burkhardt et al.[7] noted that tunneling was not observed when the pressure was more slowly increased and decreased.

Peeling losses are affected by a number of factors, the most obvious of which are peeling time and root size (Figure 1). It should be noted that longer peeling times have a greater effect on peeling depth but root size has little effect. Peeling time may be slightly reduced if the roots are preheated. The polyphenol oxidase enzyme substrate complex, which causes darkening, is concentrated in the cambial areas of the root. In the absence of preheating, peeling is often continued until the cambial area is removed in order to reduce darkening. Preheating permits the heat inactivation of these enzymes during peeling without removal of the cambium and results in less peeling loss without discoloration.

Peeling time may affect the uniformity of color among various roots. The location of the cambium layer is affected by cultivar, root size and curing (Table 2 and 3). The color of the area of the cambium may be slightly different than the internal part of the root. If the peeling time can be adjusted so that acceptable peeling is obtained without peeling to the depth of the cambium, peeling losses can be held to a minimum and a product of more uniform color can be obtained.

Peeling losses and trimming losses are inversely related. The longer peeling times that

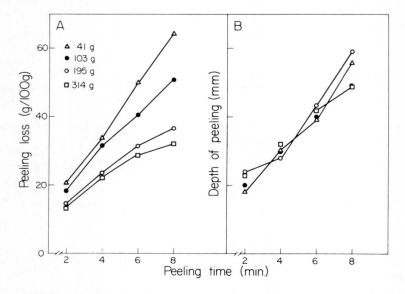

FIGURE 1. Peeling loss and depth of peeling as influenced by peeling time and root size.

Table 2
DEPTH OF CAMBIUM OF NEMAGOLD ROOTS OF VARIOUS SIZE CLASSES ROOTS CURED AND STORED 2 MONTHS

Root diameter (cm)	Depth of cambium from surface (mm)
>7.0	3.0
6—7	2.8
5—6	2.5
4—5	2.4
3—4	2.2
<3.0	1.7

Table 3
DEPTH OF CAMBIUM OF FRESH AND CURED PORTO RICO SWEET POTATOES

Root diameter (cm)	Depth of cambium from surface (mm)	
	Not cured	Cured
>10	4.25	3.25
5—10	2.75	2.69
< 5	2.58	

From Woodroof, J. G., DuPree, W. E., and Cecil, S. R., *Ga. Agric. Exp. Stn. Bull.*, N.S. 12, 1955.

result in greater peeling losses will require less hand trimming to remove surface imperfections. The best combination will depend on raw product cost and quality and labor costs. Peeling losses and the amount of alkali required are directly related.[9] Greater amounts of lye are required for the actual peeling. In addition, the peeling solution may become unusable more quickly due to the increasingly greater buffering capacity of dissolved organic material. Woodroof et al.[9] reported that the use of surfactants reduced peeling time by as much as 10%.

Following treatment with lye or steam, the peeling is completed by washing and brushing to remove the softened material. The amount of trimming required depends on the final product. Surface imperfections, whether caused from mechanical injury, disease or insects are removed. Fibrous ends or side roots are removed unless the roots are to be pureed or pulped by a process which will remove excessive fibers. As previously noted the amount of trimming required will be affected by the quality of the raw product and the peeling time. Unusable roots are discarded.

Table 4
BACTERIOLOGICAL STANDARDS FOR SUGAR FOR USE IN CANNING LOW-ACID PRODUCTS

1. Total Thermophilic Spore Count
   For the five samples examined, there shall be a maximum of not more than 150 spored and an average of not more than 125 spores per 10 g of sugar.
2. Flat Sour Spores
   For the five samples examined, there shall be a maximum of not more than 75 spores and an average of not more than 50 spores per 10 g of sugar.
3. Thermophilic Anaerobic Spores
   These shall be present in not more than three of the five samples and in any one sample to the extent of not more than four of six tubes inoculated by the standard procedure.
4. Sulfide Spoilage Spores
   These shall be present in not more than two of the five samples and in any one sample to the extent of not more than 5 spores per 10 g.

III. CANNING

A. Procedures

1. Sizing and Cutting

It is generally considered that a fairly uniform size and shape of the canned product is important in consumer acceptance. Producers may produce a variety of types to include small whole roots, cut roots in 2 to 3 cm chunks, small whole roots and halved or quartered (lengthwise) cut roots, etc. Accurate sizing is required in order to produce similar sized roots or pieces. In the U.S. this is usually accomplished with mechanical sizers but could be done as well by hand labor. Consumer preference for size and shape of the finished product should be determined among populations for which this information does not exist. In addition to the importance for consumer acceptance, a uniform size and shape will permit easier filling of cans with better control of fill weight. Sizing is not required if the roots are to be pureed or pulped before canning.

2. Blanching, Filling, Syruping

Blanching prior to canning helps to maintain can vacuum by driving out gasses and increasing the initial temperature of the contents of the can. One to three minutes blanching time in water at 77°C is usually sufficient. Immediately after blanching, the product should be placed in clean, washed containers. Filling may be by hand or by machine but cans should be inspected before syruping to be sure all cans are properly filled. Unless a vacuum pack is to be produced, sufficient liquid is added to completely fill the can except for a small headspace. Sugar syrups of various concentrations, usually 20 to 40% (w:w), or water may be used depending on consumer preferences. The liquid should be held at 95 +2°C. The sugar used for syrup preparation should be of good bacteriological quality as outlined by the National Food Processors Association[11] (Table 4).

3. Exhausting, Closing, Retorting, Cooling

Cans larger than 303 should be exhausted long enough for the internal temperature to reach 77°C[5] in order to insure a good vacuum of the finished cans. Size 303 or smaller cans do not have to be exhausted if a steam flow closing machine is used. Vacuum packed sweet potatoes are closed in a vacuum closer with a minimum machine vacuum of 23 in. (584 mm). After closing, the cans should be retorted according to one of the schedules presented in Tables 5 to 8. Cans should be quickly cooled after retorting to an internal temperature of 35°C. Undercooling may result in "stackburn" and overcooling may result in slow drying of the can and rusting. It is important that the cooling water is of drinking quality since some cans may take in small amounts of the cooling water as a vacuum develops.

Table 5
RETORT SCHEDULE FOR SOLID PACK SWEET POTATOES

Can size		Minimum initial temperature (°C)	Minutes at retort temperature (116°C)
U.S.	Metric (mm)		
211 × 400	68.3 × 101.6	49	73
		66	68
		72	61
307 × 409	87.3 × 115.9	49	105
		66	100
		82	84
401 × 411	103.2 × 119.1	49	130
		66	120
		82	105

Table 6
RETORT SCHEDULES FOR SYRUP PACK (20° BRIX OR LESS), FRESHLY DUG OR STORED, WHOLE, CUT, OR SLICED SWEET POTATOES

Can size		Minimum initial temperature (°C)	Minutes at retort temperature		
U.S.	Metric (mm)		(116°C)	(118°C)	(121°C)
303 × 406	81 × 111.1	21	36	28	23
		38	35	27	22
		60	32	24	20
401 × 411	103.2 × 119.1	21	52	42	37
404 × 307	108.0 × 87.3	38	49	40	35
401 × 602	103.2 × 155.6	60	45	37	32
603 × 700	157.2 × 177.8	21	57	46	40
		38	54	44	38
		60	50	40	35

B. Factors Affecting Firmness

1. Preprocessing Factors

Differences among cultivars in the firmness in the processed roots are frequently observed[12-16] and are an important criterion in the determination of whether or not a cultivar is acceptable in the U.S. for processing. A considerable effort to study other variables affecting firmness may be partially attributed to the fact that several popularly grown cultivars were less firm than the processors desired.

It was demonstrated early[12] and repeatedly[13-19] that delay in processing resulted in a less firm product. As little as 2 days delay resulted in a less firm product when the roots were stored at 30°C[17] but storage at low temperatures (2 to 5°C) resulted in a firmer processed product.[14,17]

Kattan and Littrell[14] and Scott and Bouwkamp[20] found that sweet potatoes harvested later

Table 7
SCHEDULES FOR RETORTING SYRUP PACK (20 TO 40° BRIX), FRESHLY DUG OR STORED, WHOLE, CUT, OR SLICED SWEET POTATOES

Can size		Minimum initial temperature (°C)	Minutes at retort temperatures		
U.S.	Metric (mm)		(116°C)	(118°C)	(121°C)
303 × 406	81.0 × 111.1	21	42	33	26
		38	49	30	24
		60	35	27	21
401 × 411	103.2 × 119.1	21	57	47	39
404 × 307	108.0 × 87.3	38	53	43	35
401 × 602	103.2 × 155.6	60	47	37	30
603 × 600	157.2 × 177.8	21	77	59	46
		38	69	52	41
		60	57	43	34

Table 8
SCHEDULING FOR RETORTING VACUUM PACK SWEET POTATOES

Can size		Minimum initial temperature (°C)	Minutes at retort temperature (116°C)
U.S.	Metric (mm)		
404 × 307	108.0 × 87.3	21	45

Note: This process is dependent on closing the cans at a minimum machine vacuum of 23 in.

in the season tended to be softer than those harvested early. Scott and Bouwkamp[20] concluded that these differences were due to age of roots rather than time of season since roots from earlier planted plots were softer than from later planted plots when harvested on the same date. Kattan and Littrell[4] found that irrigation increased firmness while Constantin et al.[21] noted a decrease in firmness in irrigated plots. They reported slight increase in firmness with higher levels of nitrogen fertilization but no effect on firmness due to levels of phosphorus application.[22]

2. Processing Factors

It is tempting to imagine that firmness can be increased by reducing processing time. The results of a series of experiments conducted by Scott on several cultivars over a two-year period (6 tests) are presented in Table 9. Since the effects of processing time on firmness were very similar regardless of cultivar, style of pack or storage prior to processing, the average firmness of the 6 tests are presented. It is clear that no benefits on firmness can be expected by processing less than the recommended time. Kattan and Littrell[14] also reported no significant differences in firmness of samples processed for 35 or 55 min at 116°C.

Sucrose concentration of the syrup had a slight but consistent effect on firmness in a series of tests conducted by Scott (Table 10) on 3 cultivars processed the day after harvest and 8 days after harvest, with and without curing. The firmest product was obtained the

Table 9
EFFECT OF PROCESSING TIME AT 116°C ON FIRMNESS OF CANNED SWEET POTATOES

Time (min)	20	30	40[a]	50	60
Firmness[b]	7.9	7.7	8.0	8.2	8.2

[a] Recommended processing time (check)
[b] Evaluated subjectively by a trained panel of judges. Firmness rated 1 (very soft) to 10 (very firm).

Table 10
EFFECTS OF SUCROSE CONCENTRATION OF THE SYRUP ON FIRMNESS OF PROCESSED SWEET POTATOES

"Fill in" syrup conc.	"Cut out"[a] syrup conc.	Firmness[b] of samples processed		
		1 Day after harvest	8 Days after harvest	8 Days after harvest and cured
0	13.2	8.4	8.2	7.3
10	16.2	8.9	7.9	7.9
20	19.9	8.7	8.3	7.4
30	24.1	9.3	8.3	8.1
40	28.1	9.1	8.5	7.9
50	32.8	9.4	8.9	8.3
60	37.6	9.5	9.0	8.9
LSD (0.05)	1.0	0.4	0.4	0.3

[a] Sucrose equivalent as measured by refractive index.
[b] Firmness evaluated subjectively by a trained panel of judges rated 1 (very soft) to 10 (very firm).

day after harvest at the highest sucrose concentrations. Woodroof et al.[9] reported similar effects of sucrose concentrations on firmness.

The work of Baumgardner and Scott[17,18] indicated that pectin levels were related to firmness. This suggests that calcium additions may result in a firmer processed product. Scott and Twigg[23] in a series of studies confirmed this hypothesis. They soaked peeled and trimmed roots in a solution of 2% $CaCl_2 \cdot 2H_2O$ (w:w) at various temperatures for various times. They found that 60°C was the optimum temperature and processed roots were firmer with increased soaking time up to 30 min. Calcium concentration of the roots increased more or less in proportion to the firmness (Table 11). Since the optimum temperature was considered risky from the standpoint of microbiological contamination, it was decided to lower the pH of the soaking solution to 2.6 by adding citric acid at the rate of 11 g/ℓ. The lower pH was found to enhance the effect of calcium (Table 12). Calcium concentration of the can contents was increased by the higher temperatures and low pH soaking solutions. The authors state that the low pH soak had no noticeable effect on flavor of the processed product.

Further studies indicated that addition of 10 mℓ of a solution containing 81.6 g/ℓ of $CaCl_2 \cdot 2H_2O$ to each can before syruping was nearly as effective as a 10 min soak at 60°C. Rao and Ammerman[19] reported an increase in firmness after soaking peeled roots in a solution of $Ca(OH)_2$ for 3 to 6 min. Sistrunk[24] found firmness to be greatly improved by vacuum infiltration of low pH solutions of citric acid prior to processing.

Table 11
EFFECT OF TEMPERATURE AND DURATION OF CALCIUM CHLORIDE TREATMENT ON FIRMNESS AND CALCIUM CONTENT OF PROCESSED CENTENNIAL SWEET POTATOES

Treatment[a]			Duration of treatment in minutes			
			Shear press (kg force)		Ca content (%)[b]	
			10	30	10	30
Control	20° No	Ca	12.7	11.8	0.018	0.018
	20° +	Ca	15.9	15.9	0.042	0.092
	55° +	Ca	17.7	25.9	0.076	0.105
	60° +	Ca	21.8	29.1	0.124	0.222
	65° +	Ca	23.2	22.7	0.139	0.260
	70° +	Ca	17.7	20.9	0.097	0.263
	75° +	Ca	17.3	18.2	0.205	0.171
L.S.D.	(5%)		2.3		0.056	

[a] Sweet potatoes were held in 2% calcium chloride solutions at temperatures indicated; control held in tap water.
[b] On basis of wet weight of total can contents.

Table 12
EFFECT OF ACIDIFICATION OF CALCIUM CHLORIDE SOLUTION ON FIRMNESS AND CALCIUM CONTENT OF PROCESSED SWEET POTATOES

Treatment[a]		Shear press (kg force)		Shear press (%)[c]	
		Control	Acidified[b]	Control	Acidified
Control	68° No Ca	16.3	17.7	0.008	0.008
	68° + Ca	20.4	23.2	0.048	0.045
	122° + Ca	29.5	31.3	0.064	0.089
	140° + Ca	28.1	36.8	0.079	0.096
	148° + Ca	28.6	34.1	0.102	0.092
L.S.D.	(5%)	2.3		0.010	

[a] Sweet potatoes were held in 2% calcium chloride solutions for 10 min at temperatures indicated; control held in tap water.
[b] Citric acid at rate of 11 g/ℓ was used to acidify treatment solutions. Resulting pH was 2.6.
[c] On basis of wet weight of total can contents.

C. Factors Affecting Color
1. Preprocessing Factors

It is obvious that the color of the raw unprocessed roots has a profound effect on the color of the processed product. It is less obvious, and of greater concern, that a variety of preprocessing factors may have an effect on the amount of darkening exhibited by the processed product. Smittle and Scott[25] proposed that the darkening reaction is a two step process, both requiring oxygen. The first step is the enzymatic oxydation of phenolic compounds to quinones. The second step is the nonenzymatic polymerization of quinones to melanin-like compounds. These melanin-like compounds are black in the presence of iron ions, and yellow when chelated with tin ions. The nonenzymatic step is responsible for the darkening often observed after the cans have been opened and the product exposed to air.

Cultivars vary greatly in their propensity to darken as demonstrated by Scott and Kattan.[10] They also noted that cultivars with low levels of carotene were more likely to darken than those with higher levels of carotene. The polyphenol oxidase activity may be estimated by blending root samples in an equal weight of water in the presence of air and an equivalent sample in an equal weight of an aqueous solution containing 1 g/ℓ of thiourea. The thiourea inhibits the action of the enzyme so that the color difference between the thiourea blend and the water blend can be taken as a measure of the enzyme activity. Since the water blend will continue to darken for a period of time, the time between blending and color evaluation must be standardized.

Kattan et al.[26] found that enzymatic discoloration was increased by irrigation. Rao and Ammerman[19] noted that roots processed after storage of 1 or 3 months were slightly darker than roots that had been processed soon after harvest.

2. Processing Factors

Preheating the roots before peeling has been found to reduce the darkening of the processed product.[4,6] This effect is thought to be due to a more rapid heat inactivation of the enzyme during peeling, and deeper penetration of inactivating temperature.[27] The preheating also drives out some internal gases and probably increases the rate of respiration, both of which would serve to reduce oxygen levels and limit enzyme activity.

Canning in plain, unlined tin-plated cans has been shown to produce a brighter product with less discoloration.[28] This effect is thought to be due to the fact that the final condensation products of the quinones are yellow when chelated with tin.[27] Although unenameled tin lined cans produce a higher product, enameled cans are often used to eliminate problems of internal can corrosion (detinning). Smittle and Scott[25] have shown that internal can corrosion is related to polyphenol oxidase activity and nitrate concentration of the roots. They suggested that internal can corrosion could be greatly reduced by selecting cultivars which do not contain excessive levels of nitrate and by controlling polyphenol oxidase activity by preheating and limiting exposure of peeled roots to oxygen.

Several additives have been shown to result in a brighter canned product. Sistrunk[24] found that darkening was reduced as the pH was decreased by the use of citric acid in a preprocessing soak. Twigg et al.[30] found EDTA, $SnCl_2$, citric acid and ascorbic acid to improve the color and reduce the amount of darkening.

The sucrose concentration of the syrup has an effect on the color of the processed product. The data presented in Table 13 show the greatest improvement in attractiveness was between 40 and 50% (w:w) sucrose. Similar results were obtained by Jenkins and Anderson[31] and Woodroof et al.[9] Replacement of sucrose by corn sweeteners resulted in a slightly darker processed product.[9]

C. Factors Affecting Nutritional Quality

1. Preprocessing Factors

Differences among cultivars have been noted for most nutrients and would certainly affect the nutritional quality of the canned product. Differences in carotene levels among cultivars are easily observed and often noted. Sweet potatoes contain mostly all trans beta carotene with small amounts of neo-beta carotene B and neo-beta carotene U and have an estimated biopotency of 94.9.[32] Differences among cultivars have been noted for ascorbic acid,[33,34] protein content,[34,35] B vitamins,[36] and various minerals.[37]

Constantin et al.[21] reported that irrigation decreased protein and carotene concentrations. Nitrogen applications increased protein while decreasing carotene concentrations and differences among years were significant for both carotene and protein concentrations. Scott

Table 13
EFFECT OF SUCROSE CONCENTRATION (w:w) ON ATTRACTIVENESS OF PROCESSED SWEET POTATOES

Concentration (%)	0	10	20	30	40	50	60
Attractiveness[a]	7.4	7.8	7.9	8.0	8.3	9.0	9.2

[a] Attractiveness evaluated subjectively by a trained panel of judges, rated 1 (very unattractive) to 10 (very attractive).

and Bouwkamp[20] found a slight increase in carotene content, as measured by Hunter Color Difference meter, as the growing season progressed. The data of Ezell et al.[39,40] indicate about a 50% reduction of ascorbic acid after curing and storing.

2. Processing Factors

Elkins[41] found that 86% of the carotene content was retained after processing and there was no additional loss of carotene after storage of the processed product for 18 months. This is very similar to the reported values of Panalaks and Murray[32] and Watt and Merrill.[42] Panalaks and Murray[32] found that canning resulted in the isomerization of carotene, that neo beta-carotene B was the main isomer produced and that calculated biopotency values were reduced from 95 to 91%. Lee and Ammerman[43] reported similar results and noted that longer cooking times or higher cooking temperatures without agitation resulted in greater carotene destruction and more isomerization. Retort cooking at 127°C for 13 min or 132°C for 12.6 min with agitation resulted in similar carotene retention and isomerization to still cooking at 116°C for 34 min. Arthur and McLemore[44] found no effect on carotene content due to differences in syrup concentrations (0 to 35% sucrose), cooking time (50 to 90 min) or peeling conditions.

Elkins[41] reported that riboflavin and niacin were completely retained during processing and were retained during 18 months of storage of the canned product. Thiamin retention was 73% of raw product concentration after processing and 44% after 18 months storage. Lanier and Sistrunk[36] found that canning resulted in the lowest concentration of niacin, riboflavin and pantothenic acid expressed on a fresh weight basis. Since the dry matter percentage was much less in the canned samples, these nutrients expressed on a dry weight basis were not less than the other treatments.

Lopez et al.[45] determined that essential elements from the roots were leached into the syrup, especially potassium, copper and magnesium. Elkins[41] found that the mineral elements were completely retained in vacuum packed sweet potatoes and that iron concentration increased nearly threefold after 18 months of storage of the processed product. The increased iron levels were obviously derived from the metal can.

Elkins[41] noted complete retention of ascorbic acid after canning but only 66% retention, as compared to raw product, offer 18 month of storage after processing. Arthur and McLemore[44] found no effect of syrup concentrations (0 to 35% sucrose) on ascorbic acid and that longer processing times resulted in greater reductions.

Amino acids may be leached from the sweet potato roots into the syrup. Meredith and Dull[46] found that leaching was greater in sweet potatoes packed in water than when packed in 30% sucrose. On a solids basis, canned sweet potatoes, as compared to fresh contained 55% of the amino acids when processed in syrup and 51% when canned in water. Essential amino acids were 70 and 58% of fresh; aromatic amino acids 69 and 49% of fresh, and sulfur containing amino acids 86 and 60% of fresh when canned in 30% sucrose syrup or water, respectively. Purcell and Walter[41] noted a reduction of 26% (presumably leached into the syrup) and significant reductions in methionine and lysine. This suggests that the canning liquor should be consumed for optimum nutritional benefit. It is likely that vacuum-packed sweet potatoes would retain more amino acids although this has not been reported.

IV. FREEZING

A. Procedures
1. Sizing and Cutting

If the roots are to be frozen whole, sizing is important to assure appropriate blanching or cooking and freezing time. If the roots are to be communited, sliced, or diced prior to further processing, sizing is probably not necessary. Sweet potatoes may be frozen in a variety of styles: as puree, slices (either transverse or longitudinal), dices, french fry strips, halves or quarters (cut longitudinal), or as whole roots. Cutting machines are available for the various cutting procedures.

2. Blanching or Cooking

If the product is intended to be cooked by the consumer, the sweet potatoes may be blanched. The internal temperature should reach 88°C in order to inactivate the various enzymes. Woodroof and Atkinson[8] studied several methods of blanching including boiling in water, exposure to free and pressurized steam and baking. Blanching with steam at 10 psi pressure (116°C) was considered the best method. It was fairly quick (7 min for a root of 5 cm diameter), resulted in little discoloration and the products were of high quality with regard to flavor, color, and aroma. Higher temperatures were unsatisfactory because the outer portions of the roots became overcooked before internal temperatures reached 88°C. Boiling and free steam blanching resulted in products that were soggy and of poorer quality with regard to flavor. However, they noted that blanching in free steam may be appropriate for the production of frozen slices. Baking resulted in well-flavored products but was slow and required excessive hand labor to remove an objectional crust which formed on the root surface. Harris and Barber[48] found that good quality strips could be produced if blanched in a 60% sucrose solution for 4.5 min. Hoover and Pope[49] reported that 1.25 cm slices absorbed sucrose from the media when blanched 10 min in solutions containing 30 to 60% sucrose. Hoover[50] found that sodium acid pyrophosphate and tetrasodium pyrophosphate added to the blanching solution would reduce discoloration of the frozen product.

Kelley et al.[51] reported that french fry pieces (9.5 mm square strips) could be blanched/precooked in cooking oil at 135°C for 3 to 7 min. The shorter the blanching times required longer reheating times by the consumer to produce an acceptable product. They also produced dices, chips, and julienne strips by cooking in oil. Frozen french fries may also be produced by blanching the strips in boiling water and quick freezing. The frying in hot oil is done by the consumer.

Sweet potatoes may be pulped after blanching as suggested by Woodroof and Atkinson,[8] or after 1 hr cooking in a sucrose solution or water as reported by Harris and Barker.[48] It is also likely that a good quality frozen puree may be obtained by the enzyme activation processes developed by Hoover for the production of dehydrated flakes. Citric acid added to puree at the rate of 0.2% was found by Woodroof and Atkinson[8] to preserve the bright color of the puree and of pies produced from the frozen puree.

3. Freezing

It is good practice to cool and freeze the product as soon as possible although Woodroof and Atkinson[8] found no quality difference in quick-frozen puree and puree frozen more slowly in large containers. Packaging may precede freezing or follow freezing if whole roots or cut pieces are individually quick frozen. Since the product is not exposed to heat after blanching, sanitation is of paramount importance during these steps. The nutritional value of frozen sweet potatoes has not received much study. Processing procedures for freezing are relatively gentle as compared to other methods of preservation so that heat destruction of vitamins would not be expected to be greater than from home preparation. Since the product is not packed in liquid, leaching of vitamins and minerals would probably be less than experienced by syrup-packed canned or home preparation by boiling.

V. DEHYDRATION

A. Production of Slices, Dices, Strips

Lease and Mitchell,[53] quoting a report by Newman, stated that dehydrated sweet potatoes, produced in 1899 without blanching, were difficult to rehydrate and did not resemble the fresh product. The following year, cooking preceded dehydration and a complete success was reported. Hand and Cockerham[54] described a dehydration procedure consisting of slicing to a uniform size, blanching 6 to 10 min in boiling water and dehydration at 63 to 66°C at the early stages of drying. The temperature was gradually increased to about 71°C as drying progressed.

Caldwell et al.[55] reported that the most successful procedures for producing dehydrated pieces included lye-peeling and slicing longitudinally into triangular strips (as an orange may be divided into sections) about 2 cm thick. The strips were transferred immediately after slicing to a solution containing 2% citric acid for 3 to 5 min to prevent discoloration, placed on trays and steamed at 93 to 97°C for 20 to 30 min for cured and stored roots and 35 to 40 min for freshly dug stock. Drying was accomplished in a forced air dryer. Dryer temperatures of 50 to 73°C were found to produce acceptable dehydrated products. Higher temperatures resulted in scorching and at lower temperatures, microbiological contamination occurred. Strips dehydrated to 5 to 15% residual moisture could be stored in fairly humid conditions in nonairtight containers for up to 2.5 years without apparent loss in quality. Strips dehydrated to more than 15% moisture slowly absorbed moisture and became moldy. The authors concluded that dehydration to 12 to 15% moisture was optimum since considerably longer drying time was required to further reduced the moisture content without an increase in storage life and the strips dehydrated to lower moisture contents were rehydrated with difficulty. Samples, stored in nonairtight containers, were not infested with Indian meal moth even though the insect was present in dried fruits stored in the same room, but were occasionally infested with tobacco beetle, drug store beetle or weevils. The authors recommend cellophane, waxed cardboard or metal containers to prevent insect contamination. Burkhardt[56] described a two-stage tunnel dehydrator. The first stage used air velocities of 355 linear m/min, the second stage 55 linear m/min. Mahoney et al.[57] reported a procedure of slicing the peeled roots into 4.8 mm slices, blanching in steam at 102°C for 6 min and dehydrating in the dehydrator described by Burkhardt.[56] They found that first stage temperatures of 104°C for 2.5 hr and a second stage temperature of 71°C for 2 hr resulted in the best colored product. Increasing the second stage temperature to 79°C resulted in much poorer color. First stage temperatures of 60 to 120°C had little effect on rehydration but a first stage temperature of 127°C resulted in poorer rehydration. Arthur et al.[58] reported a procedure for producing dices (9.5 mm on a side) similar to the procedure of Mahoney et al.[7] except that the dices were treated after blanching with a spray containing 2.1 parts of sodium sulfite and 0.7 parts of sodium bisulfite per 1000 parts of water. They found that carotene was relatively stable but that decreases in carotene concentration could be detected at higher storage temperatures (24 and 38°C) with increasing time of storing the dehydrated dices (9 to 27 weeks). Ascorbic acid seemed to be less stable showing greater losses at all time and temperature combinations. Time of storage and storage temperatures had no effect on rehydration properties. Dehydrated dices were held in airtight containers in nitrogen atmosphere.

Lambou[59] found that the temperature of storage of roots prior to dehydration had considerable effects on the quality of the rehydrated product. Storage of roots at 10°C resulted in an umpalatable product but storage at 15 to 20°C for 2.5 months produced an acceptable product. Temperatures of storage was more important than time. Flavor and odor were most affected, followed by color. Texture was least affected. Time and temperature of storage of the dehydrated product had no affect on the quality of the rehydrated dice. Arthur and McLemore[60] reported the effects of preheating treatments (54 to 60°C for 30 to 40 min or 74° for 10 to 20 min) and initial drying temperatures (93, 79, and 66°C) "do not indicate any one combination of conditions to be materially superior." Drying at 66°C resulted in less destruction of ascorbic acid. Molaison et al.[61] found that dehydrated dice stored for up to 6 years at room temperature in airtight containers in nitrogen atmospheres retained their original quality.

Woodroof and Cecil[62] reported a procedure for dehydration by cooking thin slices (0.6 to 0.8 mm) in cooking oil heated to 175°C. They preheated the roots 15 to 20 min at 77 to 88°C, peeled in boiling lye and cooked in oil for 1.5 to 2 min. Oil temperature should not be less than 150°C. If the chips are too hard, preheating time should be increased or the roots should be sliced more thinly. If the roots are too greasy the temperature of the cooking oil should be increased. Shelf life of the chips could be increased by the additions of antioxidants to the oil such as butylated hydroxyanisole (BHA) at the rate of 0.03%, butylated hydroxyoluene (BHT) at the rate of 0.03% OR 0.015% BHA + 0.015% BHT.

Kelley et al.[51] found that chips (0.8 mm thick) julienne strips (1.6 × 1.6 mm) or dice (1.6 mm on a side) could be dehydrated in 135°C cooking oil at cooking times of 3.25, 5, and 5 min, respectively. Sugars can be leached from the chips prior to cooking in order to prevent the darkening associated with carmelization of sugars.[63]

B. Production of Dehydrated Flakes

Taubenhaus[64] described a procedure for producing flakes consisting of washing, cooking, mashing, and drying on steam-heated drum dryers. The production of flakes which produced an acceptable product on rehydration was difficult to achieve consistently. The use of uncured roots resulted in a product which, when rehydrated, was too starchy and not sweet enough.[65]

Deobald et al.[66] found that under some conditions, holding the cured roots at 74°C for 30 min prior to peeling resulted in improved quality. Hoover[67] successfully produced acceptable flakes from uncured sweet potatoes by adding starch-hydrolyzing enzymes after cooking and pulping the cut roots. The amount of starch hydrolysis could be controlled by controlling the time the enzyme was allowed to act before heat inactivation and by regulating the amount of enzyme treated puree added back to the nontreated puree before drying.

Spadaro et al.[2] reported on modifications of an earlier procedure[1] which had been developed for the production of flakes from the cured roots of the cultivar 'Goldrush'. They found that additions of amylase and/or sucrose after cooking and pureeing would result in flakes which, on rehydration, were acceptable. The specific combinations depended on the cultivar used and probably on the length of storage after curing. Bertonier et al.[68] found that curing and storing affected the optimum amount of amylase required for acceptable flakes but had no effect on the properties of the starch.

A second, although related, procedure for controlling the quality of the processed product was reported by Hoover.[69] The process includes peeling, communiting to 0.7 mm, raising the temperature rapidly to 70 to 85°C, holding for specific time before heating to 103°C to inactivate enzymes, then drying on a drum dryer. Holding temperatures of 79 to 85°C were found to be optimum for the activity of the native amylase.[70] Lower or higher temperatures resulted in reduced conversion of starches to maltose. Approximately 90% of the conversion occurred within the first 10 min although the quality of the flakes continued to increase with holding times up to 60 min, due possibly to changes in viscosity of the remaining starch.[71]

Deobald et al.[72,73] found that differences in alpha amylase activity of the raw roots as measured by a procedure developed by Ikemiya and Deobald[74] could be related to processing characteristics and quality of the dehydrated flakes produced. Amylase activity could be enhanced in freshly harvested roots by additions of calcium ions and by increasing the holding time for conversion. Cured and stored roots may have higher than optimum levels of amylase. These could be reduced by longer preheating at higher temperatures and reducing the conversion time.

Hoover[75] found that additions to the puree or dried flakes of sodium acid pyrophosphate (SAPP) and tetrasodium pyrophosphate (TSPP) were most effective in preventing discoloration followed by sodium tripolyphosphate, hexametaphosphate, and monosodium phosphate. When flavor was considered, the best treatment was a mixture of SAPP and TSPP in a ratio of 3:1. Spadaro et al.[2] reported that a concentration of 0.2% of the 3:1 mixture received slightly higher flavor ratings than a 0.4% concentrations. A 0.05% concentration of sodium sulfite and sodium bisulfite in a 2:2 ratio did not give adequate color preservation and received low color scores for flavor. Citric acid at 0.2% concentration gave good color in the product but slightly lower taste ratings although Woodroof and Atkinson[8] reported that this concentration could not be detected in the flavor of purees.

The rheological properties of puree have been associated with drying characteristics[3,76] and quality.[77,78] Ice et al.[79] and Creamer et al.[80] have studied the properties of acidified purees.

Walter et al.[81] described a procedure for fortifying sweet potato flakes with soy flour, cottonseed flour and wheat gluten flour. The protein levels were higher in the fortified flakes but water-binding capacity of the reconstituted flakes was reduced.

Sweet potato flakes, in contrast to other dried forms, seem to be rather subject to deterioration on storage. When stored in air at room temperature, development of off flavors and autoxidation of carotene and lipid fractions has been noted.[82-85]

It has been suggested that the "hay-like" off flavors may be related to the formation of monocarbonyls, particularly aldehydes[86] or a decrease in the unsaturation ratio of fatty acid.[87] Deobald and McLemore[88] and Deobald et al.[89] found that addition of antioxidants to sweet potatoe puree prior to dehydration would extend shelf life of the dried flakes but not sufficiently to warrant their use. Haralampu and Karel[90] found that in the presence of oxygen, water activity of dried uncooked sweet potato flour affected the rate of carotene and ascorbic acid destruction. Carotene destruction was greatest at low water activities while ascorbic acid destruction was greatest at higher water activity levels.

VI. CONCLUSIONS

The development of the sweet potato processing industry occurred in the U.S. in the context of several important considerations.

1. The fresh commodity was available on a seasonal basis but consumers were willing to use the product the year around.
2. Fresh market requirements and grades were based on cosmetic criteria rather than edibility (cosmetic defined as size, shape, or slight surface blemishes).
3. The off grade product was priced at a level such that the processed product was economically competitive with comparable commodities.
4. The raw product was subjected to losses during storage and marketing not experienced by the processed product; thus products that would have been wasted were converted into a stable useful product.
5. The convenience of the processed product contributed to its desirability.

The relative importance of the various factors is not known but is likely that cultural and economic limitations are of equal or greater importance than technical limitations in the successful development of a processing industry. It is also likely that the successful development of a processing industry would greatly increase the utilization of sweet potatoes. It remains for scientists to determine which products are most likely to be accepted, which attributes are most important to quality as judged by consumers and finally to modify existing technology and develop new technology in order to produce those products.

REFERENCES

1. **Spadaro, J. J. and Patton, E. L.,** Precooked dehydrated sweet potato flakes, *Food Eng.*, 33(6), 46, 1961.
2. **Spadaro, J. J., Wadsworth, J. I., Zeigler, G. M., Gallo, A. S., and Kilter, S. P.,** Instant sweet potato flakes — processing modifications necessitated by varietal differences, *Food Technol. (Chicago)*, 21, 326, 1967.
3. **Wadsworth, J. I., Kolter, S. P., Gallo, A. S., Zeigler, G. M., and Spadaro, J. J.,** Instant sweet potato flakes: factors affecting drying rates on a double drum drier, *Food Technol. (Chicago)*, 20, 815, 1966.
4. **Scott, L. E., Appleman, C. O., and Wilson, M.,** The discoloration of sweet potatoes during preparation for processing and the oxidase in the root, *Md. Agric. Exp. Stn. Tech. Bull.*, A33, 1944.
5. **Edmond, J. B. and Ammerman, G. R.,,.** *Sweet Potatoes: Production, Processing, Marketing*, AVI Publishing, Westport, Conn., 1971, 264.
6. **Scott, L. E.,** Sweet potato discoloration in relation to HP steam peeling, *Food Packer*, 33(3), 38, 1952.
7. **Burkhardt, G. J., Merkel, J. A., and Scott, L. E.,** Time and pressure regulated steam for peeling fruits and vegetables, *HortScience*, 8, 485, 1973.
8. **Woodroof, J. G. and Atkinson, I. S.,** Preserving sweet potatoes by freezing, *Ga. Agric. Exp. Stn. Bull.*, 232, 1944.
9. **Woodroof, J. G., DuPree, W. E., and Cecil, S. R.,** Canning sweet potatoes, *Ga. Agric. Exp. Stn. Bull.*, N.S.12, 1955.
10. **Scott, L. E. and Kattan, A. A.,** Varietal differences in the catechol oxidase content of the sweet potato root, *Proc. Am. Soc. Hortic. Sci.*, 69, 436, 1957.
11. National Food Processors Association, Thermal Processes for Low Acid Foods in Metal Containers, Bull. 26L, 12th ed., NFPA, Washington, D.C., 1982.
12. **Magoon, C. A. and Culpepper, C. W.,** A Study of Sweet Potato Varieties with Special Reference to their Canning Quality, *U.S. Bull.* 1041, U.S. Department of Agriculture, Washington, D.C., 1922.
13. **McConnell, E. R. and Gottschall, P. B.,** Effects of canning on new and familiar sweet potato varieties with particular emphasis on breakdown and firmness, *Food Technol. (Chicago)*, 11, 209, 1957.
14. **Kattan, A. A. and Littrell, D. L.,** Pre- and post-harvest factors affecting firmness of canned sweet potatoes, *Proc. Am. Soc. Hortic. Sci.*, 83, 641, 1963.
15. **McConnell, E. R., Gottschall, P. B., Jr., and Huffington, J. M.,** Influence of variety and storage on the quality of canned Louisiana sweet potatoes, *Proc. Am. Soc. Hortic. Sci.*, 67, 493, 1956.
16. **Huffington, J. M., McConnell, E. R., and Gottschall, P. B., Jr.,** Sweet potato varieties for canning: influence of storage on quality, *Proc. Am. Soc. Hortic. Sci.*, 67, 504, 1956.
17. **Baumgardner, R. A. and Scott, L. E.,** Firmness of processed sweet potatoes *(Ipomoea batatas)* as affected by temperature and duration of the post-harvest holding period, *Proc. Am. Soc. Hortic. Sci.*, 80, 507, 1962..
18. **Baumgardner, R. A. and Scott, L. E.,** The relation of pectin substances to firmness of processed sweet potatoes, *Proc. Am. Soc. Hortic. Sci.*, 83, 629, 1963.
19. **Rao, C. S. and Ammerman, G. R.,** Canning studies on sweet potatoes, *J. Food Sci. Technol.*, 11, 105, 1974.
20. **Scott, L. E. and Bouwkamp, J. C.,** Effect of chronological age on composition and firmness of raw and processed sweet potatoes, *HortScience*, 10, 165, 1975.
21. **Constantin, R. J., Hernandez, T. P., and Jones, L. G.,** Effects of irrigation and nitrogen fertilization on quality of sweet potatoes, *J. Am. Soc. Hortic. Sci.*, 99, 308, 1974.
22. **Hammett, H. L., Constantin, R. J., Jones, L. G., and Hernandez, T. P.,** The effect of phosphorus and soil moisture levels on yield and processing quality of 'Centennial' sweet potatoes, *J. Am. Soc. Hortic. Sci.*, 107, 119, 1982.

23. **Scott, L. E. and Twigg, B. A.**, The effect of temperature of treatment and other factors on calcium firming of processed sweet potatoes, *Maryland Processors Report*, 15(2), 1, 1969.
24. **Sistrunk, W. A.**, Carbohydrate transformations, color and firmness of canned sweet potatoes as influenced by variety, storage, pH and treatment, *J. Food Sci.*, 36, 38, 1971.
25. **Smittle, D. A. and Scott, L. E.**, Internal can corrosion by processed sweet potatoes as affected by phenolase activity and nitrate concentration, *J. Am. Soc. Hortic. Sci.*, 94, 649, 1969.
26. **Kattan, A. A., Horton, B. D., and Moore, J. N.**, Effect of supplemental irrigation on yield and quality of two vegetable crops, *Ark. Farm Res.*, 7(3), 3, 1958.
27. **Walter, W. M., Jr. and Schadel, W. E.**, Effect of eye peeling conditions on sweet potato tissue, *J. Food Sci.*, 47, 813, 1982.
28. **Scott, L. E., Twigg, B. A., and Bouwkamp, J. C.**, Color of processed sweet potatoes: effects of can type, *J. Food Sci.*, 39, 563, 1974.
29. **Hernandez, H. H. and Vosti, D. C.**, Dark-colored discoloration of canned all-green asparagus. I. Chemistry and related factors, *Food Technol. (Chicago)*, 17, 95, 1963.
30. **Twigg, B. A., Scott, L. E., and Bouwkamp, J. C.**, Color of processed sweet potatoes: Effect of additives, *J. Food Sci.*, 39, 565, 1974.
31. **Jenkins, W. F. and Anderson, W. S.**, Quality improvement in canned sweet potatoes, *Miss. Farm. Res.*, 20(6), 2, 1957.
32. **Panalaks, T. and Murray, T. K.**, The effect of processing on the content of carotene isomers in vegetables and peaches, *Can. Inst. Food Technol. J.*, 3, 145, 1970.
33. **Massey, P. H., Jr., Eheart, J. F., Young, R. W., and Camper, H. M.**, The effects of variety on the yield and vitamin content of sweet potatoes, *Proc. Am. Soc. Hortic. Sci.*, 69, 431, 1957.
34. **Collins, W. W. and Walter, W. M., Jr.**, Potential for increasing nutritional value of sweet potatoes, in *Sweet Potato: Proc. First Int. Symp.*, Villareal, R. L. and Griggs, T. D., Eds., Asian Vegetable Research and Development Center, Shanhua, Tainan, Taiwan, 1982, 355.
35. **Li, L.**, Breeding for increased protein content in sweet potatoes, in *Sweet Potato Proc. First Int. Symp.*, Villareal, R. L. and Griggs, T. D., Eds., Asian Vegetable Research and Development Center, Shanhua, Tainan, Taiwan, 1982, 345.
36. **Lanier, J. J. and Sistrunk, W. A.**, Influence of cooking method on quality attributes and vitamin content of sweet potatoes, *J. Food Sci.*, 44, 374, 1979.
37. **Scott, L. E. and Bouwkamp, J. C.**, Seasonal mineral accumulation by the sweet potato, *HortScience*, 9, 233, 1974.
38. **Hammett, H. L.**, Total carbohydrate and carotenoid content of sweet potato as affected by cultivar and area of production, *HortScience*, 9, 467, 1974.
39. **Ezell, B. D. and Wilcox, M. S.**, Influence of storage temperature on carotene, total carotenoids and ascorbic acid content of sweet potatoes, *Plant Physiol.*, 27, 81, 1951.
40. **Ezell, B. D., Wilcox, M. S., and Crowder, J. N.**, Pre- and post-harvest changes in carotene, total carotenoids and ascorbic acid content of sweet potatoes, *Plant Physiol.*, 27, 355, 1951.
41. **Elkins, E. R.**, Nutrient content of raw and canned green beans, peaches and sweet potatoes, *Food Technol. (Chicago)*, 33(8), 66, 1979.
42. **Watt, B. K. and Merrill, A. L.**, Composition of Foods, Handbook 8, U.S. Department of Agriculture, Washington, D.C., 1963.
43. **Lee, W. G. and Ammerman, G. R.**, Carotene stereoisomerization in sweet potatoes as affected by rotating and still retort canning processes, *J. Food Sci.*, 39, 1188, 1974.
44. **Arthur, J. C., Jr. and McLemore, T. A.**, Effects of processing conditions on the chemical properties of canned sweet potatoes, *Agric. Food Chem.*, 5, 863, 1957.
45. **Lopez, A., Williams, H. L., and Cooler, F. W.**, Essential elements in fresh and in canned sweet potatoes, *J. Food Sci.*, 45, 675, 1980.
46. **Meredith, F. and Dull, G.**, Amino acid levels in canned sweet potatoes and snap beans, *Food Technol. (Chicago)*, 33(2), 55, 1979.
47. **Purcell, A. E. and Walter, W. M., Jr.**, Stability of amino acids during cooking and processing of sweet potatoes, *J. Agric. Food Chem.*, 30, 443, 1982.
48. **Harris, H. and Barber, J. M.**, New uses for low grade sweet potatoes, *Highlights of Agric Res.*, 5(3), 1958.
49. **Hoover, M. W. and Pope, D. T.**, Effect of variety, curing and processing on carbohydrate content of precooked frozen sweet potatoes, *Food Technol. (Chicago)*, 14, 227, 1960.
50. **Hoover, M. W.**, Preservation of the natural color in processed sweet potato products. II. Precooked frozen, *Food Technol. (Chicago)*, 18, 1793, 1964.
51. **Kelley, E. G., Baum, R. R., and Woodward, C. F.**, Preparation of new and improved products from eastern (dry-type) sweet potatoes: chips, dice, julienne strips, and frozen fresh fries, *Food Technol. (Chicago)*, 12, 510, 1958.

52. **Smith, D. A., McCaskey, T. A., Harris, H., and Rymal, K. S.,** Improved aseptically filled sweet potato purees, *J. Food Sci.,* 47, 1130, 1982.
53. **Lease, E. J. and Mitchell, J. H.,** Biochemical and nutritional studies of dehydrated sweet potato, *S.C. Agric. Exp. Stn. Bull.,* 329, 1940.
54. **Hand, T. E. and Cockerham, K. L.,** *The Sweet Potato,* Macmillan, New York, 1921.
55. **Caldwell, J. S., Moon, H. H., and Culpepper, C. W.,** Comparative study of suitability for drying purposes in forty varieties of the sweet potato, *U.S. Dep. Agric. Circ.,* 499, 1938.
56. **Burkhardt, G. J.,** Basic engineering principles of dehydration, in Proc. Maryland Dehydration Conf., *Md. Agric. Exp. Stn. Misc. Publ.,* 18, 3, 1943.
57. **Mahoney, C. H., Burkhardt, G. J., Hunter, H. A., and Walls, E. P.,** The effects of drying temperatures on the drying rate and quality of sweet potatoes, in Proc. Maryland Dehydration Conf., *Md. Agric. Exp. Stn. Misc. Publ.,* 18, 65, 1943.
58. **Arthur, J. C., Jr., McLemore, T. A., Miller, J. C., Jones, L. G., and Sistrunk, W. A.,** Effects of conditions of storage of raw materials on chemical properties of dehydrated products, *Agric. Food Chem.,* 3, 151, 1955.
59. **Lambou, M. G..,** Sweet potato dehydration: time and temperature of storage related to organoleptic evaluations, *Food Technol. (Chicago),* 10, 258, 1956.
60. **Arthur, J. C., Jr. and McLemore, T. A.,** Effects of processing conditions and variety on properties of dehydrated products, *Agric. Biol. Chem.,* 3, 784, 1955.
61. **Molaison, L. J., Spadaro, J. J., Roby, M. T., and Lee, F. H.,** Dehydrated diced sweet potatoes — a pilot plant process and product evaluation, *Food Technol. (Chicago),* 16(11), 101, 1962.
62. **Woodroff, J. G. and Cecil, S. R.,** Sweet potato chips, *Ga. Agric. Exp. Stn. Leaflet,* 6, 1955.
63. **Wiley, R. C. and Bouwkamp, J.,** Sweet potato chips, *Food Eng. Inter.,* 4(5), 20, 1979.
64. **Taubenhaus, J. J.,** *The Culture and Diseases of the Sweet Potato,* E.P. Dutton, New York, 1923, 179.
65. **Hoover, M. W. and Kushman, L. J.,** Influence of raw product storage treatments on the quality of sweet potato flakes, *Proc. Am. Soc. Hortic. Sci.,* 88, 501, 1966.
66. **Deobald, J. J., McLemore, T. A., McFarlane, V. H., Roby, M. T., Peryam, D. R., and Heiligman, F.,** Precooked dehydrated sweet potato flakes, *U.S. Dept. Agric. Bull., Agric. Res. Serv. Circ.,* 72—23, 1962.
67. **Hoover, M. W.,** An enzyme process for producing sweet potato flakes from starchy and uncured roots, *Food Technol. (Chicago),* 20, 84, 1966.
68. **Bertoniere, N. R., McLemore, T. A., Hasling, V. C., Catalano, E. A., and Deobald, J. J.,** Effect of environmental variables on the processing of sweet potatoes into flakes and on some properties of their isolates starches, *J. Food Sci.,* 31, 574, 1966.
69. **Hoover, M. W.,** An enzyme activation process for producing sweet potato flakes, *Food Technol. (Chicago),* 21, 322, 1967.
70. **Hoover, M. W. and Harmon, S. J.,** Carbohydrate changes in sweet potato flakes made by enzyme activation technique, *Food Technol. (Chicago),* 21, 1529, 1967.
71. **Walter, W. M., Jr., Purcess, A. E., and Hoover, M. W.,** Changes in amyloid carbohydrates during preparation of sweet potato flakes, *J. Food Sci.,* 41, 1374, 1976.
72. **Deobald, H. J., McLemore, T. A., Hasling, V. C., and Catalano, E. A.,** Control of sweet potato alpha-amylase for producing optimum quality precooked dehydrated flakes, *Food Technol. (Chicago),* 22, 627, 1968.
73. **Deobald, J. H., Hasling, V. C., Catalano, E. A., and McLemore, T. A.,** Relationship of sugar formation and sweet potato alpha-amylase activity during processing for flake production, *Food Technol. (Chicago),* 23, 826, 1969.
74. **Ikemiya, M. and Deobald, H. J.,** New characteristic alpha-amylase in sweet potatoes, *J. Agric. Food Chem.,* 14, 237, 1966.
75. **Hoover, M. W.,** Preservation of the natural color in processed sweet potato products. I. Flakes, *Food Technol. (Chicago),* 17, 636, 1963.
76. **Waldsworth, J. I., Ziegler, G. M., Jr., Gallo, A. S., and Spadaro, J. J.,** Factors affecting film thickness and uniformity on double-drum dryer during dehydration of sweet potato flakes, *Food Technol. (Chicago),* 21, 668, 1967.
77. **Gross, M. O. and Rao, V. N. M.,** Flow characteristics of sweet potato puree as indicators of dehydrated flake quality, *J. Food Sci.,* 42, 924, 1977.
78. **Rao, V. N. M. and Graham, L. R.,** Rheological, chemical and textural characteristics of sweet potato flakes, *Trans. Am. Soc. Agric. Engin.,* 25, 1792, 1982.
79. **Ice, J. R., Hamann, D. D., and Purcell, A. E.,** Effects of pH, enzymes, and storage time on the rheology of sweet potato puree, *J. Food Sci.,* 45, 1614, 1980.
80. **Creamer, G., Young, C. T., and Hamann, D. D.,** Changes in amino acid content of acidified sweet potato puree, *J. Food Sci.,* 48, 382, 1982.

81. **Walter, W. M., Jr., Purcell, A. E., Hoover, M. W., and White, A. G.**, Preparation and storage of sweet potato flakes fortified with plant protein concentrates and isolates, *J. Food Sci.*, 43, 407, 1978.
82. **Purcell, A. E. and Walter, W. M., Jr.**, Autoxidation of carotenes in dehydrated sweet potato flakes using ^{14}C-B-carotene, *J. Agric. Food Chem.*, 16, 650, 1968.
83. **Walter, W. M., Jr., Purcell, A. E., and Hansen, A. P.**, Autoxidation of dehydrated sweet potato flakes. The effect of solvent extraction on flake stability, *J. Agric. Food Chem.*, 20, 1060, 1972.
84. **Walter, W. M., Jr. and Purcell, A. E.**, Lipid autoxidation in precooked dehydrated sweet potato flakes stored in air, *J. Agric. Food Chem.*, 22, 298, 1974.
85. **Walter, W. M., Jr., Purcell, A. E., Hoover, M. W., and White, A. G.**, Lipid autoxidation and amino acid changes in protein-enriched sweet potato flakes, *J. Food Sci.*, 43, 1242, 1978.
86. **Lopez, A., Wood, C. B., and Boyd, E. N.**, Changes in some monocarbonly classes during processing and storage of sweet potato flakes, *J. Food Sci.*, 41, 524, 1976.
87. **Alexandridis, N. and Lopez, A.**, Lipid changes during processing and storage of sweet potato flakes, *J. Food Sci.*, 44, 1186, 1979.
88. **Deobald, H. J. and McLemore, T. A.**, The effect of temperature, antioxidant and oxygen on the stability of precooked dehydrated sweet potato flakes, *Food Technol. (Chicago)*, 18, 739, 1964.
89. **Deobald, H. J., McLemore, T. A., Bertoniere, N. R., and Martinez, J. A.**, The effect of antioxidants and synergists on the stability of precooked dehydrated sweet potato flakes, *Food Technol. (Chicago)*, 18, 1970, 1964.
90. **Haralampu, S. G. and Karel, M.**, Kinetic models for moisture dependence of ascorbic and B-carotene degradation in dehydrated sweet potato, *J. Food Sci.*, 48, 1872, 1983.

Chapter 10

FORMULATED SWEET POTATO PRODUCTS

Stanley J. Kays

TABLE OF CONTENTS

I.	Introduction	206
II.	Types of Processed Products	206
	A. Sweet Potato Puree	206
	1. Preparation	206
	2. Measurement of Puree Quality	208
	B. Dehydrated Sweet Potatoes	208
	1. Preparation	208
	2. Causes of Quality Loss	209
	C. Chips	210
	1. Preparation	210
	2. Production Problems	211
	D. Patties and Formed Potatoes	211
	E. Baked Products	211
	F. Specialty Items	212
III.	Problems with Development of New and Improved Processed Products	212
	A. Flavor	213
	B. Color	214
	C. Texture	214
IV.	Conclusions	215
References		215

I. INTRODUCTION

The world population is expanding rapidly at a rate of approximately 70 to 80 million people per year.[1] At this rate, the number of inhabitants of the earth would double within approximately 38 years. Because of the resulting increase in demand for food, there is a need for high yielding, nutritious staple crops that have a wide range of production adaptability. The sweet potato would appear to be an obvious choice, however, world production figures do not support this position. The production of sweet potatoes has not increased over the past 15 years.[2]

In the U.S., per capita consumption of the product has progressively declined each year. In contrast, the white potato has experienced a tremendous increase in production, a large portion of which in the U.S. has been due to increased use of processed products [e.g., frozen products and chips (crisps)]. It would appear, therefore, that development of appealing processed products from sweet potatoes represents perhaps one of the most important keys to expanded utilization of the crop.

This chapter will focus upon existing processed food products from the sweet potato, the current state of the art, and inherent problems associated with developing wider utilization of sweet potato products as a food or food ingredient.

II. TYPES OF PROCESSED PRODUCTS

A. Sweet Potato Puree

Sweet potatoes can be pureed to produce a product that may be used for baby foods, pie fillings, reconstituted potatoes and with the addition of starch, frozen sweet potato patties. A puree is also utilized in the production of sweet potato flakes and numerous other products.

One of the major advantages of the use of a puree for processed products is that there are no size or shape criteria for the roots. A high quality puree can be produced from virtually any size or shape of storage roots, thus the entire crop can be utilized. Since an aseptically processed puree can be stored at ambient temperatures in bulk containers, there is a minimal storage energy requirement after initial processing. Other advantages include a year-round supply and a substantially reduced product volume for storage.

1. Preparation

Several techniques have been developed for the production of sweet potato purees over the past 20 years. Initially, the process involved simply cooking the peeled roots then pureeing.[3] However, due to differences among cultivars, curing, handling, storage, and other parameters, the properties of the puree obtained were often highly variable. This in turn affected the product produced from it. A second technique minimized this problem through the addition of limited amounts of α- and β-amylase after cooking and pureeing, and gave a consistent level of starch conversion. This, however, constituted the introduction of a food additive. The problem was circumvented using an enzyme activation process that utilized native amylolytic enzymes.[4] This process is now widely used by the food processing industry. The following is the general sequence of steps for puree preparation using this process.

a. Washing

The storage roots are soaked briefly in cold water and then washed using a revolving drum washer. Pressurized water (6.3 kg cm^{-2}) is sprayed on the roots to remove any remaining soil or debris.

b. Peeling

The washed roots can be peeled by hand paring, abrasive rollers, high pressure steam or hot lye solutions. A 5- to 6-min exposure to 20 to 22% lye solution at 104°C is commonly used for cured roots while a 10% solution (104°C) for 3 to 6 min is sufficient for freshly harvested roots. Exposure to high pressure steam (6.3 kg cm^{-2}) for approximately 90 sec will also do an excellent job;[5] however, superior results have been obtained using flash in-chamber cooling with cold water.[6] This decreases peeling losses and gives a product with better color and appearance.

c. Rewashing

After peeling, the roots should be rewashed. This is especially important after lye and steam peeling. The roots are moved through a revolving drum washer with the residual lye and/or skin removed with high pressure water.

d. Trimming and Sorting

The sweet potatoes are then conveyed along a sorting line where diseased or otherwise unacceptable roots are removed. Surface blemishes are cut away as are the fibrous ends of the storage roots.

e. Pureeing

The roots are pureed using a pulper in which the blades force the puree through a 0.8 mm screen. The puree is then placed in a surge tank.

f. Enzyme Activation

The natural amylolytic enzymes within the roots are activated and the starch gelatinized by rapidly heating the puree to 74 to 85°C using steam injection.[4]

g. Partial Starch Hydrolysis

Starch, the major carbohydrate component of the storage roots (16 to 25% on a fresh weight basis), is allowed to be partially degraded by the activated amylolytic enzymes. β-Amylase, the prevalent form in the sweet potato,[7] hydrolyzes starch and starch fragments to maltose, α-amylase produces short chain-length polymers called dextrins. Maltose appears to be the only sugar produced during conversion,[8] and its production is essentially completed (≤90%) during the first 10 min of conversion.[9] Additional changes in the macromolecular components (starch and dextrins), however, occur for up to 60 min and have been shown to strongly affect the texture of the finished product.

Since the concentration of soluble solids in the puree varies widely due to cultivar[4,10] and length of storage of the roots,[11,12] adjusting the length of time the puree is hydrolyzed can be used to produce a more consistent product. As a consequence, conversion time is extremely important with regard to the quality. Typical conversion times range from 2 to 60 min.

h. Enzyme Inactivation

The partially hydrolyzed puree is then pumped through a flash heat exchanger where the product temperature is rapidly raised to 88 to 100°C. This inactivates the enzymes and completes the cooking of the puree. Harris[13] has described a specially designed heat exchanger which allows very rapid uniform heating with precise temperature control.

i. Packaging

The puree is then either placed in sterilized containers (e.g., cans or jars), sealed and sterilized, or in suitable containers for frozen storage. The high temperatures involved in retorting results in significantly poor quality product; as a consequence, freezing has become increasingly popular. With aseptic processing, bulk storage of the puree could be accomplished with reusable containers such as bulk tanks or glass lined silos.

2. Measurement of Puree Quality

In that there are a number of products that utilize the puree for their production, some variation in the physical and chemical properties of the puree is required. As a consequence, it is essential to monitor specific chemical and/or physical changes occurring in the puree during starch hydrolysis. Knowing the relationship between these parameters and the quality of the final product allows precise formulation of the puree by deactivation of the amylolytic enzymes at the appropriate time. Two commonly used measurements are the concentration of soluble solids (Brix) in the puree and analysis of viscometric properties of the puree. Measurements of puree flow index, coefficient of shear rate, and apparent viscosity have been shown to correlate with taste panel analysis of finished products.[14,15]

Several studies have focused on the storage stability of sweet potato puree. Techniques tested to increase the microbiological stability, such as radical shifts in the pH of the product, also change the viscosity of the puree.[16] Generally the puree, if sterilized and aseptically packaged, is relatively stable at ambient temperatures for at least 9 months. There appears to be little change in the component amino acids and protein during normal storage.[17] Smith et al.[18] found that the best quality was produced with flash heating at 123°C. The length of time the quality of the puree can be maintained, however, is substantially extended when stored as a frozen product.

B. Dehydrated Sweet Potatoes

Dehydration of sweet potatoes for human consumption was reported during the late nineteenth century; however, an acceptable product was not produced until it was found that the roots must first be cooked. Initially, the cooked roots were cut into small slices or cubes and dried. As drying technology progressed, sweet potatoes began to be pureed and then dried on revolving drum driers producing a thin flake. This gave a much more consistent product that could be used for mashed potatoes, pies, and other products.

The advantages of producing a dehydrated flake include: (1) a substantial decrease in the weight of the product, (2) elimination of the need for refrigeration during storage, and (3) a product that could be quickly reconstituted and prepared. Interest in dehydrated sweet potatoes increased dramatically during World War II. In 1943 and 1944, approximately 23 million kilograms of dehydrated sweet potatoes were produced per year in the U.S. After the end of the war, however, interest subsided.

1. Preparation

The initial steps for the preparation of the puree are as previously outlined (Section II.A.1). The puree is then fed into a double drum dryer.

a. Drying

The puree is dried to 2 to 3% moisture using heated revolving drums. Puree is picked up by the drums which are 2.5 to 3.3 mm apart and heated with steam (5.3 kg cm^{-2}) (Figure 1). The film of puree adhering to the revolving drum is composed of a burn-on layer and a layer formed from the shear forces of the drum and the hydrostatic force of the puree on the dryer.[19] Drum speed is to a large extent determined by the puree viscosity and diameter of the drums. Film thickness of the puree on the drum is affected by the space between the drums, drum velocity, puree viscosity, and level of puree in the dryer.[19] The rate of drying is affected by the thickness of the film on the drum, drum speed and temperature, and the amount of soluble solids in the puree.

b. Flaking

The thin sheet of dried sweet potato puree must be reduced to flakes of 1.6 to 6.4 mm diameters. A suitable grinder such as that used for coconut can be utilized.[20]

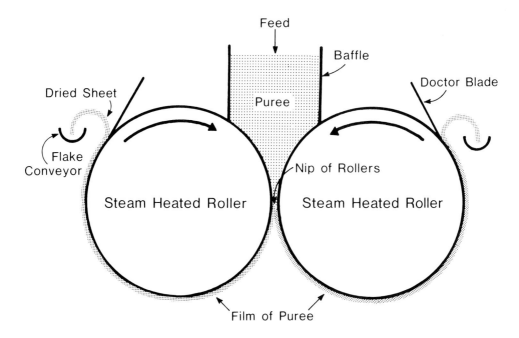

FIGURE 1. A schematic diagram of a double drum drier for producing sweet potato flakes.

c. Packaging

The flakes, when sufficiently cool, should be packaged immediately to prevent the uptake of moisture and oxidation of components lipids. Metal, glass, or film containers that have very low water vapor and oxygen transfer rates should be used. The package should be purged with an atmosphere containing less than 2% oxygen at closing.[10] This has a tremendous effect on the length of time the product will remain in good condition.[21]

2. Causes of Quality Loss

Precooked dehydrated sweet potato flakes undergo rapid deterioration if stored at an ambient oxygen concentration. Typically, quality losses include a relatively rapid loss of part of the carotene present and the development of off-flavors. Storage in a nitrogen or low oxygen ($\leq 2\%$) environment appears, at present, to be the best means of long-term preservation in that antioxidants have not been effective.[22]

The oxidation of carotene in sweet potato flakes is seen as a loss of color. When stored in air, 20 to 40% of the carotene is destroyed within approximately the first 30 days.[22] The predominant form of carotene in the sweet potato storage root is β-carotene, although other carotenes and xanthophylls are present. Studies with pure β-carotene indicate that autoxidation initially involves the formation of epoxides and 3- or 3′-hydroxycarotenes with subsequent oxidation of the epoxides to carbonyls. Work by Purcell and Walter,[23] however, suggests that the autoxidation of carotenoids in cooked food products may be more complex. For example, xanthophylls, such as leutin, which are much more resistant to autoxidation in vitro than carotenes, are only slightly more so in sweet potato flakes.

The carotene present in sweet potato flakes appears to be found in both a water-dispersible and non-water-dispersible fraction. The water dispersible fraction, formed when the sweet potatoes are cooked, consists of starch and lipid-carotene components. A major portion of the water dispersible fraction is made up of starch or dextrins of fairly high molecular weight which in some way protects the carotene from oxidation. Extraction of surface carotene from the flakes with organic solvents (e.g., chloroform-methanol, 2:1) has been shown to decrease the rate of autoxidation and color loss. These surface pigments appear to be more susceptible to attack by oxygen.[24]

Concurrent with loss of carotene is the autoxidation of lipids and the development of unpleasant off-flavors. Breakdown products of lipids are known to cause off-flavors in many products. Although the lipid content of sweet potato flakes is low (i.e., 1.5 to 2.5% on a dry weight basis) their component fatty acids are highly unsaturated and, as a consequence, very susceptible to oxidation. The unsaturated fatty acids linoleic (18:2) and linolenic (18:3) represent approximately 55% of the total of the fatty acids present.[25] The saturated fatty acids palmitic (16:0) and stearic (18:0) are also broken down, however, the rate is substantially slower than linoleic and linolenic.[26] These four fatty acids together make up about 95% of the total fatty acids in the cultivar Centennial.

The lipids from the sweet potato flakes can be separated into two general classes: surface and bound. Surface fatty acids are oxidized at a much more rapid rate than are the bound lipids.[26] This appears to occur via a dual reaction mechanism. During processing, most of the lipids present are trapped in a dense carbohydrate matrix that impedes their rate of autoxidation. The remaining lipids, however, are spread over the surface of the flake and are much more susceptible to oxidation. As with surface carotenes, these lipids can be removed with organic solvents, which increases the stability of the flakes.[24]

During storage of the flakes there is an increase in the total monocarbonyls and saturated aldehydes present and a decrease in the concentration of methyl ketones.[27] The aldehydes, in particular, may be important in the development of the characteristic off-flavor of rancid sweet potato flakes.

C. Chips

The sweet potato can be processed into chips (crisps) that, while differing distinctly in taste from those derived from *Solanum tuberosum,* are enjoyed by many people. The final product can be grouped into three general types: sweet, salted, and spicy hot chips. Sugar-coated chips are popular to some extent in the Republic of China where they are commercially produced, although on a limited scale. Salted chips have been tested on several occasions during the past century in the United States and elsewhere, but have not proved to be commercially viable. The acceptability of the natural sweetness of the cooked chip after salting appears to vary widely among individuals. Some prefer sweet-salty chips and, as a consequence, processing techniques have been tested to enhance the natural sweetness of the chip.[28] More commonly, the sweetness was found to be objectionable and techniques to minimize the development of maltose have been explored. In addition, spicy hot chips produced using cayenne pepper and citric acid have been tested in Bangladesh with relatively favorable results.[29]

1. Preparation

Sweet potato chips are produced by either hand, steam, or lye peeling the cured roots and then slicing them into thin chips. The thickness of the chip is important since it affects the length of cooking and the quality of the finished product. Generally, slices with an initial thickness of 0.8 mm are preferred.[30] Because of the relatively high concentration of reducing sugars in the roots, discoloration during cooking is often a problem. The addition of 0.5 to 0.75% sodium acid pyrophosphate to the blanch water helps to minimize this.[31] The chips are blanched for 2 min at 93°C then drained and partially dehydrated using heated forced air (119°C). Partial drying has a pronounced effect on the quality (appearance, flavor, and texture) of the finished product.[31] The higher the moisture content, the greater the quantity of fat retained after cooking. The length of frying is influenced by the moisture content of the chips and the temperature of the cooking oil. Chips that were approximately 50% moisture required 4.5 min at 138°C.[31] Optimum cooking temperature was between 143 and 154°C.

Following cooking, the chips are drained and salted with 2.9 to 3.7 kg/100 kg of chips (or sugared). After cooling, the chips should be packaged immediately to exclude water and oxygen. The yield of chips is typically around 40% of the weight of the peeled roots prior to cooking.[32]

2. Production Problems

Several problems are associated with the production of high quality sweet potato chips. These include discoloration, hardness, leatheriness, and variation in quality due to cultivar.[33] Discoloration of chips during processing is perhaps one of the most critical problems. This appears to be largely due to the relatively high concentration of reducing sugars in the chips. For example, the concentration of maltose increases from a trace amount to over 2% during cooking.[34] The concentration of sugars in the chip can be altered significantly by both preprocessing and processing conditions.[31,35] A diffuser-extractor can be used to lower the sugar content of the slices.[36] Discoloration is also influenced by the temperature of the cooking oil and the length of cooking. Both excessively high temperatures and prolonged exposures decrease color quality.

Under certain conditions the chips produced are extremely hard. This appears to be minimized by blanching[31] or preheating[28] the chips or roots prior to cooking and by dehydration prior to cooking.[31] This helps to decrease blister formation and oil uptake. Proper thickness of the chip is also important, with thicker chips being more prone to case hardening.

Leatheriness or loss of crispness of the finished product after holding is caused by the uptake of moisture from the atmosphere. Partial dehydration prior to cooking helps to decrease this,[31] however, storage in moisture-proof packages is essential.

D. Patties and Formed Potatoes

Considerable interest has been displayed in the use of sweet potato puree to produce patties and formed potatoes for the frozen convenience food market. Since the products are precooked, preparation time prior to serving is minimal.

The production of sweet potato patties requires the incorporation of sufficient starch into the puree to allow molding into a thin (1.5 to 2.0 cm) patty. The outer surface of the patty can be coated with various batters or seasoned to add variety to the product. Frozen sweet potato patties are currently produced and marketed by several companies in the United States.

Formed sweet potatoes are produced by placing the puree in aluminum boats resembling the general shape of a potato. This also requires the addition of starch to increase the firmness of the product. Crumpton[37] tested several concentrations of starch and found that firmness increased substantially at the higher levels. Starch-free purees were extremely soft. After filling, the containers were placed in an air-blast freezer and quickly frozen.

A wide variety of ingredients have been added to formed sweet potato products to diversify the flavor of the product.[38,39] These include raisins, pineapple, diced ham, pineapple and diced ham, orange flavor with coconut, and imitation butter. All of the products were judged by taste panels to be acceptable. Frozen samples were also stored at $-25°C$ for 45, 90, or 135 days and evaluated after warming (177°C). No decline in quality was found with storage.

E. Baked Products

Raised white breads made from wheat flour have become increasingly popular in many tropical and subtropical countries. For the most part, wheat is not grown in these areas and must be imported, unfavorably altering import-export balance of the importing countries. As a consequence, considerable interest has been displayed in the possibility of using locally grown sweet potatoes as a substitute for a portion of the wheat flour needed.

The possible use of sweet potato flour in bread making has been known for some time,[40] however, recently interest has expanded considerably.[41-46] Most research has utilized sweet potato flour, although sweet potato puree also represents a viable possibility.[47]

The substitution of sweet potato flour for 10 to 20% of the wheat flour generally alters to some extent the properties of the baked product. For example, dough strength, extensibility, and resistance are decreased as is the specific volume of many of the baked products[42,43,46] With pan-de-sal, a Philippine bread, decreases in volume could be off-set with changes in the formulation of the bread.[46] Increases in water and shortening compensated for the decreased volume. For most breads, addition of 10 to 20% sweet potato flour gave a readily acceptable product.

Up to 30% sweet potato flour could be added to cake mixes without affecting the volume of the cake. The mixes remained acceptable during storage in simulated shelf-life experiments for up to 165 days.[48,49]

Incorporation of sweet potato flour or puree (up to 21%) into raised doughnuts altered a number of the physical and chemical properties of the doughnuts.[47] For example, there were significant changes in the internal color of the doughnuts in that an orange-fleshed cultivar was utilized. Texture and specific volume were also affected, however, for the most part, overall quality of the doughnuts was not reduced.

F. Specialty Products

A number of specialty products have been studied using sweet potatoes as a major ingredient. These include: baby foods, candies, sauces, breakfast foods, snacks, crackers, reconstituted chips, and other products.

The use of sweet potato puree as an infant food represents the most successful processed sweet potato product. These foods are typically available in 133 or 222 cc glass jars as either a plain puree or with a butter sauce added. Two consistencies are produced, a finely strained product and a slightly coarser puree. Dried sweet potato flakes may also be used as an infant food.[50]

Candies, icings, and other confectionary products may also be prepared from sweet potatoes. These are made from both cooked and pureed sweet potatoes and from processed (toasted) sweet potato flour. Considerable research was directed toward developing sweet potato candies during the 1940s and early 1950s by the Alabama Agricultural Experiment Station.[51-53] Brittle type candies (e.g., coconut brittle) were found to have good consumer acceptance[52,54] although they were never commercially developed. Several candy products, however, are produced and marketed in Taiwan on a commercial scale.

The use of sweet potato flour as a partial substitute for wheat flour in soy sauce gives a product with generally acceptable quality attributes except for taste.[55,56] Sweet potato flour has also been tested as a component of a dry powdered mix for lumpia sauce.[57]

Sweet potato pies[58] and turnovers[59] produced from frozen or canned puree represent an attractive commercial possibility. A frozen pie mix containing milk, egg white, margarine, powdered milk, salt, and sugar added to the puree has been test marketed under the name Presto-Pi.[60] Turnovers consist of sweet potato puree completely enclosed within a pie crust. Marketed frozen, the turnover is cooked by deep fat frying at 193°C for 2.5 min. Several variations in recipes (e.g., with raisins or crushed pineapple) have been tested.[59]

Other processed sweet potato products include breakfast foods,[61] crackers,[62] snacks,[54] a potato chip-like product called "leather",[63,85] ice cream,[51,64] noodles,[65] and milkshakes.[51]

III. PROBLEMS WITH DEVELOPMENT OF NEW AND IMPROVED PROCESSED PRODUCTS

In the introduction in this chapter, it was suggested that expanded utilization of the sweet potato as a human food will be enhanced by the development of processed products (other than canned sweet potatoes). Producing a new or improved food product that achieves widespread popularity and commercial success is a difficult task with any staple. With the sweet potato this is especially true. Of the products reviewed above, only baby food can be considered a commercial success. Several products have displayed very limited acceptance in specific geographical regions of the world. However, for the most part, processed sweet potato products have not succeeded.

The success of a new product, assuming it is economically viable and adequately promoted, lies to a large extent in it's flavor (taste and aroma), visual appeal (color and shape), and texture (mouthfeel). We presented the hypothesis in 1980 at the Sweet Potato Collaborators Meeting that the primary reason the sweet potato does not enjoy wider popularity as a food is that the flavor of the cooked product is extremely dominant and, as a consequence, can not be sufficiently altered.

Most staple crops (e.g., rice, wheat, cassava, white potato, etc.) act to a large degree as flavor carriers (products with low flavor impact) to which flavor is added during or after preparation. This greatly increases the flexibility of the product for preparation. Thus, a wide variety of distinctly different flavored dishes can be prepared. In addition, this flexibility increases the products potential for introduction into new geographical areas, since distinctive "flavor principles" of the culture can be successfully added to increase acceptability.

Because of this, we proposed that at least one alternate flavor type of sweet potato for human consumption be developed. A sweet potato with a much lower flavor impact (taste and aroma) would greatly expand the number of potential direct and processed uses of the crop in the various cuisines of the world. This does not imply abandoning our existing sweet potato, rather complementing it with a second flavor type.

Interest in selection for a lower flavor impact sweet potato has increased substantially. The United States Department of Agriculture currently plans the release of a very low sugar white-fleshed selection.[66] Nonsweet, low flavor impact selections have also been found by other breeders.[67] In some cases, the selection for low flavor impact sweet potatoes represents a formal part of the breeding program.[68] These new selections should greatly increase our potential to develop new and improved processed sweet potato products.

To explore the potential and problems of new processed sweet potato products, the following section will review our current understanding of sweet potato flavor, color, and texture.

A. Flavor

The flavor of a processed sweet potato product is to a large extent a function of the base flavor of the sweet potato. This flavor is comprised of both taste (perceived by the mouth) and aroma (perceived by the olfactory epithelia). The principal taste sensation of most existing cultivars is sweetness, which is due largely to maltose[28] and sucrose. Smaller concentrations of glucose and fructose are also present.[69] The increase in sweetness upon cooking is due to the disaccharide, maltose, which is formed by the hydrolysis of starch. The quantitative and qualitative range in sugars for a wide range of sweet potato germplasm has not been assessed. A very limited sample, however, representing Asian, Central American, and North American centers of selection did not display qualitative differences in component sugars;[69] quantitative differences were substantial, however.

A better understanding of the operative mechanism(s) giving low starch conversion during cooking of the nonsweet selections is needed. This would allow the development of new cultivars with maximum phenotypic stability for low sugar.

Compounds other than sugars that are important in the taste of the sweet potato have not been identified. The presence of off-flavors in certain selections, however, is well known although little documented. The chemical identity of these compounds and their relative importance are presently unknown.

Volatile compounds which make up the aroma of food products are often extremely important components of the overall flavor perceived. With some products (e.g., boiled white potatoes, apples), the characteristic aroma is due predominately to only one compound. These are commonly referred to as "character impact compounds". The presence or absence of a character impact compound in the sweet potato has not yet been ascertained.

Cooked sweet potatoes give off a large number of volatile compounds.[69-71] For example, pentene-2, acetone, 2,3-butanedione, hexane, pentane-2-one, methylbenzene, 2-pentanol, methylacetate, dimethylbenzene, pyridine, 2-phenyl-2-methylbutane, furfural, 2-acetylfuran, benzaldehyde, 2-propenylfuran, 2-methyl-6-ethylpyridine, α-terpineol, heptylbenzene (isomer), hexadecanol, heptadecanol, 1-isopropyl-4-isopropenyl benzene, tetradecanol, a monoterpene, and 4-sesquiterpenes have been isolated from the cultivar Morado.[69]

The volatiles and flavors of cultivars from different centers of selection are known to differ widely both quantitatively and qualitatively.[69] Most of the volatile compounds probably do not represent critical components of the characteristic aroma of the sweet potato. Identification of the critical compounds is, however, an essential step in taking a more analytical and pragmatic approach to breeding for specific flavor types.[72]

B. Color

The sweet potato exhibits a wide range in flesh color of the storage roots. Typical cultivars exhibit white, yellow and orange pigmentation[69] although red and purple types have been found.[73] While all of the pigments existing in the sweet potato genepool have not been identified, the most prevalent pigment in the yellow and orange types are carotenoids, predominately β-carotene.[74-77] Other carotenes and xanthophylls (e.g., violaxanthin) may also be present, to a much lower extent.[76] White or buff coloration is the dominant trait. Interest in increasing the concentration of β-carotene, the precursor of vitamin A, has resulted in a high selection pressure being placed on this character in some production areas. Skin color also varies widely but is of little importance with processed products.

Discoloration can be a serious problem during processing of sweet potatoes and storage of their products. Significant color changes are not limited to the highly pigmented types. The peeled surface of the roots can turn dark brown due to oxidation. High temperatures encountered during lye peeling also results in an oxidative discoloration. The degree of discoloration is correlated with the activity of catechol oxidase in the roots.[78] Discoloration has also been related to the concentration of phenolics[79] in the roots, the highest concentration of which and polyphenol oxidase are associated with laticifer cells.[80] During preparation, some lines develop a green or gray coloration which is particularly noticeable in white-fleshed roots. Exposure of the cut surface of the roots to iron gives an intense black discoloration.[64] In addition, cooking at an excessively high temperature results in caramelization. This results in browning of the product, the degree of which is dependent upon the absolute temperature and length of exposure.

Loss of color in processed products such as sweet potato flakes may also be a serious problem. This is due to the oxidation and decomposition of the carotenoids present (see Section II.B.2). Storage under appropriate conditions, however, can greatly reduce this problem.

The sweet potato genepool contains a wide diversity of color types which can be selected relatively easily. As a consequence, incorporation of specific pigments needed to enhance the attractiveness of a product would not appear to be a formidable obstacle in developing new and improved processed products.

C. Texture

Cooked sweet potatoes can be characterized on the basis of sensory "mouthfeel" into two primary classes. Dry fleshed roots are firm, dry, and mealy while moist fleshed roots are soft, moist, and somewhat slick.[81,82] Moist root types are found primarily in the U.S. Drier fleshed types are found throughout much of the world's sweet potato production zone and represent the predominate form. The differences between the two types are independent of the actual moisture content of the root.[83]

While mouthfeel can be determined using sensory panels, a more rapid and consistent means of measurement is desirable. Mouthfeel quality can be characterized by uniaxial compression, shear[81] and by determining flow behavior of purees.[15] Measurements of puree texture can be used to accurately predict the textural properties of processed products, e.g., sweet potato flakes.[84] In addition, textural characteristics of a processed sweet potato product can be altered by a number of preprocessing and processing factors. For example, the length of hydrolysis of a puree has a substantial affect on the finished product. It is evident, therefore, that the sweet potato genepool includes a range in textural properties and these properties can be further altered during processing to achieve desired results.

IV. CONCLUSIONS

Development of new and improved processed products from the sweet potato appears to represent an excellent means of increasing the utilization of this high yielding, nutritious species. The existing gene pool contains a wide range of flavor, color, and textural types which, coupled with our potential to modify the product, offers exciting new possibilities for processed products. Rapid advances in this area would be facilitated by close integration of breeding, biochemical, and processing programs.

REFERENCES

1. Population Research Bureau, World Population Data Sheet, Population Research Bureau, Washington, D.C., 1976.
2. FAO, *FAO Production Yearbook*, Food and Agriculture Organization of the United Nations, Rome, 35, 296 pp, 1981.
3. **Deobald, H. J. and McLemore, T. A.**, A Process for Preparing a Precooked Dehydrated Sweet Potato Product, U.S. Patent 3046145, 1962.
4. **Hoover, M. W.**, An enzyme-activation process for producing sweet potato flakes, *Food Technol. (Chicago)*, 21, 4, 1967.
5. **Hoover, M. W.**, Preservation of the natural color in processed sweet potato products. I, *Flakes Food Technol.*, 17, 128, 1963.
6. **Smith, D. A., Harris, H., and Rymal, K. S.**, Effect of cold water injection during high pressure steam peeling of sweet potatoes, *J. Food Sci.*, 45, 750, 1980.
7. **Balls, A. K., Walden, M. K., and Thompson, R. R.**, A crystalline β-amylase from sweet potato, *J. Biol. Chem.*, 173, 9, 1948.
8. **Hoover, M. W. and Harmon, S. J.**, Carbohydrate changes in sweet potato flakes made by the enzyme activation techniques, *Food Technol. (Chicago)*, 21, 115, 1967.
9. **Walter, W. W., Jr. and Purcell, A. E.**, Changes in amyloid carbohydrates during preparation of sweet potato flakes, *J. Food Sci.*, 41, 1374, 1976.
10. **Spadaro, J. J., Wadsworth, J. I., Ziegler, G. M., Gallo, A. S., and Koltun, S. P.**, Instant sweet potato flakes — processing modifications necessitated by varietal differences, *Food Technol. (Chicago)*, 21, 8, 1967.
11. **Bertoniere, N. R., McLemore, T. A., Hasling, V. C., Catalano, E. A., and Deobald, H. J.**, The effect of environmental variables on the processing of sweet potato into flakes and on some properties of their starches, *Food Technol. (Chicago)*, 20, 574, 1966.
12. **Deobald, H. J., McLemore, T. A., Hasling, V. C., and Catalano, E. A.**, Control of sweet potato α-amylase for the production of optimum quality precooked dehydrated flakes, *Food Technol. (Chicago)*, 22, 627, 1968.
13. **Harris, H.**, Process and Apparatus for Continuous High Temperature Processing of Food Products, U.S. Patent 4302111, 1981.

14. **Gross, M. O. and Rao, V. N. M.,** Flow chacteristics of sweet potato puree as indicators of dehydrated flake quality, *J. Food Sci.,* 42, 924, 1977.
15. **Rao, V. N. M., Hamann, D. D., and Humphries, E. G.,** Flow behavior of sweet potato puree and its relation to mouth feel quality, *J. Texture Studies,* 3, 319, 1975.
16. **Ice, J. R.,** Effects of pH, Enzymes, and Storage Time on Carbohydrate Changes and Rheology of Sweet Potato Puree, Ph.D. Thesis, North Carolina State University, Raleigh, 1979.
17. **Creamer, G., Young, C. T., and Hamann, D. D.,** Changes in amino acid content of acidified sweet potato puree, *J. Food Sci.,* 48, 382, 1983.
18. **Smith, D. A., McCaskey, T. A., Harris, H., and Rymal, K. S.,** Improved aseptically filled sweet potato purees, *J. Food Sci.,* 47, 1130, 1982.
19. **Wadsworth, J. I., Kolter, S. P., Gallo, A. S., Ziegler, G. M., and Spadaro, J. J.,** Instant sweet potato flakes: factors affecting drying rate on double drum drier, *Food Technol. (Chicago),* 20, 815, 1966.
20. **Spadaro, J. J. and Patton, E. L.,** Precooked dehydrated sweet potato flakes, *Food Eng.,* 33, 46, 1961.
21. **Molaison, L. T., Spadaro, J. J., Roby, M. T., and Lee, F. H.,** Dehydrated diced sweet potatoes — a pilot-plant process and product evaluation, *Food Technol. (Chicago),* 16, 101, 1962.
22. **Deobald, H. J. and McLemore, T. A.,** The effect of temperature, antioxidant and oxygen on the stability of precooked dehydrated sweet potato flakes, *Food Technol. (Chicago),* 18, 739, 1964.
23. **Purcell, A. E. and Walter, W. M., Jr.,** Autoxidation of carotenes in dehydrated sweet potato flakes using ^{14}C-β-carotene, *J. Agric. Food Chem.,* 16, 650, 1968.
24. **Walter, M. W., Jr., Purcell, A. E., and Hansen, A. P.,** Autoxidation of dehydrated sweet potato flakes. The effect of solvent extraction on flake stability, *J. Agric. Food Chem.,* 20, 1060, 1972.
25. **Walter, M. W., Jr. and Purcell, A. E.,** Characterization of a stable water-dispensible carotene fraction from sweet potato flakes, *J. Agric. Food Chem.,* 19, 175, 1971.
26. **Walter W. M., Jr. and Purcell, A. E.,** Lipid autoxidation in precooked dehydrated sweet potato flakes stored in air, *J. Agric. Food Chem.,* 22, 298, 1974.
27. **Lopez, A., Wood, C. B., and Boyd, E. N.,** Changes in some monocarbonyl classes during processing and storage of sweet potato flakes, *J. Food Sci.,* 41, 524, 1976.
28. **Sistrunk, W. A., Miller, J. C., and Jones, L. G.,** Carbohydrate changes during storage and cooking of sweet potatoes, *Food Technol. (Chicago),* 8, 223, 1954.
29. **Molla, M. R. I.,** Sweet potato chips, a possible product for urban consumers in Bangladesh, *Bangladesh Hortic.,* 1, 77, 1973.
30. **Kelly, E. G. and Baum, R. R.,** Preparation of tasty vegetable products by deep-fat frying, *Food Technol. (Chicago),* 9, 388, 1955.
31. **Hoover, M. W. and Mills, N. C.,** Process for producing sweetpotato chips, *Food Technol. (Chicago),* 27(5), 74, 1973.
32. **Kelly, E. G., Baum, R. R., and Woodward, C. F.,** Preparation of new and improved products from eastern (dry type) sweet potatoes: chips, dice, julienne strips and frozen french fries, *Food Technol. (Chicago),* 12, 510, 1958.
33. **Baba, T., Kouno, T., and Yamamura, E.,** Development of snack foods produced from sweet potatoes. I. Factors affecting hardness and discoloration of sweet potato chips, *J. Jpn. Soc. Food Sci. Technol.,* 28, 318, 1981.
34. **Baba, T. and Yamamura, E.,** Development of snack foods produced from sweet potatoes. II. Optimization of processing conditions in blanching and freezing for sweet potato chips, *J. Jpn. Soc. Food Sci. Technol.,* 18, 355, 1981.
35. **Olurunda, A. O. and Kitson, J. A.,** Controlling storage and processing conditions helps produce light colored chips from sweet potatoes, *Food Prod. Dev.,* 11, 44, 1977.
36. **Hannigar, K.,** Sweet potato chips, *Food Eng. Intern.,* 4(5), 28, 1979.
37. **Crompton, W. R.,** Thermal Diffusivity of Sweet Potatoes, Ph.D. Thesis, Mississippi State University, Starkeville, 1974.
38. **Collins, J. L., Sanders, G. G., Hill, A. R., and Swingle, H. D.,** Sweet potato products made from baked, cured roots, *Tenn. Farm Home Sci., Prog. Rep.,* 91, 30, 1974.
39. **Collins, J. L., Hill, A. R., Sanders, G. G., and Gaines, C. S.,** Color and taste evaluations of a frozen sweet potato product, *Food Prod. Dev.,* 10(3), 53, 1976.
40. **Gore, H. C.,** The value of sweet potato flour in breadmaking, *Ind. Eng. Chem., Ind. Ed.,* 15, 1238, 1923.
41. **De La Fuente, L. R.,** Experiments in breadmaking from flour mixtures of wheat and three varieties of sweet potatoes, *Bolivian Trim. Exp. Agropecmar,* 9, 10, 1960.
42. **Hamed, M. G. E., Refai, F. Y., Hussein, M. F., and El-Samahy, S. K.,** Effect of adding sweet potato flour to wheat flour on physical dough properties and baking, *Cereal Chem.,* 50, 140, 1973.
43. **Hamed, M. G. E., Refai, F. Y., Hussein, M. F., and El-Samahy, S. K.,** Effect of dehydrated sweet potato flour on the rheological behavior of wheat flour dough, *Egypt. J. Food Sci.,* 1, 215, 1973.

44. **Lehmann, G. and Münzberg, F.,** Lebensmittel-und Ernährungsprobleme in Enturcklungsländern. II. Uber die Mehl-und Brotherstellung, *Ernaährungs-Umschau,* 19, 316, 1972.
45. **Madamba, L. S. P., Barba, C. V. C., Burgos, M., and Domingo, T. B.,** Formulation, development and evaluation of sweet potato products, *NSDB Tech. J.,* 2, 60, 1977.
46. **Tapang, N. P. and Del Rosario, R. R.,** Composite flours. I.. The use of sweet potato, Irish potato and wheat flour mixtures in breadmaking, *Philipp. Agric.,* 61, 124, 1977.
47. **Collins, J. L. and Abdul Aziz, N. A.,** Sweet potato as an ingredient of yeast-raised doughnuts, *J. Food Sci.,* 47, 1133, 1982.
48. **El-Samahy, S. K., Morad, M. M., Seleha, H., and Abdel-Baki, M. M.,** Cake-mix supplementation with soybean, sweet potato or peanut flours. II. Effect on cake quality, *Baker's Dig.,* 54(5), 32, 1980.
49. **Morad, M. M., Abdel-Baki, M. M., Seleha, H., and El-Samahy, S. K.,** Cake-mix supplementation with soybean, sweet potato or peanut flours. III. Effect on storability and the role of packaging materials, *Baker's Digest,* 54, 34, 1980.
50. **Lee, S. R.,** Preparation of drum-dried weaning food based on sweet potato and soybean, *Korean J. Food Sci. Technol.,* 2, 1, 1970.
51. **Van de Mark, M. S.,** Alamalt — Its properties and uses, *Ala. Exp. Stn. Mimeo,* 23, 5, 1945.
52. **Van de Mark, M. S. and Ware, L. M.,** Candies from sweet potatoes, *Sweet Potato J.,* 3, 1947.
53. **Van de Mark, M. S., Ware, L. M., and Harris, H.,** Candies from sweet potatoes pack health promoting values, *Food Industries,* Sept., 1947.
54. **Lanham, B. T.,** Consumer reaction to Alayam candy, *Ala. Agric. Exp. Stn. Bull.,* 271, 1950.
55. **Lee, J. M., Hong, W. S., Kim, Y. S., Hong, Y. M., and Yu, J. H.,** Studies on the substitution of raw materials in soy sauce manufacture. II. Use of potatoes and sweet potatoes, *Korean J. Microbiol.,* 10, 79, 1972.
56. **Han, P. J., Choi, K. S., and Lee, S. J.,** Studies on the effect of various cooking conditions of soybean on the quality of prepared Meju, *Res. Rep. Off. Rural Dev. (Korea),* 12(6), 63, 1969.
57. **Briones, P. R., Banzon, E. A., Librea, M. C., Gonzalez, O. N., and Alabastro, V. Q.,** The effect of storage on the acceptability of instant lumpia sauce packed in tin containers, bottles, and poly film bags, *Philipp. J. Nutr.,* 21, 172, 1968.
58. **Turner, J. L. and Danner, M. J.,** Acceptance of an improved frozen sweet potato puree, *Ala. Agric. Exp. Stn. Circ.,* 121, 14, 1957.
59. **Jaynes, H. O. and Corley, D. T.,** Sweet potato turnovers, *Tenn. Farm Home Sci.,* 120, 11, 1981.
60. **Marshall, W. W. and Danner, M. J.,** Consumer reaction to Presto-Pi, *Ala. Agric. Exp. Stn. Circ.,* 134, 1959.
61. **Ware, L. M.,** Nuggets, curls, waves, shreds: how you'll have your potatoes for breakfast this morning?, *Sweet Potato J.,* November 1947.
62. **Popova, I. N. and Babichenko, L. V.,** Changes in starch grains during cracker manufacture, *Izvestiya Vysshikh Uchebnykh Zavedenii, Pishchevaya Tekhnologiya,* 5, 128, 1977.
63. **Yaakob, C. M. and Raya, S.,** A preliminary study on processing of sweet potato leather, *Pertanika,* 6, 17, 1983.
64. **Woodroof, J. G. and Atkinson, I. S.,** Preserving sweet potatoes by freezing, *Ga. Agric. Exp. Stn. Bull.,* 232, 26, 1944.
65. **Lü, C. Y., Chen, C. Y., and Wang, H. H.,** Studies on the processings and qualities of starch noodles from various starches, *Am. Chem. Soc. Abstr.,* 177, 87, 1979.
66. **Jones, A.,** personal communication, 1984.
67. **Martin, F. W.,** personal communication, 1983.
68. **Martin, F. W.,** Goals in breeding the sweet potato for the Caribbean and Latin America, *Breeding New Sweet Potatoes for the Tropics,* Martin, F. W., Ed., *Proc. Am. Soc. Hortic. Sci. Trop. Region,* 27(B), 61, 1983.
69. **Kays, S. J. and Horvat, R. J.,** A comparison of the volatile constituents and sugars of representative Asian, Central American and North American sweet potatoes, *6th Intern. Symp. Trop. Root and Tuber Crops,* International Potato Center, Lima, Peru, 577, 1984.
70. **Nagahama, T., Inoue, K., Nobori, Y., Fujimoto, S., and Kanie, M.,** On some components in steam distillate of sweet potato, *J. Agric. Chem. Soc. Jpn.,* 51, 597, 1977.
71. **Purcell, A. E., Later, D. W., and Lee, M. L.,** Analysis of the volatile constituents of baked, Jewel sweet potatoes, *J. Agric. Food Chem.,* 28, 939, 1980.
72. **Kays, S. J. and Horvat, R. J.,** Insect resistance and flavor chemistry: Integration into future breeding programs, *In: Breeding New Sweet Potatoes for the Tropics,* Martin, F. M., Ed., *Proc. Amer. Soc. Hort. Sci. Trop. Region,* 27(B), 97, 1983.
73. **Yen, D. E.,** *The Sweet Potato and Oceania,* B. P. Bishop Museum Press, Honolulu, 1974, 236.
74. **Ezell, B. B. and Wilcox, M. S.,** The ratio of carotene to carotenoid pigments in sweet potato varieties, *Science,* 103, 193, 1946.

75. **Lease, E. J.,** Sweet potato as a source of vitamin A for man and domestic animals, *Proc. Assoc. Southern Agric. Workers,* 42, 162, 1941.
76. **Matlach, M. B.,** The carotenoid pigments of the sweet potato (*Ipomoea batatas* Poirs), *J. Wash. Acad. Sci.,* 27, 493, 1937.
77. **Villers, J. F., Heinzelman, D. C., Pominoki, J., and Wareham, H. R. R.,** Isolation of carotene from sweet potatoes, *Food Ind.,* 16, 76, 1944.
78. **Scott, L. E., Appleman, G. O., and Wilson, M.,** The discoloration of sweet potatoes during preparation for processing and the oxidase in the root, *Md. Agric. Exp. Stn. Tech. Bull.,* 33, 11, 1944.
79. **Schadel, W. E. and Walter, W. M., Jr.,** Localization of phenols and polyphenol oxidase in 'Jewel' sweet potatoes (*Ipomoea batatas* 'Jewel'), *Can. J. Bot.,* 59, 1961, 1981.
80. **Walter, W. M., Jr. and Purcell, A. E.,** Effects of substrate levels and polyphenol oxidase activity on darkening in sweet potato cultivars, *J. Agric. Food Chem.,* 28, 941, 1980.
81. **Hayward, H. E.,** *The Structure of Economic Plants,* Macmillan, New York, 1938.
82. **Rao, V. N. M., Hamann, D. D., and Humphries, E. G.,** Mechanical testing as a measure of kinesthetic quality of raw baked sweet potatoes, *Trans. Am. Soc. Agric. Eng.,* 17, 1187, 1974.
83. **Nelson, A. M.,** Shear Press Testing to Define Mouthfeel Characteristics of Baked Sweet Potato, M.S. Thesis, North Carolina State University, Raleigh, 1973.
84. **Rao, V. N. M. and Graham, L. R.,** Rheological, chemical and textural characteristics of sweet potato flakes, *Trans. Am. Soc. Agric. Eng.,* 25, 1792, 1982.
85. **Washam-Hutsell, L. J.,** Evaluation of Sweet Potato Leather as Food Product, M.S. Thesis, University of Tennessee, Knoxville, 1983.

Chapter 11

INDUSTRIAL PRODUCTS FROM SWEET POTATOES

Satoshi Sakamoto and John C. Bouwkamp

TABLE OF CONTENTS

I.	Introduction		220
II.	Utilization as a Raw Material for Starch Manufacturing		220
	A.	Processes to Produce Starch	221
		1. Washing	222
		2. Grinding	223
		3. Sieving	223
		4. Refining	223
		5. Drying	223
	B.	Uses of Starch	223
	C.	Measurement of Starch Properties	224
		1. Water Content	224
		2. Whiteness	224
		3. Viscosity	224
		4. Size Distribution of Starch Grains and Average Grain Size	224
	D.	Properties of Starch	224
	E.	Measurement of Properties of Raw Materials	225
		1. Starch Content	225
		2. Dry Matter Content	225
III.	Fermented Products		226
	A.	Processes to Produce Distilled Spirit "Shochu"	226
		1. Production of "Koji"	226
		2. Production of the First Stock	226
		3. Preparation of the Raw Material	226
		4. Production of the Second Stock	226
		5. Distillation	227
	B.	Alcohol Production	227
IV.	By-Product Utilization or Disposal		229
	A.	Use of By-Products	229
	B.	Disposal of Waste Water	229
		1. Sedimentation Tank	229
		2. Dehydration	230
		3. Rotary Disk	230
		4. Surface Aeration	230
V.	Cultivation of Sweet Potatoes for Industrial Use		230
	A.	Cultivars	230
		1. "Koganesengan"	230
		2. "Minamiyutaka"	230
	B.	Cultivation Method	231

VI. Conclusions 231

References 232

I. INTRODUCTION

Sweet potato is one of the crops capable of producing high dry matter yields from solar radiation through photosynthesis. Solar energy reaching the earth is estimated at 7.2×10^{20} kcal a year, and plant photosynthesis utilizes only a small part of this.[1] The energy fixed by plant photosynthesis is useful for human life not only for food but also for fiber and fuel. The energy productivity of representative crops under average yield are shown in Table 1. From the table it is observed that the calorie per unit area of sweet potato is superior to cereals and comes after sugar cane, which requires a much longer growing period.

At present, sweet potato is cultivated on about 12 million ha in more than 90 countries,[2] and mainly the consumption of products are as human food and animal feed (Table 2). In some countries, sweet potatoes are used as raw material of industries, e.g., 38, 34, and 16% of the product in Korea, Japan, and Taiwan, respectively. The items of industrial use include starch, distilled spirits, alcohol production, etc. These industries, in many cases, bring a chance of economic prosperity and employment to the area, though they need a reasonable investment in equipment before the industries start.

In Indonesia, for example there is a scheme to develop the transmigration and resettlement project by establishing 4,000 factories for alcohol production, which will have a capacity of 5,000 kℓ ethanol a year from cassava and sweet potato in each factory. The aim of this project is to reduce the overpopulation of Java island by transmigration to outer islands such as Sumatra, Kalimantan, Sulawesi, and Irian Java and to give jobs and increase the income of transmigrated farmers through sales of raw material feedstocks to alcohol factories.

This chapter reviews the technologies and practices for the utilization of sweet potato for industries of starch, distilled spirit "Shochu" and alcohol production. Utilization or disposal of by-product are investigated also.

II. UTILIZATION AS A RAW MATERIAL FOR STARCH MANUFACTURING

Roots of sweet potatoes contain 20 to 30% of starch and are one of the important starch-producing crops in the world. They can be grown throughout the year in the tropics, though production is limited to about 5 to 6 months in the temperate areas. About 8, 16, and 28% of production is used as raw material of starch in Korea, Taiwan, and Japan, respectively.[3-5]

The starch plants in the U.S. closed down due to high production costs, poor quality of starch and low returns per unit area.[6] Consumption of sweet potato starch decreased drastically after the relaxation of import restrictions on sugar and raw material of starch in 1963 in Japan[5] (Figure 1). It is difficult to surpass the competition of other cheap starchy crops such as corn, because of production costs in temperate areas. The reasons for high production cost of sweet potato starch in temperate areas are short the duration of factory operation, e.g., about 3 to 5 months a year, high labor costs and high price of raw material.

To establish starch industries of sweet potato in the tropics, it is important to provide supplies of the raw material to the factory the year round and to use raw material of high quality (such as high starch content) immediately after harvest so as to maintain good efficiency in the operation.

Table 1
ENERGY PRODUCTION OF CROPS

Name of crops	Yield[a] (kg/ha)	Calorie per unit[b] (kcal/kg)	Calorie per unit area (10^3 kcal/ha)
Barley	1,987	3,305	6,567
Corn	3,370	3,001	10,113
Potato	14,387	566	8,143
Rice	2,855	3,062	8,742
Sugar cane	56,102	318	17,840
Sweet potato	12,384	972	12,037

[a] FAO.[2]
[b] Working group of energy utilization. Unit energy price in Agriculture, Forestry and Fisheries. GEP 1-2, Series of the result No. 3, M.A.F.F. Japan 1981.

Table 2
CONSUMPTION OF PRODUCT

Name of country	Consumption ratio (%)					
	Human food	Animal feed	Industry	Loss	Seed	Ref.
Indonesia	90			10		31
Japan	37	17	34	7	5	5
Korea	59		38		3	3
Taiwan	11	73	16			4

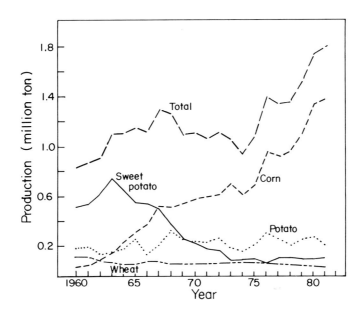

FIGURE 1. Trend of starch production in Japan.

A. Processes to Produce Starch

The processes to produce starch from sweet potato are grinding the root tissue, washing out the starch grains from the tissue, separating the starch from cellulose and protein and drying, etc. These processes are shown in Figure 2. The details of these steps are as follows:

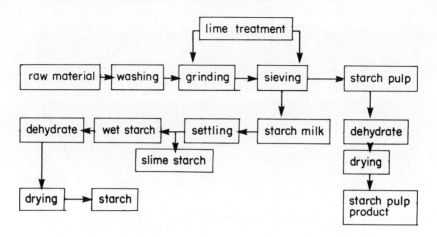

FIGURE 2. Process of starch production.

FIGURE 3. Transporting through floating channel.

1. Washing

After receiving the roots in the yard, they are transported by a stream of water through a floating channel to the washer. During transporting by the stream of water, part of the dust and mud on the roots is washed off and settled and the roots are carried into a washer by a belt-conveyor. The washer consists of a water tank and turning paddles which carry roots to the grinder after washing (Figure 3).

FIGURE 4. Shaker sieve.

2. Grinding

For grinding, a metallic roll of about 60 cm diameter with a rasp on the surface is used. The turning cylinder grinds through the roots and starch grains of the tissue are washed out by the water. The size of the rasp, speed of turning, amount of fiber, starch content and hardness of the roots affect the grinding efficiency. Grinding efficiency is expressed by the following:

3. Sieving

There are two steps in sieving, course sieving and the refining sieve, and the former is to remove course fibers and the latter for refining. There are many kinds of sieves such as rotary sieves, shaker sieves, jet extractor, sieve bent, and so on, and they are employed singly or jointly (Figure 4).

Saturated lime water is used to keep the process of grinding and sieving alkaline (pH 8.6 to 9.2). This treatment is very useful for sweet potato starch production and is effective in increasing the whiteness of starch and increasing the sieving efficiency. By this treatment, CaO content of starch increases a little but does not cause any trouble.

4. Refining

Refining is the most important step in order to produce starch of high quality and the most modernized step in the process. The sieved suspension of starch is a brownish color due to melanin pigment which is caused by oxidized polyphenols. The starch suspension is settled in a settling tank so as to separate slime starch, protein and polyphenol from the starch. Recently nozzle separators have been used for pigment separation.

5. Drying

Using a centrifugal dehydrator, the water content of starch is reduced to about 40%. Then, starch is dried by a flash drier to about 18% moisture.

B. Uses of Starch

At present, more than 90% of sweet potato starch is used for manufacturing starch syrup, glucose, and isomerized glucose syrup in Japan.

Starch syrup is used as a raw material of candy drops, ice cream, sausage, jams, and so on. Main uses of sweet potato starch are similar to the use of starch syrup and the starch is also ingredients of bread, biscuit, cake, juice, ice cream, and so on. The glucose of isomerized syrup changes to fructose and increases the sweetness of glucose by isomerization. This procedure uses the technique of immobilized enzymes. The isomerized glucose syrup contains fructose, glucose, and other sugars. The uses of isomerized syrup are for the production of lactic acid beverages, bread, juice, and so on. The price of isomerized syrup is cheaper than sugar and the consumption will increase rapidly in the future.[7]

Only a small part of sweet potato starch is consumed for production of food such as noodles or in the production of chemicals, medicines, and textiles.

C. Measurement of Starch Properties

Following are fundamental properties to estimate the quality of starch, although evaluation of quality is dependent on the intended purpose of the starch.

1. Water Content

There are many ways to measure the water content such as the drier method, infrared drier method and so on. Using the drier method, a definite amount of starch sample (e.g., 5 or 10 g) is placed in a weighing bottle which is weighed and kept 4 hr in an oven at 105°C and weighed. Drying and weighing repeated until the weight is constant. Difference in the weight of the starch before and after drying is the water content of the starch.

2. Whiteness

Using electric spectro photometer, whiteness of starch is measured by ratio of reflected light compared with a whiteness standard. Before the measurement, the starch should be refined and sieved through a 200 mesh sieve. The result of measurement coincides with the result of other methods such as visual observation or the toruen method.

3. Viscosity

Viscosity of starch is measured by amylograph. Variation of viscosity according to change of temperature is shown by amylogram using a 5 to 10% suspension of starch. The following important measurements are taken from the result of the amylogram: Maximum viscosity, retrogradation and stability.

4. Size Distribution of Starch Grains and Average Grain Size

Using sedimeter or particle size analyzer, the distribution of starch particle and average particle diameter are measured. These properties are important for starch industries.

D. Properties of Starch

Madamba et al.[8] reported amylose contents of sweet potato starches to be from 29.4 to 32.2% with small but statistically significant differences among the six cultivars under study. Uehara[9] found an amylose content of 21.6% in sweet potato starch. It was also determined that treatment of starch with 7 M urea for 1 hr at 30°C resulted in solubilization of 93.6% of the amylose fraction while 99.8% of the amylo pectin remained in the insoluble fraction. Watanabe et al.[10] reported that the amylose content of sweet potato starch was 20.9%. Doremus et al,[11] found a range of amylose content of 17.5 to 21.7% in 22 cultivars. Others[12,13] have reported about 18% amylose from individual cultivars.

Starch grains are of variable shape (oval, round, faceted-round, and polygonal) and are generally nonaggregated.[8,14] Granule size ranges from 4 to 43 microns,[8] depending on the cultivar, with cultivar mean granule size ranging from 12.3 to 21.5 microns. Barham et al.[15] found starch grains in the five cultivars studied ranging from 2 to 26 microns with cultivar means ranging from 8.49 to 9.50 microns. Curing the roots resulted in starch grains averaging about 1 micron smaller than those extracted from freshly harvested roots. Barham and Wagoner[16] reported that the easily extracted starch grains were generally smaller than the more difficult to extract grains. Madamba et al.[8] found a large and statistically significant range of phosphorus content from 9 to 22 mg%. The intrinsic viscosity of starches from six cultivars was found to be 120 to 155 mℓ/g.[8] This indicates that sweet potato starches are not as highly polymerized as potato starch and are more like cereal starches. They found no relation between intrinsic viscosity and particle size. Rasper[17] found a maximum viscosity of sweet potato starch of 590 Brabender units, slightly less viscous than gelatinized cassava starch but more viscous than corn starch. Barhem et al.[15] studied the starch from five cultivars of sweet potatoes and reported a range of about 675 to 810 g required to keep a cylinder from turning at maximum viscosity.

Average gelatinization temperature for starch from six cultivars of sweet potatoes was found to range from 63.6 to 70.7°C by Madamba et al.[8] A significant positive correlation was found between average gelatinization temperature and amylose content of the starches. Gelatinization occurred over a range of 12° to 17°C of temperature change. Barham et al.[15] found that studies from five cultivars had average gelatinization temperatures from 69.0 to 75.5°C and that the average gelatinization temperature was reduced after curing the roots. Rasper[17] reported that sweet potato starch began to gelatinize at 77°C and continued to increase in viscosity until a temperature of 85°C was attained. Sweet potato starches exhibit a single stage swelling pattern.[8,17] Swelling potential of six cultivars was found to be in the range of 18 to 26 g of water absorbed per gram of residual starch.[8] The swelling power curves followed the same pattern as the solubility curves indicating a direct relation of these functions. Rasper[17] reported the swelling power of the sweet potato starch tested to be 27.2 g of water absorbed per gram of original starch.

The susceptibility of sweet potato starch to alpha amylase after 1 day incubation was found to range from 35.7 to 65.5% weight loss among the six cultivars tested.[8] A highly significant negative correlation (-0.91) between amylosis loss and average particle size was reported. Rasper[17] also noted among ''root'' starches smaller starch grains were more susceptible to enzymatic degradation. The susceptibility of sweet potato starches to acid corrosion (2.2 N HCl) showed highly significant differences among cultivars but was only slightly correlated (0.22) with particle size.

E. Measurement of Properties of Raw Materials

Desirable properties of raw material for starch production are high starch content and good quantity. There is a positive high correlation ($r = 0.848$) between starch content and dry matter content.[18] Starch content and dry matter content are the main properties of raw material.

1. Starch Content

Starch content of sweet potato roots increases until 4 months after planting and remains constant after reaching the maximum value. To measure the starch content in small scale, about 2 kg of samples are washed, sliced by a slicer and mixed well, sampling 100 g of sliced fresh root with two replications. The root slices are ground with an electric blender at 10,000 rpm for 1.5 min with 250 mℓ of water to pulverize. Then, it is sieved through 200 mesh sieves using about 5 ℓ of water into a 6 ℓ vessel. The top water is drained after settling overnight. Starch is collected using small brush after drying one day in the sun and weighing after complete drying using electric oven for 8 hr at 80°C and 6 hr at 105°C. Starch contents of recommended cultivars in Japan are between 23 to 28%.

2. Dry Matter Content

Estimating dry matter content is almost the same as the first portion of starch content

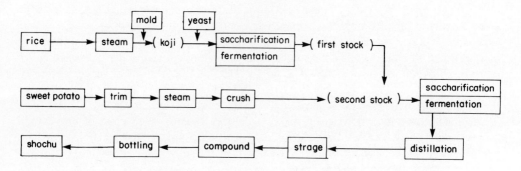

FIGURE 5. Process of shochu production.

measurement. After weighing the sample of 100 g of sliced roots, the sample is dried by an electric oven for 8 hr at 80°C and 6 hr at 105°C. Dry matter content of cultivars is between 35 to 40%.

III. FERMENTED PRODUCTS

A. Processes to Produce Distilled Spirit "Shochu"

Many kinds of starchy crops, such as rice, barley, corn, sweet potatoes and potatoes are used as a raw material to produce a distilled spirit called "Shochu" in Japan. Among them, sweet potatoes are the most famous raw material of "Shochu" production and about 5% of the total production is used for "Shochu" production in Japan.[5] Besides this, there are fermented local alcoholic beverages using sweet potatoes as raw material in Africa and in other Asian countries.[6]

The process to produce "Shochu" is similar to processes to produce whiskey and is shown in Figure 5.

The details of the process are as follows:

1. Production of "Koji"

"Koji" is same as malt in the case of whiskey production, and rice is used to produce "Koji". Selected rice is steamed after soaking in water overnight and conidium of Kawachi mold (*Aspergillus kawachii* Kitahara) is inoculated on rice for 2 days at 35 to 37°C. This mold was discovered from a natural mutant of awamori mold (*Aspergillus awamori* Nakazawa).

2. Production of the First Stock

Above mentioned "Koji" is mixed with water and yeast (*Saccharomyces cerevisiae* Hanson) so as to promote the activity of saccharification and fermentation. They are held for one day at 30°C.

3. Preparation of the Raw Material

To get high quality "Shochu", it is important to carefully select the raw material. It is required that diseased (rotted) roots are discarded. After washing by running water, roots are trimmed, steamed, cooled, and crushed (Figure 6).

4. Production of the Second Stock

First stock and steamed sweet potato are mixed for saccharification and fermentation, and fermented for 8 days at 30°C. Starch is changed to hexose by the action of amylase contained in "Koji" and the hexose is converted into ethanol and carbon dioxide gas by the alcohol fermentation of yeast. The ratio of first stock and steamed roots is about 1:3 by weight.

FIGURE 6. Trimming of raw material of "Shochu" production.

5. Distillation

Ripened second stock is distilled by ordinary methods. At first the distilled liquid contains about 70% alcohol, and it contains 35 to 36% alcohol in average. About 800 ℓ of "Shochu" which contains 25% alcohol can be manufactured from 1 ton of sweet potato, though the amount of product depends on the starch content of raw material and the efficiencies of the systems.

B. Alcohol Production

After the oil crisis, energy farming policies which include production of alcohol fuels from farm products have been discussed in many countries in the world. In energy farming, there are two types of utilization, one is produce ethanol using farm products and another produce methanol from farm waste material. In Brazil, the project to supply ethanol from sugar cane, cassava and Babasu palms started already.[1] In U.S., whey, juice processing waste, corn and wheat are included as a raw material of ethanol production.[19] In Thailand, it is planned to produce ethanol as a new application for surplus sugar cane and cassava. A plan of alcohol fuel production using sugar cane is being discussed in the Philippines. In Indonesia, there are factories to manufacture ethanol using cane molasses. At present, governmental pilot plants to produce ethanol using cassava and sweet potato are under construction at Tulang Bawang and Sulusban, Lampung Province. In these projects it is planned to produce 5,000 kℓ of ethanol a year in each place.[20] In Japan, there are seven semigovernmental alcohol factories and many private factories. Figure 7 shows one of the semigovernmental alcohol factories.

The process to produce alcohol is almost the same as distilled spirit production. Concerning the energy balance of ethanol production from sweet potato, there are different results of estimation and these are presented in Table 3. The estimation of IEE shows more output than input and a small positive balance, though the estimate of NIR shows more input than output and no benefit to produce the ethanol with regard to the energy balance. The difference of these estimations depends on the difference of estimation of input energy for production of raw material. Therefore, it is important to improve the energy balance by increased efficiency of production. There are many directions to improve the energy balance such as improvement of fermentation techniques, and the development of a technique of nonsmothering liquefaction and saccarification.[1]

FIGURE 7. Ethanol factory.

Table 3
ENERGY BALANCE OF ETHANOL PRODUCTION

	Input			Output			Output/Input	Source
	Farming	Production	Total	Ethanol	By-product	Total		
kcal/ℓ ethanol	1,000	3,500	4,500	5,260		5,260	1.17	a
10 kcal/ha	36,820	34,690	71,510	24,430	6,170	30,600	0.43	b

a: The Institute of Energy Economics, *The Alcohol Fuels from Energy Farming*, Tokyo, Japan, 1980, chap. 2.
b: National Institute of Resources, Science and Technology Agency, Tokyo, Japan, 1980.

The use of sweet potatoes as a feedstock for ethanol was suggested as early as 1909.[21] A study of the effects of nutrient addition indicated that up to 0.97 ℓ of 100% ethanol could be produced per kilogram of sweet potato flour.[22]

Badger et al.[23] reported an experimental procedure consisting of grinding the sweet potatoes to pass through a 0.32 cm screen, adding water and heating by steam injection to a temperature of 78°C. Alpha amylase and a mixture of enzymes and vitamins were added as in maize fermentation, and later, yeast. Additional water was added as required in order to permit the pumping of the mash from the cooker to fermentation tanks. Conversion efficiencies of 59 to 87% were obtained based on actual yields as a fraction of theoretical yields. Production estimates ranged from 2319 to 2833 ℓ/ha. A smaller batch resulted in a conversion efficiency of 92% and an estimated production of 3684 ℓ/ha.

Sreekantiah and Rao[24] found that fermentation of cooked sweet potatoes with *Rhizopus niveus* resulted in production of up to 37.5 mℓ of ethanol per 100 g of starch, a fermentation efficiency of 75%. Svendsby et al.[25] incubated minced sweet potato roots at room temperature with glucoamylase, pectin depolymerase and yeast for five days and obtained nearly 100 g of ethanol per kilogram of roots.

McArdle[26] reported an experimental procedure which included grinding the roots, adding water and heating to 70 or 80°C. The starch was presumably gelatinized at these temperatures and native amylase activity resulted in soluble solids levels of up to 90% of the total solids.

Table 4
POLLUTION REGULATION IN JAPAN

pH		BOD(ppm)		COD(ppm)		ss(ppm)	
River	Sea	Avg.	Max	Avg.	Max	Avg.	Max
5.8—8.6	5.0—9.0	120	160	120	160	150	200

Addition of pectinase and cellulase resulted in higher fermentation efficiency. Conversion efficiencies of up to 80% of theoretical were reported.

It is important to reduce the cost of production. At present, production cost of ethanol from sweet potato is about 2 times the gasoline price in Japan. It is necessary to reduce the price of raw material by production practices which will increase the production per unit area of labor saving.

IV. BY-PRODUCT UTILIZATION OR DISPOSAL

A. Use of By-Products

Starch pulp of 5 to 6% of the raw material is produced as a by-product of starch. It contains 55 to 65% starch, 2 to 4% protein, 0.8 to 1.5% fat, 5 to 14% fiber, 2 to 10% ash and 13 to 15% water.

Dehydrators and driers are used for drying the starch pulp. It is utilized as a raw material for citric acid production and for animal feed. To produce citric acid, either dry or wet pulp may be used. The pulp is steamed, cooled and inoculated by spraying of a conidium suspension of *Aspergillus niger*. It is fermented for 72 to 96 hr at 28 to 30°C and citric acid is extracted. Then it is neutralized by lime water to approximately pH 6.0 and dehydrated by centrifugal dryers until about 10% water content.

Starch pulp as an animal feed is palatable and digestible especially for pigs, but it is necessary to use as only part of formula feed for fattening of livestock because it contains too much carbohydrate in proportion to protein. Dawson et al.[27] describes a system of nutrient recovery and waste water cleanup consisting of precipitation of the proteins by reducing the pH to 3.8 and heating the waste water to 80°C. After cooling, the waste water is innoculated with *Torulopsis utilis* and the pH raised to 6.5 with ammonium hydroxide. About 2 tons of by-products containing approximately 50% protein could be recovered per 100 tons of sweet potatoes processed. Waste water of distilled spirit "Shochu" production contains 95% water and only 5% solid matter. It is utilized as animal feed, or fertilizer to return the organic matter to upland fields or pasture. Recently, investigations to collect methane gas from waste water have succeeded. In this case, waste water is sealed up with methane mold for 20 days in a tank and then 25 m^3 methane gas is produced per kiloliter waste water. It takes several years to depreciate the considerable investment in required equipment.

B. Disposal of Waste Water

Pollution due to waste water of starch factories or shochu factories is regulated by governmental regulation in Japan and it is shown in Table 4. For this purpose, following treatments are used to meet the situation.

1. Sedimentation Tank

Waste water is discharged after holding 2 or 3 days in the sedimentation tank and the effect of sedimentation of suspended solids is fairly good though removal of BOD is less than 30%.

2. Dehydration

Waste water is ground again before dehydration by centrifugal dehydrator, and then the solid matter is sieved. The dehydrated fraction is rich in protein and fat and suitable for animal feed or fertilizer.

3. Rotary Disk

This method was developed in West Germany for sewerage. Slowly turning (1.5 to 2.5 rpm) disks are kept in the sedimentation tank, as the result organic matter decomposition and the efficiency of the separator is very good.

4. Surface Aeration

Waste water is held for a long time (about 10 days) in the sedimentation pond so as to aerate from the surface. A large area is required for effective aeration and activity.

V. CULTIVATION OF SWEET POTATOES FOR INDUSTRIAL USE

A. Cultivars

It is necessary to grow a cultivar with high starch content and high yield for industrial use. Starch content affects the efficiency of not only starch factories but also distilled spirit and ethanol factories. In addition, it is desirable for cultivars to possess the characteristics of pest and disease resistance and good storage ability.

For this purpose, breeding for cultivars for industrial use has been developed and has produced some excellent results. The following are recommended varieties for industrial use in Japan.

1. "Koganesengan"

It was selected from the cross Kakei 7 - 120 and Pelican Processor, registered as Norin 31 and released in 1966. It shows high dry matter content (38.3%), high starch content (26.7%) and high yield (32 t/ha) and high adaptability for heavy fertilizing. Further, it is well adapted for early harvesting, an advantage for industrial use because the duration of factory operation is extended. "Koganesengan" has moderate resistance to root-knot nematode (*Meloidogyne* spp.), and root lesion nematode (*Pratylenchus* spp.), but is susceptible to black rot (*Ceratocystis fimbriata*).[28]

2. "Minamiyutaka"

It was selected from the cross Koganesengan and Kyushu 58 and released in 1975. The pedigree of Minamiyutaka is shown in Figure 8. It is the first cultivar developed by the use of germplasm from wild relatives in Japan and has high dry matter content (35.9%) and high starch content (23.8%).[29]

The yield is higher than Koganesengan in the southern part of Japan but less in the northern part. The difference in the yield performance among regions may be due to the varietal response to temperature.

The development of high starch and high yield cultivars, required selections for dry matter content or starch content and root yield over a long time. Figure 9 shows efficacy of high starch content and high yield breeding. In this figure, Norin 2 was used as the standard cultivar and the yield of materials in 1957 was about 20% higher than Norin 2 but starch content was almost the same as Norin 2.

On the contrary, starch content of 1966 was 3 to 5% higher than Norin 2 due to breeding efficacy, but the yield was almost the same as that of Norin 2.

The yield in 1976 increased remarkably and starch content is higher than Norin 2. The strain of the highest starch content was 7% higher than Norin 2.[30]

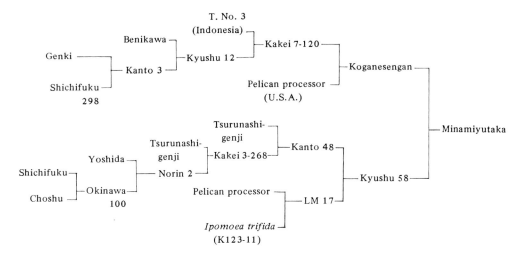

FIGURE 8. Pedigree of Minamiyutaka.

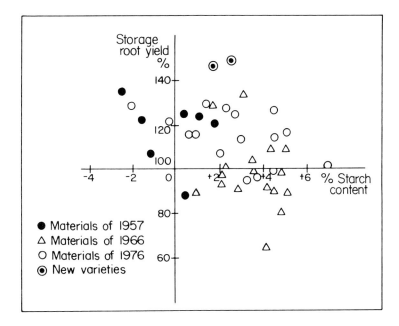

FIGURE 9. Comparison of starch content and tuberous yield using the materials of performance test in 1957, 1966 and 1976 with Norin 2.

B. Cultivation Method

There is a positive correlation between duration of growth and root yield. In tropics, it can be grown all the year and it is possible to be harvested two or three times a year. Year round production may be limited by rainfall and might be possible only in irrigable areas. Industrial use of sweet potato may be a growth industry, if the raw material supplies are constant and efficiency of production is stabilized.

VI. CONCLUSIONS

The utilization of sweet potato for industries has not been well developed thus far. But in the future, it may be useful in development and employment policy of the area, and effective to stabilize the agricultural management of upland fields. It produces almost same starch yield in shorter duration than cassava, and may be superior crop except it is less drought resistant. Damages due to weevil and viruses may be overcome by the improvement of cultivar and cultivating methods. It may be possible to double the production of sweet potato in the tropics and to fulfill the role that corn and other cereals have realized in the temperate regions of the world.

REFERENCES

1. Institute of Energy Economics, *The Alcohol Fuels fron Energy Farming*, Tokyo, Japan, 1980, chap. 2.
2. FAO, Production Yearbook, Food and Agriculture Organization of the United Nations, Rome, 1981.
3. **Hong, E. H.**, The storage, marketing and utilization of sweet potatoes in Korea, *Sweet Potatoes, Proc. First Int. Symp.*, Villareal, R. L. and Griggs, T. P., Eds., Asian Vegetable Research and Development Center, Shanhua, Tainan, Taiwan, 1982, 405.
4. **Yeh, T. P.** Utilization of sweet potato for animal feed and industrial use; Potential and problems, *Sweet Potatoes, Proc. First Int. Symp.*, Villareal, R. L. and Griggs, T. P., Eds., Asian Vegetable Research and Development Center, Shanhua, Tainan, Taiwan, 1982, 385.
5. **M.A.F.F.**, Information Papers on the Production and Distribution of Root Crops, Tokyo, Japan, 1983, Chap. 6.
6. **Winarno, F. G.**, Sweet potato processing and by-product utilization in the tropics, *Sweet Potatoes, Proc. First Int. Symp.*, T. P., Eds., Asian Vegetable Research and Development Center, Shanhau, Tainan, Taiwan, Villareal R. L. and Griggs, 1982, 373.
7. **Kainuma, K.**, Isomerized glucose syrup-history, chemistry and industrial aspects, *J. Jpn. Soc. Starch Sci.*, 27, 139, 1980.
8. **Madamba, L. S. P., Bustrillos, A. R., and San Pedro, E. L.**, Sweet potato starch: physiochemical properties of the whole starch, *Philipp. Agric.*, 58, 338, 1975.
9. **Uehara, S.**, Amylase — amylopection ratio of soluble and insoluble fractions of sweet potato starch granules treated with urea, *J. Agric. Chem. Soc. Jpn.*, 57, 529, 1983.
10. **Watanabe, T., Akiyama, Y., Takahashi, H., Adachi, T., Matsumoto, A., and Matsuda, K.**, Structural features and properties of Nageli amylodextrin from waxy-maize, sweet potato and potato starches, *Carbohydrate Res.*, 109, 221, 1982.
11. **Doremus, G. L., Crenshaw, F. A., and Thurber, F. H.**, Amylose content of sweet potato starch, *Cereal Chem.*, 28, 308, 1951.
12. **Takahashi, K.**, Calorimetric studies on alpha — 1,4 glucosidic linkage content in sweet potato starches at two different stages of development *Agric. Biol. Chem.*, 30, 629, 1966.
13. **Bertoniere, N. R., McLemore, T. A., Hasling, V. C., Castalano, E. A., and Deobald, H. J.**, Effect of environmental variables on the processing of sweet potatoes into flakes and on some properties of their isolated starches, *J. Food Sci.*, 31, 574, 1966.
14. **Thurber, F. H., Gastrock, E. A., and Guilbeau, W. F.**, Production of sweet potato starch, in *Crops in War and Peace*, Yearbook of Agriculture, 1950-1951, U.S. Department of Agriculture, Washington, D.C., 163.
15. **Barham, H. N., Wagoner, J. A., Williams, B. M., and Reed, G. N.**, A comparison of the viscosity and certain microscopical properties of some Kansas starches, *J. Agric. Res.*, 68, 331, 1944.
16. **Barham, H. N. and Wagoner, J. A.**, Effect of time and conditions of care on the carbohydrate composition of sweet potatoes and the composition of their starches, *J. Agric. Res.*, 73, 255, 1946.
17. **Rasper, V.**, Investigations on starches from major starch crops grown in Ghana. II. Swelling and Solability patterns: amyloclastic susceptibility, *J. Sci. Food Agric.*, 20, 642, 1969.
18. **Sakai, K., Hirosaki, S., Shirasaka, S., and Marumine, S.**, Studies on the measuring method of starch content in sweet potato (2), *Rep. Kyushu Branch Crop. Sci. Soc. Jpn.*, 15, 64, 1960.
19. USDA The report of the Alcohol Fuels Policy Review, U.S. Department of Agriculture, Washington, D.C., 1979.
20. J.I.C.A., Report of Japanese implementation survey team, Indonesia Biomass Energy Research and Development Center, 1982.
21. **Keitt, J. E.**, Sweet potato work in 1908, *S.C. Agric. Exp. Stn. Bull.*, 146, 1909.

22. **Hao, L. C. and Delahanty, M. E.,** Ethanol fermentation of dehydrated sweet potatoes — nutrient study, *J. Bacteriol.,* 49, 524, 1945.
23. **Badger, P. C., Pile, R. S., Bandi, D. W., Mays, D. A., and Lewis, J. W.,** TVA/DOE integrated on farm alcohol production system progress report, Phase II, October 1981 — February 1982, Circular Z-134, Tennessee Valley Authority, NFDC, Muscle Shoals, Ala., 1983.
24. **Sreekantiah, K. R. and Rao, B. A. S.,** Production of ethanol from tubers, *J. Food Sci. Technol.,* 17, 194, 1980.
25. **Svendsby, O., Kakutani, K., Matsumura, Y., Izuka, M., and Yamamoto, T.,** Ethanol fermentation of uncooked sweet potato with application of enzymes, *J. Ferment. Technol.,* 59, 485, 1981.
26. **McArdle, R. N.,** The Use of Endogenous Amylase to Produce Fermentable Saccharides in *Ipomoea batatas* (L.) Lamand. *Zea mays* L., Ph.D. Dissertation, University of Maryland, College Park, Md., 1983.
27. **Dawson, P. R., Greathouse, L. H., and Gordon, W. O.,** Sweet potatoes: More than starch, in Crops in Peace and War, Yearbook of Agriculture 1950-1951, U.S. Department of Agriculture, Washington, D.C., 195.
28. **Sakai, K. et al.,** On the new variety of sweet potato "Koganesengan", *Bull. Kyushu Agric. Exp. Stn.,* 13, 55, 1967.
29. **Ono, T., Marumine, S., Yamakawa, O., Hirosaki, S., Sakamoto, S., and Ide, Y.,** A new sweet potato variety "Minamiyutaka" *Bull. Kyushu Agric. Exp. Stn.,* 19, 133, 1977.
30. **Sakamoto, S.,** Breeding of sweet potato varieties for high starch content and high yield, in *Proc. 5th Int. Symp. Trop. Root and Tuber Crops,* Manila, Philippines, 1979, 33.
31. **Anonymous,** Statistical Yearbook of Indonesia 1977-1978, Biro Pusat Statistik, Jakarta, Indonesia.

Chapter 12

ROOTS AND VINES AS ANIMAL FEEDS

T. P. Yeh and John C. Bouwkamp

TABLE OF CONTENTS

I.	Nutritional Quality of Sweet Potatoes		236
	A.	Nutrient Composition	236
	B.	Digestible Nutrients of Sweet Potatoes	236
		1. Protein Quality and Amino Acid Contents	236
		2. Digestibility of Sweet Potatoes by Various Livestock	236
II.	Trypsin Inhibitors in Sweet Potatoes		236
III.	Processing for Feed		238
	A.	Dry-Heating Process	238
	B.	Pelleting	240
	C.	Starch Urea (Starea) for Cattle	243
	D.	Silage	244
IV.	Sweet Potatoes (Roots and Tops) in Livestock Rations		244
	A.	Swine Rations	244
		1. Feeding Efficiency and Growth Rates	244
		2. Carcass Quality of Pigs Fed Sweet Potato Chips	245
		3. Recommended Swine Rations	245
		4. Swine Grazing Systems	245
	B.	Cattle Diets	247
	C.	Sheep Production	248
	D.	Poultry Rations	248
	E.	Aquaculture Feeds	250
	F.	Microbiological Feedstocks	250
V.	Conclusions		250
References			251

I. NUTRITIONAL QUALITY OF SWEET POTATOES

A. Nutrient Composition

Sweet potatoes are a high moisture feed (about 68% moisture) in their fresh state, but the dry matter yield per hectare of high yield cultivars is competitive with many other feed sources (Table 1).[1] The nutrient composition of sweet potatoes grown in Taiwan and elsewhere is presented in Table 2[2,3] and Table 3. Roots have rather low levels of crude protein, fat and fiber and generally high levels of NFE; while the above ground portions contain higher levels of crude protein, fat and fiber and considerable amounts of NFE. The roots and especially the vines contain relatively high calcium concentrations with lower levels of phosphorus in contrast with feed grains of corn, barley, wheat, and sorghum. Starch content in roots among 316 sweet potato cultivars grown in Taiwan ranged from 7.0 to 22.0% (fresh weight basis) with an average of 13.5%.

B. Digestible Nutrients of Sweet Potatoes

1. Protein Quality and Amino Acid Contents

Yang et al.[15] reported that the crude protein content of sweet potatoes varied from 3.5 to 7.1% on dry matter basis, and that the protein and lysine contents are significantly influenced by variation among cultivars. One example of data analysis of amino acids for sweet potato chips is shown in Table 4[2,16] indicating comparatively lower contents of methionine, cystine, tryptophan, and tyrosine in Tainung 57 than HP-18 (high-protein cultivar). Also Li[17] has reported that the average protein in roots of native cultivars in Taiwan was 4.4%, ranging from 1.3 to 7.2%, and that of improved cultivars was 4.7%, ranging from 2.5 to 8.0%, while in foreign introductions the mean value was 5.4%, ranging from 1.9 to 10.1%. The variation of protein content in the tops was large in native cultivars (12.1 to 25.7%) and in foreign introductions (13.2 to 25.3%), but less in improved cultivars (15.4 to 23.3%). Thus, it seems clear that cultivars vary to a considerable extent in their protein levels and that it may be possible through breeding programs to select new cultivars for both high yield and high protein. Among the different crop seasons, highly significant differences in protein contents of roots and tops were found, influenced by cultivars and nitrogen levels. Protein content of roots was increased from 4.3 to 5.3% by applications of 50 and 150 kg/ha of nitrogen, respectively. The protein content of tops was likewise increased from 17.1 to 19.3%, respectively. There was a tendency for protein content of the roots to increase with the 160 kg N/ha application and dry matter to decrease at the heaviest N dosage applied.[18] Yeh et al.[19] (Table 5) reported that increasing the rate of nitrogen application resulted in a tendency to decrease yields of roots, and increase the protein and lysine contents of both roots and tops.

2. Digestibility of Sweet Potatoes by Various Livestock

Wu[20] reported that the average energy values for sweet potato chips consumed by pigs, in kilocalories per gram of dry matter, were: gross energy, 4.024; digestible energy (DE), 3.497; metabolizable energy (ME), 3.332, and net energy, 2.020. Comparison of the net energy (NE) values of corn and sweet potato chips showed that the net energy of sweet potato is about 78.8% that of corn (Table 6). Digestibility of the various fractions by other livestock is presented in Table 7. Digestibility of the crude protein fraction of dehydrated uncooked roots appears to be low. Since the crude protein fraction is low, total digestible nutrient levels remain high.

The feeding value of sweet potatoes is frequently compared to maize. A summary of results expressed as a comparison to maize is presented in Table 8.

II. TRYPSIN INHIBITORS IN SWEET POTATOES

The presence of trypsin inhibitors in sweet potatoes was first reported by Sohonie and

Table 1
SELECTED HIGH-YIELD CULTIVARS OF SWEET POTATOES IN TAIWAN

Varieties	Fresh root (kg/ha)		Sun-cured chips (%)	Crude protein[a] (%)
	Autumn	Spring		
Tainung 57	43,000	49,000	30—37	3.2—4.5
Tainung 62	36,000	44,000	31	4.0—4.5
Tainung 63	32,000	—	27	4.5—5.3
Tainung 64	34,000	43,100	27—29	4.5—5.2
Tainan 15	—	33,920	33	—
Tainan 17	30,530	—	31.4	—

[a] On dry basis.

Table 2
NUTRIENT COMPOSITION OF SWEET POTATOES GROWN IN TAIWAN

		Roots, fresh	Roots, sun-cured chips	Tops, fresh	Tops, dehydrated
Dry matter	(%)	17.8—40.7	83.3—85.0	9.1—19.5	80.2—88.2
Crude protein	(%)	1.1—1.9	2.5—3.5	2.5—3.8	11.7—18.6
Crude fat	(%)	0.3—0.7	2.1	0.2—1.4	2.8—5.6
Crude fiber	(%)	0.4—1.0	2.9—3.0	1.7—4.3	10.3—17.0
Crude ash	(%)	0.4—1.0	3.2—3.3	1.4—2.8	7.4—9.9
NFE	(%)	14.7—36.1	72.5—73.1	1.7—7.6	38.5—49.3
Calcium	(%)	0.2—0.3	0.2—0.3	0.2	1.0—1.78
Phosphorus	(%)	0.06—0.07	0.02—0.015	0.05—0.07	0.03—0.27
Copper	(ppm)	—	1.9	—	23.1
Manganese	(ppm)	—	14.1	—	45.2
Zinc	(ppm)	—	8.5—222	—	35.5
Iron	(ppm)	—	790	—	1548

Bhandarker.[29] Further studies[30] revealed the presence of two fractions with trypsin inhibitor activity after fractionation with $(NH_4)_2SO_4$ and acetone. Sugiura et al.[31] described a procedure for isolation and fractionation of trypsin inhibitors by column chromatography on Sephadex G-50, DEAE - and CM-cellulose. They reported three fractions and described the properties of the two main inhibitors. The inhibitors had molecular weights of 23,000 and 24,000 and were shown to be arginine inhibitors according to the definitions of Liu et al.[32]

Several reports have indicated a positive correlation between trypsin inhibitor activity and protein concentration in the roots.[33-35] Trypsin inhibitor activity has also been correlated with dry matter content in roots. Notwithstanding, it is likely that lines may be selected which are low in trypsin inhibitor activity but with good levels of dry matter and protein.[35] Differences among cultivars in the heat stability of cultivars has also been noted.[33,34] Yeh et al.[36] demonstrated that trypsin inhibitor activity could not be completely eliminated when the sun cured chips of high-protein cultivar (HP-18) was treated with dry heat at 100°C for 5 min. One example of the effect of trypsin inhibitor on the performance of growing-finishing pigs is shown in Table 9.[36] When the diets contain sweet potato chips, days needed to 90 kg body weight were increased, daily feed intake and daily gain were decreased, resulting

Table 3
PROXIMATE ANALYSIS OF SWEET POTATO ROOTS AND VINES

Plant part	DM	CP	EE	CF	Ash	NFE	Ca	P	Ref.
Fresh roots	35.0	8.8	0.5	2.5	2.6	85.7			4
Fresh roots	28.1	5.4	0.5	0.3	3.2	90.6			5
Dried roots	88.4	2.8	1.9	1.0	3.5	90.8			6
Dried roots	91.0	3.3	0.8			88.4	0.11	0.14	7
Dried roots	91.3	5.5	0.4	4.1	3.2	86.8			8
Dried roots		2.8	1.9	1.2			0.60	0.15	9
Dried roots		4.9	0.8	3.3	3.3	87.6			10
Dried roots		7.2	1.4	3.7		81.5			11
Dried roots		4.3	0.6	1.7	2.1	91.2			12
Dried roots		3.3					0.09	0.19	13
Dried roots	90.8	6.2	1.4	4.0	6.3	81.8	0.30	0.22	55
Leaves + vines	12.0	17.2	2.8	13.2	13.6	53.2			4
Vines	8.7	21.9	3.4	15.0	18.0	41.7			4
Leaves	10.8	19.4	3.7	10.2	25.9	40.8			5
Vines, hay	86.6	16.4	5.2	27.4	12.6	38.4			5
Vines		13.5					1.24	0.27	13
Stems		6.6					3.30	0.36	14
Leaves		14.3					4.20	0.34	14

Table 4
COMPARISON OF SELECTED AMINO ACIDS BETWEEN TAINUNG 57 AND HP-18 (CHIPS, % AIR DRY SAMPLE)

	Sweet potato chips[a]		
Amino acids	Tainung 57	HP-18	Yellow corn[b]
Dry matter	90.50	90.40	89.00
Crude protein	2.90	6.35	8.90
Cystine	0.03	0.10	0.09
Isoleucine	0.12	0.24	0.45
Lysine	0.13	0.26	0.18
Methionine	0.04	0.12	0.09
Phenylalanine	0.13	0.31	0.45
Threonine	0.11	0.29	0.36
Tryptophan	0.41	—	0.09
Tyrosine	0.05	0.27	0.39

[a] From Yeh et al.[36]
[b] From Livestock Research Institute.[2]

in inferior feed efficiency as compared to a corn-soybean meal ration that did not include sweet potato chips.

III. PROCESSING FOR FEED

The effort to find out the most effective process to enhance the nutritive value of sweet potatoes has been made by some Taiwanese workers. It is also important to make the materials easy to handle, store, transport, and formulate.

A. Dry-Heating Process

The starch availability and nutritive value of dried sweet potato chips are not improved

Table 5
EFFECTS OF DIFFERENT LEVELS OF NITROGEN APPLICATION ON THE YIELDS IN kd/ha OF CRUDE PROTEIN AND LYSINE OF SWEET POTATOES

	Nitrogen (kg/ha)					
	N_0		N_{100}		N_{200}	
	CP	Lysine	CP	Lysine	CP	Lysine
HP-18 (High-protein cultivar for roots)						
Roots	463	20	589	24	477	21
Tops	585	29	724	38	870	44
Total	1048	50	1313	62	1347	65
HP-14 (cultivar for vine production)						
Roots	62	3	58	3	69	3
Tops	754	34	1132	49	1453	64
Total	816	37	1190	52	1522	67

Table 6
ENERGY VALUES FOR SWEET POTATO CHIPS AS COMPARED TO YELLOW CORN

	Sweet potato chips (Tainung)		Yellow corn (U.S. No. 2)	
	kcal/kg	DM (%)	kcal/kg	DM (%)
DE	3497	91.1	3837	100
ME	3377	90.9	3714	100
ME_n	3332	—	—	—
NE	2020	78.8	2564	100

by either microwave heating (80 to 90°C, 1.5 to 4.5 min, (wave frequency, 2450 MHz) nor by rapid high temperature drying (390 to 430°C 1 min).[37,38] But "popping" the chips (in the same way poprice is made, 6 to 8 kg/cm^2 pressure at 164 to 178°C) can improve the starch availability and nitrogen digestibility[20,37-39] (Table 10). The trypsin inhibitor activity of the chips can be completely eliminated by popping treatment, but not by the rapid high temperature drying process, which is designed for large scale production. Available lysine in the chips is reduced by high temperature heat-drying (Table 11).[38] The results of feeding trials showed that daily gain and feed efficiency of pigs fed with popped dried chips were significantly ($p<0.05$ or $p<0.01$) improved as compared to those of pigs fed unheated dried chips, and the performance of pigs was comparable to that of pigs fed corn diets. The daily gain and feed-to-gain ratio of growing-finishing pigs fed unpopped dried chips substituted for 40% of the corn in their ration were significantly ($p<0.01$) inferior to those of pigs fed corn or popped dried chips.[39,40] Also the daily gain and feed-to-gain ratio of growing-finishing pigs fed the chips treated with momentary heating at 390°C were not improved as compared to those of pigs fed nonheating dried chips (Table 12).[38] The result of the feeding trial also indicated that the performance of pigs was inferior as the amounts of dried chips was increased in the pig ration.

Table 7
DIGESTIBILITY OF SWEET POTATO FEED STUFFS AND THEIR COMPONENTS BY VARIOUS TEST ANIMALS

% Digestibility

Animal	Feed[a]	CP	EE	CF	NFE	TDN	ME	Ref.
Young chicks	DUR					46.7		21
Young chicks	DCR					64.6		21
Beef cattle	Vines					62.0	2.06	22
Beef cattle	FR					92.0	2.63	22
Dairy cattle	DUR	3.2	52.0	−51.6	90.1	80.0		10
Beef cattle	DUR	50.8			96.0			11
Lambs	DUR	19.8			90.1			11
Sheep	DUR	14.0	74.0	37.0	90.0		2.71	5
Sheep	FR	37.5	51.6	79.3	95.5		3.24	5
Dairy cattle	DUR	−25.2	71.6		87.7	77.1		23
Sheep	Leaves	80.0	84.0	55.0	86.0		2.39	5
Cattle	Hay	64.5	72.8	35.7	74.1		2.13	5

[a] DUR = Dehydrated uncooked roots; DCR = dehydrated cooked roots; FR = fresh roots.

Table 8
VALUE OF SWEET POTATO ROOTS AS COMPARED TO CORN IN VARIOUS FEEDING TRIALS

Animal	Substitution for corn	Comparative value	Parameter compared	Ref.
Young chicks	Up to 60%	nsd	Wt gain	6
Cattle	100% root Trimmings	80%	Wt gain	24
Cattle	50%	nsd	Wt gain	7
Dairy cattle	100%	nsd	Milk production	25
Dairy cattle		91%	Milk production	26
Dairy cattle		88%	Milk production	23
Dairy cattle	50%	97%	Milk production	10
Dairy cattle	100%	91%	Milk production	10
Lambs, steers	100%	92%	Digestibility	11
Chicks	10 or 20%	nsd	Wt gain	27
Dairy cattle	100%	nsd	Milk production	28

B. Pelleting

Trypsin inhibitor activity of the chips prepared from cultivar HP-18 could not be eliminated by pelleting at a temperature of 80°C, and the performance of pigs fed pellet ration with chips by pelleting at 80°C was inferior to those of pigs fed the same diets of mash ration with chips (Table 13).[39] There was still a tendency towards poor performance of pigs fed diets with dried chips of HP-18 as compared to those of pigs fed corn diets.

During 1977 to 1979, an attempt to manufacture the commercial products of pellet forms of dried roots and tops was conducted by Pu-li Byproduct plant of Taiwan Sugar Corporation.[41]

High yields cultivars (HP-4 and Tainung 31) for tops were planted to supply fresh material for momentary high-temperature dry-heating, and then to manufacture a pellet form of green feed. Also the same process was applied to manufacture a pellet form of dried chips from fresh raw roots.

Because of high production cost to produce those pellets, the attempt for industrial uses was discontinued.

Table 9
THE EFFECT OF TRYPSIN INHIBITOR ON THE PERFORMANCE OF GROWING-FINISHING PIGS

Treatments[a]	1	2	3	4
% Inhibition of TIA of mixed feed	23	44	52	49
Days needed to 90 kg[b]	118(A)	147(B)	163(C)	201(D)
Avg. daily gain (kg[b])	0.603(A)	0.482(B)	0.437(C)	0.355(D)
Avg. daily feed intake (kg[b])	1.85(A)	1.83(A)	1.78(B)	1.74(C)
Feed/gain[b]	3.08(A)	3.84(B)	4.09(B)	4.99(C)

[a] The diets offered to Treatment 1 was corn-soybean meal ration, to Treatment 2 was 1/2 corn + 1/2 sweet potato chip-soybean meal ration, to Treatment 3 was sweet potato chip-soybean meal ration, and to Treatment 4 was sweet potato chip + vines-soybean meal ration.

[b] Values in the same line followed by different small letters differ significantly ($p<0.01$ or $p<0.05$).

From Yeh, T. P., Wu, M. C., and Chen, S. Y., unpublished data, Animal Industry Research Institute, Taiwan Sugar Corporation, Chunan, Taiwan, 1979.

Table 10
STARCH AVAILABILITIES OF SWEET POTATO CHIPS TREATED WITH DRY-HEATING

Treatment	Starch availability (mℓ CO_2/g DM/hr)	% Improvement	Ref.
Microwave heating[a]			
Nontreated	20.50	—	37
For 1.5 min 80—90°C	19.87	−3.17	
For 3.0 min 80—90°C	19.90	−3.00	
For 4.5 min 80—90°C	19.85	−3.17	
Popping[b]			
Untreated	19.26	—	37
6 kg/cm^2, 164°C	52.02	170.09	
8 kg/cm^2, 175°C	53.65	178.56	
12 kg/cm^2, 191°C	46.64	142.16	
Popping[c]			
Untreated	15.21	—	39
6 kg/cm^2, 164°C	47.77	194.35	
8 kg/cm^2, 175°C	42.56	179.82	
Dry air for 1 min passing			
Untreated	10.58	—	38
At 390°C	11.35	7.28	
At 430°C	15.15	43.19	

[a] Sweet potato Tainung 57, moisture content of chips 8.75%.
[b] Sweet potato Tainung 57, moisture content of chips 8.16%.
[c] Sweet potato Tainung 57, moisture content of chips 9 to 10%.

Table 11
TRYPSIN INHIBITOR ACTIVITY, STARCH AVAILABILITY AND AVAILABLE LYSINE OF SWEET POTATO CHIPS BY DRY-HEATING

	Sweet potato chips HP-18		
	Sun-cured	Dry heat at 390°C, for 1 min	Dry heat at 430°C, for 1 min
Trypsin inhibitor activity, TIU/g dry matter	1320	1100	630
Starch availability, mℓ, CO_2/g dry matter/hr	10.58	11.35	15.15
% Improvement of starch availability as compared, to sun-cured	—	7.28	43.19
Total lysine (%)	0.20	0.17	0.17
Available lysine (%)	0.19	0.15	0.15
% Availability of lysine	95.00	88.24	88.24

From Yeh, T. P., Chen, S. Y., and Chang, Y., unpublished data, Animal Industry Research Institute, Taiwan Sugar Corporation, Chunan, Taiwan, 1980.

Table 12
THE EFFECTS OF MOMENTARY HIGH TEMPERATURE DRY-HEAT OF SWEET POTATO CHIPS ON THE PERFORMANCE OF GROWING-FINISHING PIGS
(20 to 90 kg body wt)

Treatment[a] Main carbohydrates in diet	Daily treatment	Feed/gain[b]	% of gain lean kg kg/kg
Corn (C)	No-heat	52.77	0.69(A) 2.95
Sweet potato chips	Momentary Heating at 390°C	52.92	0.61(C) 3.29(AB)
1/2 Corn + 1/2 sweet potato chips	Momentary Heating at 390°C	52.60	0.65(BC) 3.18(AB)
Sweet potato chips 1/2 Corn + sweet potato chips (BC)	No-heat	54.89	0.60(CD) 3.37(A) 54.12 0.66(AB) 3.13
Sweet potato chips + tallow (AB)	No-heat	52.37	0.61 (BC) 3.27

[a] Sun-dried chips of HP-18 were used
[b] Values on the same column followed by different letters differ significantly at ($p<0.05$) or highly significant at ($p<0.01$).

From Yeh, T. P., Chen, S. Y., and Chang, Y., unpublished data, Animal Industry Research Institute, Taiwan Sugar Corporation, Chunan, Taiwan, 1980.

Table 13
THE EFFECT OF PELLETING TEMPERATURE ON THE PERFORMANCE OF GROWING-FINISHING PIGS (19 TO 90 kg BODY WT)

Treatments			
Main dietary carbohydrates[b]	Pelleting temp	Daily gain[a] (kg)	Feed/gain[a] (kg/kg)
Corn	Room temp	0.61(A)	3.06(E)
	80°C	0.60(A)	3.10(D)
1/2 Corn +	Room temp	0.51(B)	3.62(BCD)
1/2 Sweet potato chips	80°C	0.46(C)	4.05(BC)
Sweet potato chips	Room temp	0.44(CD)	4.11(ABC)
	80°C	0.44(CD)	4.06(BC)
Sweet potato chips +	Room temp	0.34(E)	5.16(A)
Sweet potato vine	80°C	0.36(DE)	4.84(A)

[a] Values on the same column followed by different letters differ significantly at $(p<0.05)$ or highly significant at $(p<0.01)$.
[b] Sun-dried chips of HP-18 and dehydrated meal of HP-4 were used for chips and vine, respectively.

From Yeh, T. P., Wu, M. C., and Chen, S. Y., Animal Industry Research Institute, Taiwan Sugar Corporation, Chunan, Taiwan, unpublished data, 1979.

Table 14
FEEDING VALUE OF GELATINIZED UREA-SWEET POTATO MEAL (GUSP) FOR CATTLE

Treatments	Soybean meal	Urea	GUSP
No. of animals	6	6	6
Initial wt (kg)	221	221	216
Final wt (kg)	290	253	260
Avg. daily gain (kg)[a]	0.82(A)	0.37(C)	0.52(B)
Daily feed intake (kg)			
Concentrate	2.25	2.25	2.25
Hay	4.25	3.06	

[a] Values on the same line followed by different letters differ significantly $(p<0.05)$.

From Chen, M. C. and Wang, T. C., *J. Agric. Assoc. China*, NS 103, 94, 1978. With permission.

C. Starch Urea (Starea) for Cattle

A series of experiments were conducted to evaluate the effectiveness of gelatinized urea-sweet potato meal (GUSP), by Chen et al.[42] (Table 14). The cattle receiving soybean meal diets performed significantly better than on urea diets and GUSP diets, though the feeding value of GUSP diets was better than urea diets. Results from metabolism studies appeared to show that the effect of GUSP on the digestibility of dry matter, as well as on crude protein and crude fiber nitrogen retention of blood urea level was similar to the effect of soybean meal, but urea alone was inferior to soybean meal in these respects.

Table 15
FEEDING FRESH AND SUN-DRIED SWEET POTATO CHIPS FOR FATTENING PIGS (BODY WT 26-90 kg)

	Ration	Avg daily intake (kg)	Avg[a] daily gain (kg)	Backfat[a] thickness (cm)	Feeding days to 90 kg body wt
1.	Corn-soybean meal balanced diets	2.02	0.78(A)	2.89(A)	83
2.	Soybean meal	0.41	0.66(B)	2.61(AB)	98
	Corn	1.42			
	Sweet potato vines	1.03			
3.	Soybean meal	0.61	0.63(B)	2.25(B)	102
	Dried sweet potato chips	1.45			
	Sweet potato vines	1.04			
4.	Soybean meal	0.64	0.54(C)	2.46(AB)	118
	Fresh sweet potato chips	4.59			
	Sweet potato vines	1.00			

[a] Values on the same column followed by different letters differ significantly ($p<0.01$ or $p<0.05$).

From Koh, F. K., Chow, W. C., and Tai, N. L., Comparative feeding value of yellow corn and sweet potato chips for growing-finishing pigs, Farm Animal Breeding Station, Taiwan Sugar Corporation Annual Research Report, 33-36 and 37-42, Chunan, Taiwan, 1980.

D. Silage

Bennett et al.[14] found that good quality sweet potato vine silage could be produced if additives such as 75% phosphoric acid (3 to 9 kg/mt) or molasses (25 to 45 kg/mt) were added. Silage without additives became spoiled due to the presence of molds. Sweet potato vine silage contained nearly five times the calcium, nearly twice the phosphorus and 1.5 times the protein of sorghum silage. Ruiz et al.[43] found that wilted vines (leaves and stems) could produce a good quality silage if wilted 1 day before ensiling. Addition of storage roots to the silage resulted in undesirable levels of butyric acid and additions of urea resulted in excessive weight losses due to spoilage.

IV. SWEET POTATOES (ROOTS AND TOPS) IN LIVESTOCK RATIONS

A. Swine Rations
1. Feeding Efficiency and Growth Rates

Koh et al.[44] reported that a daily crude protein intake of 334 g and a daily gross energy intake of 8.5 Mcal would result in satisfactory growth performance for growing-finishing pigs fed containing raw sweet potatoes or dried sweet potato chips. Raw sweet potatoes as a feed for growing-finishing pigs requires more soybean meal or some other protein supplement to obtain satisfactory performance than does a feed of dried sweet potato chips. The cost per kilogram gain of the former is also higher than that of the later. The above results indicate that the utilization of sun-dried chips is more economical than use of raw sweet potatoes (Table 15).

The results of many feeding experiments[36,38,45-47] have confirmed that the performance of pigs fed diets with dried sweet potato chips was not comparable to that of pigs fed diets with corn, but daily gain and feed-to-gain ratio were slightly superior when the pigs were fed diets formulated with dried sweet potato chips substituting 25% of the corn in the ration (Table 16). The rations with sweet potato chips that substitute for half or all of the corn and 0.2% additional L-Lysine could result in satisfactory performance for growing pigs.[48]

Table 16
THE PERFORMANCE OF FATTENING PIGS FED DIFFERENT PROPORTIONS OF CORN AND SWEET POTATO CHIPS

Treatments		Daily[a] gain (kg)	Feed/[a] gain (kg/kg)	Ref.
Corn (% in diets)	Sweet potato chips (% in diet)			
65—83	0	0.53	3.93	45
0	56—72	0.37	4.79	
30—39	30—39	0.48	3.83	
63—81	0	0.65(AB)	3.38	46
45—58	15—20	0.66(A)	3.37	
29—37	29—37	0.62(B)	3.54	
14—18	42—54	0.58(C)	3.74	
0	54—68	0.56(D)	3.81	
72—84	0	0.60(A)	3.08(B)	36
35—41	35—41	0.48(B)	3.84(B)	
0	69—81	0.44(E)	4.08(A)	
69—75	0	0.69(A)	2.95(B)	38
0	63—68	0.60(C)	3.37(A)	
33—36	33—36	0.66(B)	3.13(B)	
72—84	0	0.56	3.14	47
35—41	35—41	0.49	3.71	
0	69—81	0.48	3.80	

[a] Values in the same column followed by different letters differ significantly ($p<0.01$ or $p<0.05$).

2. Carcass Quality of Pigs Fed Sweet Potato Chips

Tai and Lei[46] reported that the back fat thickness of pigs is thinner as the proportion of sweet potato chips is increased in the pig ration. Thinner backfat and a higher percentage of lean cut is obtained when diets are formulated with sweet potato chips to substitute half or all of the corn[36,38] (Table 17).

Porcine fat melting point and unsaturated fatty acid content are significantly ($p<0.05$ or $p<0.01$) lowered when sweet potato is fed to pigs.[49] These experimental results indicate that dried sweet potato chips could improve the carcass quality and porcine fat quality of growing-finishing pigs. Massey[50] reported that sweet potatoes in the diet resulted in firmer flesh of the butchered carcass.

3. Recommended Swine Rations

Adequate formula feeds incorporated with dried sweet potato chips for growing-finishing pigs (Table 18) are recommended by the senior author. The ration of complete balanced diets prepared for growing pigs containing 16% crude protein with 3.2 to 3.3 Mcal DE/kg, and that for finishing pigs contain 14% crude protein with 3.2 to 3.3 Mcal DE/kg are used. The additional 0.2% L-Lysine could be added to the ration when it is necessary to balance the amino acid contents in the diets.

4. Swine Grazing Systems

Massey[50] reported that a stocking rate of 25 to 30 pigs/ha averaging 55 kg/pig could be

Table 17
THE CARCASS MEASUREMENTS OF GROWING-FINISHING PIGS FED DIFFERENT PROPORTIONS OF CORN AND SWEET POTATO CHIPS

Treatment				
Corn (% in diets)	Sweet potato chips (% in diets)	Backfat thickness (cm)	% Lean cut	Ref.
63—81	0	4.47	—	46
45—58	15—20	4.34	—	
29—37	29—37	3.81	—	
14—18	42—54	3.53	—	
0	54—68	3.84	—	
72—84	0	3.57	47.19	36
35—41	35—41	3.20	47.35	
0	69—81	3.30	47.43	
69—75	0	3.25	52.77	38
0	63—68	3.22	54.12	
33—36	33—36	3.10	54.89	

Table 18
RECOMMENDED DIETS INCORPORATED WITH SWEET POTATOES FOR GROWING-FINISHING PIGS

Formula ingredient	Grower		Finisher	
	G1	G2	F1	F2
Soybean meal (44%)	24.0	21.1	26.2	20.9
Yellow corn (8%)	35.4	35.2	30.4	33.0
Sweet potato chips (sun-cured)	35.4	35.2	31.6	34.0
Sweet potato vines (dehydrated)	—	4.0	—	—
Wheat bran	—	—	5.0	5.0
Animal Fat	1.5	1.5	2.1	2.9
Dicalcium phosphate	1.8	1.2	1.2	1.2
Limestone powder	0.2	0.7	0.9	0.9
Common salt	0.5	0.5	0.5	0.5
Vitamin-mineral premix	0.7	0.6	0.7	0.6
Antibiotic premix	0.5	—	0.4	—
L-Lysine-HCl	—	—	0.2	0.2
Calculated nutrient composition				
Crude protein (%)	16.52	14.51	16.51	14.50
Crude fat (%)	4.00	3.83	4.16	4.24
Crude fiber (%)	3.25	3.37	3.45	2.98
Ca (%)	0.72	0.74	0.78	0.77
P (%)	0.60	0.52	0.58	0.56
DE (kcal/kg)	3328	3217	3261	3335

grazed for 50 to 60 days if additional grain was provided. The sweet potatoes were grown for about 4 months prior to grazing. Weight gains of 0.5 to 0.6 kg/pig/day were reported.

Rose[51] described a hog tethering system for grazing swine in Papua New Guinea. The sweet potatoes, planted in mounds about 3 m^2, are grown for 4 months before grazing begins.

A pig (10 kg) is tethered each day to a different mound and is provided 332 g of a commercial pig concentrate at the end of the day. After 4 months, the pig is marketed. Using a continuous cycle 22 pigs/ha/year could be produced. Live weight gains of 494 to 552 g/pig/day were reported.

B. Cattle Diets

The feeding value of sweet potato vines for cattle has been evaluated.[52-54] Fresh sweet vines are palatable to cattle, and a cow of 400 to 500 kg can consume 50 to 70 kg daily. An increased proportion of fresh sweet potato vines produced more milk. No ill effect on health or milk pH, acidity, or specific gravity were found to follow the feeding of fresh materials as the sole roughage. The high cost of mechanical dehydration and the reduction of palatability means that the use of dehydrated sweet potato vines would be less practical than that of fresh, sun-dried, or ensiled vines for feeding to growing heifers.

A number of trials have been conducted to determine the value of sweet potato roots for dairy cows. Replacement of corn in the concentrate ration by dehydrated sweet potato roots resulted in 91.4%,[10] 88%,[23] and 98%[25] of milk produced by cows fed the standard concentrate containing corn. Replacement of half of the corn resulted in 97%,[10] and 100%,[28] of the production of cows fed the standard concentrate. Feeding of dehydrated roots of orange-fleshed cultivars resulted in 22%,[10,26] more vitamin A in the milk and 30% more carotene.[10]

Rusoff et al.[8] found dehydrated sweet potatoes to be equal to corn-soybean silage on a dry weight basis for milk production. Bennett et al.[14] reported that fresh, chopped sweet potato roots could be substituted for sorghum silage in dairy rations at the rate of 1 kg of sweet potatoes substituted for 1.8 kg of silage with less than 1% reduction in milk production and slightly higher butterfat content. Cattle fed sweet potatoes gained 0.42 kg/day while those fed sorghum gained 0.13 kg/day. Greater weight gains for those cows fed sweet potatoes were also reported by Mather et al.[10]

Southwell and Black[7] compared the value of dehydrated sweet potato roots to corn by substituting half and all the corn with dehydrated sweet potatoes in a fatting ration for beef cattle. They found that animals receiving the standard ration gained 1.07 kg/day compared to 1.17 and 0.98 when sweet potatoes replaced half and all of the corn, respectively. Feed-per-gain ratios were 9.51, 9.31, and 9.22 for the standard, one-half replacement, and complete replacement, respectively. Price received for the finished animals was nearly identical for those animals fed on the standard ration and those fed the one-half replacement rations. Animals fed the complete replacement ration received only 92% of the price of the animals fed the standard ration. Briggs et al.[11] estimated, on the basis of digestibility trials with beef cattle, that sweet potatoes are roughly equivalent to corn in total digestible nutrients. Bond and Putnam[24] reported that sweet potato trimmings, a by-product of sweet potato canning, were not equivalent to corn in finishing rations, resulting in lower live weight gains, (1.0 kg/day vs. 1.2 kg/day for corn), higher conversion ratios (9.8 kg feed/kg gain vs. 9.4 for corn) and poorer carcass quality (low good vs. low choice for corn). Overall they estimated that sweet potato trimmings were 80% of the value of corn for fatting beef cattle and that the trimmings should be ground for best results.

Ruiz et al.[43] reported that the rate of growth of sweet potato forage (vines) is better than natural pasture under the same ecological conditions; similar to grass, legume associations; but poorer than grass receiving moderate fertilization. They estimate that "one ha could feed 12 young bulls for 100 days without protein supplements in a quantity sufficient to give high live weight gains." Ruiz et al.[22] compared various proportions of vines and storage roots in beef cattle diets. They found the lowest weight gains (0.66 kg/day) from animals fed all vines and no roots and the best weight gains (0.85 kg/day) from bulls fed 50% vines and 50% roots (fresh wt : fresh wt). Feed conversion ratios were 8.51 and 6.24 for the 100% vine and 50% vine 50% root diets, respectively. They considered the economics of chicks fed a ration low in vitamin A but with 50% sweet potato meal had slightly higher vitamin A levels in their livers after 8 weeks than those fed a low vitamin A diet with 1% cod liver oil.

E. Aquaculture Feeds

Goyert and Avaualt[64] reported on studies using various agricultural by-products as feed for crayfish (*Procambarns clarkii*). They found that dried sweet potato vines and leaves produced the highest final weight by crayfish followed by sweet potato trimmings (a by-product of sweet potato canning), rice stubble, and rye hay. Soy-bean stubble and dried sugar cane stalks were not suitable as supplemental feed for crayfish.

F. Microbiological Feedstocks

Gray and Abou-el-Seoud[65] reported that sweet potatoes could serve as a suitable substrate for a range of *Fungi imperfecti* species. Under optimum experimental conditions 100 kg of sweet potatoes contain 6.9 kg of protein could produce 81.2 kg of dried mycelium and unutilized sweet potato tissue containing 31.6 kg of protein. Thus, the protein concentration can be increased to over four times that of original concentrations. El-Ashwah et al.[66,67] reported that sweet potato homogenate alone was not suitable to support good mycelial growth. Addition of ammonium phosphate (1 g/ℓ) resulted in good mycelial growth and high total crude protein production with *Cladosporium cladosporoides*. Shaking rates were found to be optimum at 150 to 250 rpm. Analysis of amino acid composition indicated that higher levels of lysine, histidine, arginine, tryptophan, phenylalanine, and tyrosine were found in the mycelia than in casein. Leucine + isoleucine and valine were lower than in casein and threonine was very low in all cultures.

Dawson et al.[68] found that waste water from starch milling was a good medium for the production of a yeast, *Torulposis utilis*. With addition of a nitrogen source (ammonium hydroxide) sugar content of the medium was reduced from 0.75 to 1.00% to about 0.05% after 8 hr. They estimate that approximately one ton of dried yeast containing 50% protein could be produced per 100 tons of fresh sweet potatoes processed for starch.

V. CONCLUSIONS

The continuous efforts by many workers have been made to obtain better utilization of sweet potatoes as animal feed, such as (1) determining the optimum proportion of dried sweet potato chips to corn in pig and poultry diets, (2) defining the most effective process to enhance the nutritive value of sweet potatoes, (3) the feeding of a high-protein cultivar to animals, and (4) increasing the nutrient production of sweet potatoes.

It would appear that several conclusions are warranted.

1. Sun-dried sweet potato chips are seen to be most economical as a pigfeed.
2. When chips replaced 25% of corn in a pig ration the results were comparable to that of a whole corn ration.
3. The starch availability and nutritive value of chips were improved and trypsin inhibitor activity was completely eliminated by popping the chips.
4. Increasing nitrogen fertilizer application increased the protein and lysine contents of roots, stems and leaves.
5. There is potential for gelatinization of urea-sweet potato meal as a cattle diet.

It may also be concluded that orange fleshed sweet potatoes, when fed to dairy cattle or laying chickens may enhance the vitamin A value of milk or eggs produced by those animals. The enhancement of protein levels and quality in sweet potatoes through fungal fermentations would seem to be an especially promising possibility in areas where protein concentrates are unavailable or prohibitively expensive.

REFERENCES

1. **Li, L.,** Chiayi Agric. Exp. Stn., Taiwan Agric. Res. Inst., Chiayi, Taiwan, unpublished data, 1979.
2. **Chow, T. Y., Chen, M. C. Chiu, W. S., Chow, P. Y., Chung, P. O., Hsu, C. T., Lee, P. K., Lin, W. T., Peng, H. K., Shen, T. F., Tai, N. L., Yang, C. P., and Yeh, T. P.,** The Tables of Feed Composition in Taiwan, Taiwan Livestock Research Institute and Animal Industry Research Institute of Taiwan Sugar Corporation, Chunan, Taiwan, 1976, 36, 51, 52, 76, 104, 133, 155.
3. **Yeh, T. P.,** Nutrient composition tables of feed stuffs in Taiwan, *J. Chin. Agric. Chem. Soc.*, 1, 56, 1963.
4. **Devendra, C. and Gohl, B. I.,** The chemical composition of Caribbean feeding stuffs, *Trop. Agric. (Trinidad)*, 47, 335, 1970.
5. **Gohl, B.,** *Tropical Feeds,* FAO Animal Production and Health Series No. 12, Food and Agriculture Organization of the United Nations, Rome, 1981, 319.
6. **Job, T. A., Oluyemi, J. A., and Entonu, S.,** Replacing maize with sweet potato in diets for chicks, *Br. Poult. Sci.*, 20, 515, 1979.
7. **Southwell, B. L. and Black, W. H.,** Dehydrated sweet potatoes for fattening steers, *Ga. Agric. Exp. Stn. Bull.*, 45, 1948
8. **Rusoff, L. L., Miller, G. D., Burch, B. J., Jr., and Frye, J. B., Jr.,** Dehydrated sweet potatoes as a substitute for corn-soybean silage, *J. Dairy Sci.*, 33, 657, 1950.
9. **Fetuga, B. L. and Oluyemi, J. A.,** The metabolizable energy of some tropical tuber meals for chicks, *Poult. Sci.*, 55, 868, 1976.
10. **Mather, R. E., Linkous, W. N., and Eheart, J. F.,** Dehydrated sweet potatoes as a concentrate feed for dairy cattle, *J. Dairy Sci.*, 31, 569, 1948.
11. **Briggs, H. M., Gallup, W. D., Heller, V. G., Darlow, A. E., and Cross, F. B.,** The digestibility of dried sweet potatoes by steer and lambs, *Okla. Agric. Exp. Stn. Tech. Bull.*, T-28, 1947.
12. **Lease, E. J. and Mitchell, J. H.,** Biochemical and nutritional studies of dehydrated sweet potato, *S.C. Agric. Exp. Stn. Bull.*, 329, 1940.
13. **Scott, L. E. and Bouwkamp, J. C.,** Seasonal mineral accumulation by the sweet potato, *Hort. Science*, 9, 233, 1974.
14. **Bennett, H. W., Means, R. H., Cowsert, W. C., Leonard, O. A., and Gieger, M.,** Production and utilization of silage in Mississippi, *Miss. Agri. Exp. Stn. Bull.*, 425, 1945.
15. **Yang, T. H., Tai, Y. C., Shen, C. T., Ko, H. S., Chen, S. W., and Blackwell, R. O.,** Studies on protein nutrition of the main food in Taiwan, *J. Chin. Agric. Chem. Soc.*, 13, 132, 1975.
16. **Yeh, T. P., Chen, S. Y., and Wu, M. C.,** Animal Industry Research Institute, Taiwan Sugar Corporation, Chunan, Taiwan, unpublished data, 1979.
17. **Li, L.,** Breeding for increased protein content in sweet potatoes, in *Sweet Potato Proc. First Int. Symp.*, Villareal, R. L. and Griggs, T. D., Eds., Asian Vegetable Research and Development Center, Shanhua, Tainan, Taiwan, 1982, 345.
18. **Li, L.,** Studies on the influence of environment factors on protein content of sweet potatoes, *J. Agric. Assoc. China*, 92, 64, 1975.
19. **Yeh, T. P., Chen, S. Y., and Sun, C. C.,** The effect of fertilizer application on the nutrient composition of high protein cultivars of sweet potatoes, *J. Agric. Assoc. China*, 113, 33, 1981.
20. **Wu, J. F.,** Energy value of sweet potato chips for young swine, *J. Anim. Sci.*, 51, 1261, 1980.
21. **Yoshida, M., Hoshii, H., and Morimoto, H.,** Nutritive value of sweet potato as a carbohydrate source in poultry feed.IV. Biological estimation of available energy of sweet potato by starting chicks, *Agric. Biol. Chem.*, 26, 679, 1962.
22. **Backer, J., Ruiz, M. E., Munoz, H., and Pinchinat, A. M.,** The use of sweet potato (*Ipomoea batatas*, (L.) Lam) in animal feeding. II. Beef production, *Trop. Anim. Prod.*, 5, 152, 1980.
23. **Rusoff, L. L., Seath, D. M., and Miller, G. D.,** Dehydrated sweet potatoes, their feeding value and digestibility, *J. Dairy Sci.*, 30, 769, 1947.
24. **Bond, J. and Putnam, P. A.,** Nutritive value of dehydrated sweet potato trimmings fed to beef steers, *J. Agric. Food Chem.*, 15, 726, 1967.
25. **Frye, J. B., Jr., Hawkins, G. E., Jr., and Henderson, H. B.,** The value of winter pasture and sweet potato meal for lactating dairy cows, *J. Dairy Sci.*, 31, 897, 1978.

26. **Copeland, O. C.,** Dehydrated sweet potato meal vs. ground shelled corn for lactating dairy cows, *Texas Agric. Stn. Annu. Rep.,* 54, 28, 1941.
27. **Tillman, A. D. and Davis, H. J.,** Studies on the use of dehydrated sweet potato meal in chick rations, *La. Agric. Exp. Stn. Bull.,* 358, 1943.
28. **Frye, J. B., Jr., Thomason, J. H., and Henderson, H. B.,** Sweet potato meal versus ground corn in the ration of dairy cows, *J. Dairy Sci.,* 34, 341, 1948.
29. **Sohonie, K. and Bhandarker, A. P.,** Trypsin inhibitors in Indian food stuffs. I. Inhibitors in vegetables, *J. Sci. Ind. Res.* 13B, 500, 1954.
30. **Sohonie, K. and Honawar, P. M.,** Trypsin inhibitors of sweet potato *(Ipomoea batatas), Sci. Culture,* 21, 538, 1956.
31. **Sugiura, M., Ogiso, T., Takeuti, K., Tamura, S., and Ito, A.,** Studies on trypsin inhibitors in sweet potato. I. Purification and some properties, *Biochim. Biophys. Acta,* 328, 407, 1973.
32. **Liu, W. H., Feinstein, G., Osuga, D. T., Haynes, R., and Feeney, R. E.,** Modification of arginines in trypsin inhibitors by 1,2 cyclohexanedione, *Biochemistry,* 7, 2886, 1968.
33. AVRD Sweet potato report for 1975, Asian Vegetable Research and Development Center, Shanhua, Tainan, Taiwan, 1976.
34. **Lin, Y. H. and Chen, H. L.,** Level and heat stability of trypsin inhibitor activity among sweet potato (*Ipomoea batatas* L.) varieties, *Bot. Bull. Acad. Sin.,* 21, 1, 1980.
35. **Bouwkamp, J. C., Tsou, S. C. S., and Lin, S. S. M.,** Trypsin inhibitors in sweet potatoes: Genotype and environment effects, *Proc. Am. Soc. Hortic. Sci. Trop. Region,* 27B, 126, 1983.
36. **Yeh, T. P., Wu, M. C., and Chen, S. Y.,** unpublished data, Animal Industry Research Institute, Taiwan Sugar Corporation, Chunan, Taiwan, 1979.
37. **Yeh, T. P., Wung, S. C., Koh, F. K., Lee, S. Y., and Wu, J. F.,** Improvement in the Nutritive Values of Sweet Potato Chips by Different Methods of Processing, Animal Industry Research Institute, Taiwan Sugar Corporation, Annu. Res. Rep. 1977, 89.
38. **Yeh, T. P., Chen, S. Y. and Chang, Y.,** unpublished data, Animal Industry Research Institute, Taiwan Sugar Corporation, Chunan, Taiwan, 1980.
39. **Yeh, T. P., Wung, S. C., Lin, H. K., and Kuo, C. C.,** Studies on Different Methods of Processing some Local Feed Materials to Enhance their Nutritive Values for Swine, Animal Research Institute, Taiwan Sugar Corporation, Annu. Res. Rep., 1978, 25.
40. **Wu, J. F., Wu, M. C., Chen, S. Y., and Yeh, T. P.,** unpublished data, Pig Research Institute, Chunan, Taiwan, 1980.
41. Pu-li Byproduct Plant, Taiwan Sugar Corporation, unpublished data, 1977-1979.
42. **Chen, M. C. and Wang, T. C.,** Nutritional evaluation of gelatinized urea-sweet potato meal for cattle, *J. Agric. Assoc. China,* NS 103, 94, 1978.
43. **Ruiz, M. E., Lozano, E., and Ruiz, A.,** The use of sweet potato (*Ipomoea batatas* (L.) Lam) in animal feed. III. Addition of various levels of root and urea to sweet potato forage silages, *Trop. Anim. Product.,* 6, 234, 1981.
44. **Koh, F. K., Yeh, T. P., and Yen, H. T.,** Effects of feeding sweet potatoes, dried sweet potato chips and sweet potato vine on the growth performance of growing-finishing pigs, *J. Chinese Soc. Anim. Sci.,* 5, 55, 1976.
45. **Koh, F. K., Chow, W. C., and Tai, N. L.,** Comparative Feeding Value of Yellow Corn and Sweet Potato Chips for Growing-Finishing Pigs, Farm Animals Breeding Stn., Taiwan Sugar Corporation, Annu. Res. Rep. 33-36 and 37-42, Chunan, Taiwan, 1960.
46. **Tai, N. L. and Lei, T. S.,** Determination of proper amount of yellow corn and dried sweet potatoes in swine feed, *J. Agric. Assoc. China,* NS 70, 71, 1970.
47. **Lee P. K. and Lee, M. S.,** Study on hog feed fornula of using high protein sweet potato chips and dehydrated sweet potato vines as the main ingredient, *J. Taiwan Livestock Res.,* 12, 49, 1979.
48. **Chen Y., Yeh, T. P., Yang, Y. S., Chang, T. C., Wu, M. C., Siao, C. M., and Yang, W. L.,** Studies on the Formula Feeds with Sweet Potatoes as Diets for Growing-Finishing Pigs, Annu. Res. Rep. 1979-1980, 95, 1980.
49. **Kawaita, H., Fukumoto, M., Kusunoki, S., and Miyauchi, Y.,** Effects of sweet potato feeding on porcine fat melting point and fatty acid composition, *Jpn. J. Swine Husbandry Res.,* 16, 184, 1979.
50. **Massey, Z. A.,** Feeding sweet potatoes to hogs, *Ga. Agric. Exp. Stn. Press Bull.,* 306, 1929.
51. **Rose, C. J.,** A tethering system of grazing pigs on sweet potatoes in the Tari valley, in proc. *Papua New Guinea Food Crops Conf.,* Wilson, K. and Bourke, R. M., Eds., Department of Primary Industry, Port Moresby, 1976, 151.
52. **Chen, M. C., Yi, J. J., and Hsu, T. C.,** The nutritive value of sweet potato vines produced in Taiwan for cattle, *J. Agric. Assoc. China,* NS 99, 39, 1977.
53. **Chen, M. C. and Chen, C. P.,** Studies on the utilization sweet potato chips and cassava pomace for cattle, *J. Agric. Assoc. China,* NS 107, 45, 1979.

54. **Chen, M. C., Chen, C. P., and Din, S. L.,** The nutritive value of sweet potato vines for cattle, *J. Agric. Assoc. China,* NS 107, 55, 1979.
55. **Cross, F. B.,** Producing and dehydrating sweet potatoes for livestock feed, *Okla. Agric. Exp. Stn. Bull.,* B-477, 1955.
56. **Ruiz, M. E.,** Sweet potatoes (*Ipomoea batatas* (L) Lam) for beef production: agronomic and conservation aspects and animal responses, in *Sweet Potato, Proc. First Int. Int. Symp.,* Villareal, R.L. and Griggs, T. D., Eds., Asian Vegetable Research and Development Center, Shanhau, Tainan, Taiwan, 1982, 439.
57. **Massey, Z. A.,** Sweet potato meal as livestock feed, *Ga. Agric. Exp. Stn. Press Bull.,* 522, 1943.
58. **Lee, P. K. and Yang, Y. F.,** Nutritive value of high protein sweet potato chips as feed ingredient for broilers, *J. Agric. Assoc. China,* NS 106, 71, 1979.
59. **Lee, P. K. and Yang, Y. F.,** Nutritive value of high protein sweet potato meal as feed ingredient for Leghorn chicks, *J. Agric. Assoc. China,* NS 108, 56, 1979.
60. **Squibb, R. L., Mendez, J., and Serimsbaw, N. S.,** Valor de las harinas de comote y achcote en raciones para aves de corral, *Turrialba,* 3, 163.
61. **Turner, W. J., Malyhnicz, G. L., and Nad, H.,** Effect of feeding rations based on cooked sweet potato and a protein supplement to broiler and crossbreed poultry, *Papua New Guinea Agric. J.,* 27, 69, 1976.
62. **Rosenberg, M. M. and Seu, J.,** Sweet potato root meal versus yellow corn meal in the chicks diet, *World's Poult. Sci. J.,* 8, 93, 1952.
63. **Yoshida, M., Hoshii, H., and Morimoto, H.,** Nutritive value of sweet potato as a carbohydrate source in poultry feed. Part V. Reliability of available energy value of sweet potato estimated by feeding experiment, *Agric. Biol. Chem.,* 26, 683, 1962.
64. **Goyert, J. C. and Avault, J. W., Jr.,** Agricultural by-products as a supplemental feed for crayfish, *Procambarus clarkii, Trans. Am. Fish. Soc.,* 106, 629, 1977.
65. **Gray, W. D and Abou-el-Seoud, M. D.,** Fungal protein for food and feeds. II. Whole sweet potato as a substrate, *Econ. Bot.,* 20, 119, 1966.
66. **El-Ashwah, E. T., Musmar, I. T., Ismail, F. A., and Alian, A.,** *Fungi imperfecti* as a source of food protein. II. Effect of sweet potato based media on yield and protein concentration, *Egypt. J. Microbiol.,* 15, 63, 1980.
67. **El-Ashwah, E. T., Musmar, I. T., Ismail, F. A., and Alian, A.,** *Fungi imperfecti* as a source of food protein. III. Chemical composition and amino acid content in mycelium of different cultures, *Egypt. J. Microbiol.,* 15, 73, 1980.
68. **Dawson, P. R., Greathouse, L. H., and Gordon, W. O.,** Sweet potatoes. More than starch, in *Crops in Peace and War.* Yearbook of Agriculture 1950-1951, U.S. Department of Agriculture, Washington, D.C., 195.

CONCLUSIONS — PART II

It is commonly accepted that sweet potato has characteristics of high yielding potential, wide adaptability, multifunctional usage, wide range of chemical compositions available for improvement and disadvantages of high water content and less stability. The desired characteristics of sweet potato, although recognized by many scientists, are not demonstrated in tropical areas. Lin[1] summarized the present production status of sweet potato and pointed out that two groups of countries can be identified based on the yield level. Japan, Korea, and Taiwan represent the high-yielding countries. The yield level is about double that of other Asian countries. The second noticeable point is a sharp decline of harvested area in high-yielding countries after the middle of the 1960s. However, regardless of the decline in total production in these three countries, the per hectare yield improved about 10 to 15% during the same period of time. Yield improvement was not observed in low-yielding countries. The status of high-yielding countries reveals that sweet potato can respond to research and management input. Theoretically, one can double the yield of low-yielding countries simply by technology transfer, but this has not actually happened. The productivity improvement in tropical Asia on other crops such as cereal grains and other vegetables indicates that the lack of institutional mechanism of technology transfer and improvement is not the only reason accounting for present status of the sweet potato industry in tropical Asia. The real situation could be more complex than technology improvement itself.

Since 94% of world production of sweet potato is in Asia and about 80% is in China, the role of sweet potato as demonstrated in Chinese culture and Taiwan's experience can be a good reference to other tropical Asian countries due to its location in the subtropics and recent economic development.

Sweet potato is considered to be a survival crop in traditional Chinese culture. It was an important staple food during war or times of weather crises. Sweet potato contributed to 17.55% of the total food energy in Taiwan diets during 1935 to 1939. About 40% of the total production was for human consumption during that period. The production reached its peak of 3.7 million tons in 1967 at which time only 16% was used directly as food and this contributed to only 5.45% of available dietary energy. It is quite clear from these figures that the increased production of sweet potato during this period of time was not due simply to a shortage of food. The incentive behind the production was for other demands. Animal feed was the primary use of sweet potato. More than 60% of sweet potatoes produced were used as animal feed in 1967. The role of animal feed was replaced by imported corn in the early 1970s and production sharply declined. The experience of Taiwan, indicated that sweet potato can play a significant role in subsistence and post subsistence socio-economic system, but it is a rather poor competitor in a highly market-oriented society. The perspective of this crop in tropical areas depends on how well people can use it in the current socio–economic system and how well it can be improved to fit better in a market–oriented society.

POTENTIAL ROLES OF SWEET POTATO IN SUBSISTENCE SYSTEM

In general, crop yield is lower in a subsistence farming system than in market and profit-oriented systems. In a traditional subsistence economic system, farmers lack the capital to purchase materials and tools such as chemical fertilizer, pesticides, and farm machines to improve their productivity. The distribution system for modern products is also poor. This is often considered the major limiting factor for high productivity of traditional agricultural systems. This factor, however, is not necessarily the only factor limiting the productivity of sweet potato in tropical Asia. Lack of incentives for improvement may be identified as another factor contributing to the present status of sweet potato industry. Yield improvement as well as harvested areas of sweet potatoes in Asian countries is lower than for other crops such as cereal grains and popular vegetables. As a less desired food commodity of a subsistence nature, the demand for sweet potato could be a limiting factor for the development of this crop. Increasing the demand of sweet potato to fit the socio–economic status is an important task to those who are interested in the promotion of sweet potato in less developed countries. Experience obtained in the Chinese system is a case for consideration. That is, use of sweet potato as a backyard animal feed.

Although sweet potato is produced primarily for animal feed, it can be shifted to a food commodity whenever it is needed. Sweet potato is consumed in the household but its final products — animals, are market-oriented. That means the demand of sweet potato will not be limited by the need of each household, but rather by the ambitions of individual farmers to increase their income and improve the quality of their lives. This arrangement may create a much higher incentive for farmers to improve the production of sweet potato. There are, however, constraints to this type of transition and these require research effort from various disciplines.

Sweet potato as a backyard animal feed has been practiced in Chinese cultures for a long time but this system has never been developed and practiced in the Western world. Required technologies have to be developed and practiced which can be easily transferred. Technology development may be less efficient and could take a longer time. Pigs are traditionally a backyard animal among traditional Chinese farmers, but pigs may not necessarily be the best choice in other tropical countries. Culture, religion, and other factors may make pigs a less profitable commodity. Proper animal selection has to be worked out in each country as the basis for technology development.

Sweet potato, although rich in carbohydrates and several other nutrients, is deficient in many nutrients that are essential for animal production. Protein is the primary concern of sweet potato-based feed diet. By-products of the soybean industry have played an important role in traditional Chinese systems. Soybean cake of oil press process, residues of tofu manufacturing, and soybean paste of soysauce extraction are common protein supplements in sweet potato-based diets. Those materials are not readily available in many tropical Asian countries. Research to increase the protein content of the sweet potato root, and promoting acceptable high protein crops and identification of locally available low cost protein sources are important to guarantee the success of this approach. Feed formulation needs to be determined for best feeding efficiency at various growth stages of the animals.

Sweet potato-based feed diets have a lower feeding efficiency than maize, especially on protein digestibility. Trypsin inhibitor and starch properties have been identified as the two major factors effecting protein digestibility.[2-4] Attempts have been made to improve protein digestibility through breeding. Results show that this approach might be feasible. Cooking was practiced in the traditional Chinese systems to improve feeding efficiency. It is however, labor and fuel intensive. A farming system of sweet potato-pig-biogas fermentation has been suggested as a possible way to generate fuel needed from a self-sustaining system. This approach has not been tested in tropical environments.

POTENTIAL ROLES OF SWEET POTATO IN A MARKET-ORIENTED SYSTEM

Although sweet potato often fits well in subsistence agricultural systems, it can be developed into a market-oriented crop. First of all the crop yield will respond to input levels — both research and management input. An average dry matter yield of 11 tons under intermediate input level has been observed by Lin in 10 locations. High yield potential has been demonstrated also in the institutes.[2] Production-wise, sweet potato has a strong competitive potential; but as a market-oriented crop, the demand could be the major concern.

In developed countries such as the U.S. and Japan, sweet potato has been accepted as a vegetable. The consumption level of this commodity has reached a stable balanced level. Quality is the major emphasis in both of these countries although the desired characteristics are different. Sweet potato thus becomes a typical horticultural crop. Sweet potato is not yet very popular in other parts of the world such as Europe. It may be accepted in their diet if they are more available to consumers and of desired quality.

Market-oriented animal feed is a potentially feasible market of sweet potato. Dehydration and uniform quality are major constraints. Relatively large production size using the same variety is essential. Low cost dehydration processes such as solar dryers need to be developed in order to handle large volumes of products. A uniform product can be established either by a well organized extension and gathering system or through a big farm operation. Small scale production will be difficult because of quality variation due to variety and management level. Even under the present yield level of sweet potato in Taiwan, sweet potato potentially can compete with locally produced corn although it is not yet competitive with imported corn from the U.S. on a unit energy basis.

Sweet potato is a potential source of starch. The major competing crops are corn and cassava. Corn has the advantage of dry grain which is much easier for transportation and storage. One future potential of sweet potato starch is a two-step processing. Raw materials can be processed to unpurified starch at the production sites and transported to a central reprocessing site. The purified starch could then be sold in the international market in order to reduce the transportation cost. Low cost technology for starch recovery needs to be developed. Proper utilization of by-products at the tropical production sites needs to be developed in order to cover part of the processing cost. Cassava has become an international trade commodity for many reasons. One of them is it can grow well in marginal land due to its drought tolerance. The input requirement of cassava is lower than sweet potato and this makes it a very attractive commodity in developing countries. The potential of drought tolerance of sweet potato has not received enough attention. Variety and cultural practices to grow sweet potato in marginal land and dry season deserves more research attention. The advantage of a shorter growing period for sweet potato than cassava should be utilized in cropping systems.

Sweet potato starch can be used in many ways but it lacks a special property. In most cases sweet potato starch is used as a substitute raw material but it is not the first choice. Utilization based on the special properties of sweet potato starch utilization through chemical modification deserves more study.

FUTURE RESEARCH NEEDED

Unlike many other crops, sweet potato has never been an important commodity in the developed world and has received only limited attention from modern science. Technology has to be worked out by the scientists in the developing world without too much help from the experience of industrialized countries. This situation makes the progress of sweet potato improvement relatively slow and inefficient. The research needed may vary from basic research to utilization but several fields deserve particular attention.

The number and complexity of the problems remaining in the utilization of sweet potatoes in the tropics are a clear indication that coordinated efforts of multidisciplinary terms will be required to successfully attack these problems. The magnitude of the problems should be viewed in the context of the great potential of a better life for many people if the problems are resolved. The potentials for utilization as food, feed, and industrial products have been documented in the previous chapters.

Sweet potatoes are efficient producers of dry matter and calories but these are diluted in a high moisture product. Although sweet potatoes are fairly well adapted for storage and transport as compared to vegetable crops, they are poorly adapted as compared to cereal and grain legumes. As a result, most efforts at extending the utilization of sweet potatoes are conducted with the aim of increasing and storability, increasing the energy content (dry matter) or both.

The use of controlled temperature storages in temperate regions has revealed striking differences among cultivars for resistance to storage diseases and maintenance of quality. It is now possible to hold fresh sweet potatoes in commercial storages for up to 9 months. Although storage for such a duration may not be a practical goal under tropical conditions, it is likely that considerable progress can be made by breeding and selection for longer storage life in ambient tropical conditions.

Increasing the utilization of fresh roots as food will depend on the willingness of the population to increase their consumption. Tsou and Villareal[5] have outlined a number of reasons for the reluctance of people to increase their consumption of sweet potatoes. The solution to these problems will almost certainly require the efforts of social scientists as well as biological scientists.

Increased utilization of sweet potatoes does not depend on the resolution of problems surrounding consumption of fresh roots. Maize and wheat are staples in many parts of the world but are rarely consumed except as ingredients. Thorough evaluation of physical and chemical properties of the starch of a large number of cultivars would more clearly define the potential of sweet potatoes as an ingredient in various food products. The development and marketing of these products may also require substantial efforts.

Utilization of sweet potatoes as processed products would seem to depend on the determination of which products are most likely to be acceptable to consumers and which quality parameters are the most important. Economic and marketing studies may be necessary to determine if processing is feasible.

The use of sweet potatoes in animal feeds seems to have a good potential for the near term future. As economies develop, it is nearly certain that demand for animal products will increase. The presence of trypsin inhibitors requires some additional research effort to reduce the levels through breeding or treatment, especially if monogastric animals are to be fed. Research on starch digestibility may also increase feed conversion efficiency, particularly as we begin to understand the range of physical and chemical properties of the starches.

Successful utilization may depend on development and production of a variety of starch and flavor qualities. One needs only to consider the effects on starch and flavor qualities on the utilization purposes of wheat and rice to realize that utilization of a crop depends more on the development of a variety of attributes than on the limitation of a single type.

As mentioned previously, a rather extensive body of information exists, and is reasonably available. What is needed is an organized system for rapid and easy exchange of information. National and international societies and working groups have certainly assisted in the exchange of information as have various symposia and workshops. An international program, adequately funded to provide complete and up–to–date bibliographies and abstracts, would certainly increase the likelihood of developing the crop to fulfill its potential.

REFERENCES

1. **Lin, S. S. M., Peet, C. C., Chen, D. M., and Lo, H. F.**, Breeding goals for sweet potato in Asia and the Pacific: a survey on sweet potato production and utilization. *Proc. Am. Soc. Hortic. Sci., Trop. Region*, 27B, 42, 1983.
2. AVRDC Sweet Potato Report 1975, Asian Vegetable Research and Development Center, Shanhua, Tainan, Taiwan, 1976, 26.
3. AVRDC, Annu. Progress Rep. of 1983, Asian Vegetable Research and Development Center, Shanhua, Tainan, Taiwan, 1984.
4. **Sheu, C. T.**, Studies on the effect of pre-heating of sweet potato raw material for the elimination of antitryptic factor and the improvement of feed efficiency, *Mem. Coll. Agric., Nat. Taiwan Univ.*, 19(2), 33, 1979.
5. **Tsou, S. C. S. and Villareal, R. L.**, Resistance to eating sweet potato, In *Sweet Potato, Proc. First Int. Symp.*, Villareal, R. L. and Griggs, T. D., Eds., Asian Vegetable Research and Development Center, Shanhua, Tainan, Taiwan, 1982, 37.

Index

INDEX

A

Acid soils, see also Soil, 10
Adventitious roots, 82—83
Agriculture, tropical, 3
Alcidodes dentipes, 60, 61
Alcohol, production of, 227—229
Alkaline soils, see also Soil, 10
Aluminum toxicity, 21
Amino acids, see also Protein, 155, 157
Animal feed, see also Processing, 149
    back yard, 256
    in China, 255
    corn vs. sweet potato as, 240
    needed research on, 258
    processing for, 238—244
    sweet potatoes as, 250
    varieties bred for, 147
Animal manures, 19, 20
Antibiosis, 65, 66
Antinutritional factors, 159
Aphids
    cotton, 71
    green peach, 72—72
    viral pathogens transmitted by, 49—51
Aphis gossypii, 71
Aquaculture, sweet potatoes, in, 250
Architecture, sweet potato, 81
    aboveground plant parts, 88—90
    roots, 82—88
Ascorbic acid, 5, 158
    effect of canning on, 194
    effect of cooking on, 168
    nutritive value of, 151
Ash, in sweet potato tips, 181
Asia
    average yield in, 142
    desired varietal characteristics in, 144—145
    disease and insect pests in, 145, 146
    environmental constraints in, 145—146
    growing environment in, 141—144
    potential for sweet potato in, 147
    sweet potato production in, 140
    sweet potato yield in, 255
    utilization of sweet potato in, 140—141, 143
    variety improvement programs in, 146—147
Asian Vegetable Research and Development Center (AVRDC), 147
    index system of, 149
    research on sweet potato tips of, 180—182
    varieties developed for tips at, 176
    vegetable garden of, 152
Aspidomorpha spp., 60—62

B

Baked products, 211—212

Bangladesh
    sweet potato production in, 140, 142
    utilization of sweet potato in, 140, 143
Barley, see also Food crops, 4
Bemisia tabaci, 71—72
Beta-carotene, see also Vitamin A, 158, 194
Biological yield, 116, 117
Black rot, 37, 40, 41
Blue stem, 44
Boron
    deficiency, 13, 41
    requirements, 18
Branching, 89, 91
Brazil, ethanol production in, 227
Bread making, sweet potato flour in, 211
Breakfast foods, 212
By-products, use of, 229

C

Cabbage, 151
    Chinese, 152
Calcium, 5
    deficiency, 12
    nutritive value of, 151
    requirements, 18
    in sweet potato tips, 181
Calcium chloride treatment, firmness and, 193
Calorie production, see also Energy, 3, 4
Candies, 212
Canning
    factors affecting color, 193—194
    factors affecting firmness, 190—193
    factors affecting nutritional quality, 194—196
    procedures for, 189—190
    retort schedules, 190, 191
Carbofuran, 64
Carbohydrates
    composition of, 154
    effect of cooking on, 166, 167
Carbon
    deposition of, 101
    recycling of, 104
    transport, 88
β-Carotene, see also Vitamin A, 158, 194
Carotenes, 158
Carrot, nutritional value of, 152
Cassava, see also Food crops, 4, 257
Cattle
    starch urea (starea) for, 243—244
    sweet potato rations for, 247—248
Ceratocystis fimbriata, 39, 40
Character impact compounds, 214
Charcoal rot, 41, 42
China
    sweet potato as animal feed in, 256
    sweet potato production in, 140

sweet potato yield, 255
utilization of sweet potato in, 140—141, 143, 176
Chinese cabbage, 152
Chips, sweet potato
   corn chips vs., 239
   preparation of, 210—211
   production problems of, 211
Chloramben, 27
CIPC, 27
Circular spot, 42, 43
Clamps, 163
Coleoptera
   Alcidodes dentipes, 60, 61
   Aspidomorpha spp., 60—62
   Cylas spp., 61—65
   Diabrotica spp., 65, 67
   Euscepes postfasciatus, 65—68
   Plectris aliena, 68—69
   Typophorus spp., 70—71
Color
   effect of cooking on, 166
   factors affecting, 193—194
   of processed products, 214
Commodity nutrition value, 149
Confectionary products, 212
Consumer preference, determination of, 189
Cooking, effect on flavor of, 214
Copper requirements, 19
Corn
   advantages of, 257
   sweet potato vs., 143, 238
      as animal feed, 240, 247
      chips, 239
Cotton aphid, 71
Cowpea, nutritional value of, 152
Crackers, 212
Crop growth rate, 119—121
Crop rotation, to reduce weevil infestations, 64
Crops, see also Food crops
   annual vs. perennial, 6
   energy production of, 221
   fresh weight and calories for, 4
   survival, 176, 255
Cultivars
   AIS 016-1, 179
   Centennial, 167
   BNAS White, 178, 179
   Daja 380, 179
   developed in Asia, 147
   Dilaw, 178—180
   Earlyport, 179
   "Fadenawena", 115
   Garnet, 155
   "Gem", 21
   Georgia Red, 161
   Goldrush, 161, 198
   HM-16, 179
   HP-4, 240
   HP-14, 239
   HP-18, 238, 239
   Jewel, 43, 89, 90, 155, 167
   Kinangkong, 178, 179
   Koganesengan, 230
   Minamiyutaka, 230, 231
   Nancy Hall, 165
   needed research on, 258
   Nemagold, 188
   Porto Rico, 188
   resistant, 37, 43, 60, 66
   Rose Centennial, 179
   Tainan, 237
   Tainung, 237, 238, 240
   tip and root yield, 179
   Triumph, 165
Cultivation, 3
Curing, 160—162
Cylas spp., 61—65

D

Dacthal, 27
Decay, curing and, 161
Deficiency symptoms, 11
Dehydrated sweet potatoes
   causes of quality loss in, 209—210
   flakes, 198—199
   preparation of, 208—209
   slices, dices, strips, 197—198
Density, plant, 25—26, 125
Diabrotica spp., 65, 67
Digestibility, 236, 240, 256
Diphenamid, 27
Diplodia gossypina, 40—41
Diseases, effect on yield of, see also Pathogens, 126
Distilled spirit production, 226
Drainage, requirements for, 10
Dry-heating process, 238—239, 241, 242

E

Eating texture, 145
Economy
   market-oriented, 256—257
   subsistance, 255—256
Edaphic requirements
   mineral nutrition, 11—21
   soil reaction (pH), 10—11
   soil texture and drainage, 10
Electron transport pathway, 105, 106
Energy
   in nutrition value, 151
   respiratory losses of
      dark respiration, 105—107
      input-loss ratios, 109—110
      photorespiration, 104—105
      respiratory quotients, 109—110
      respiratory rate as function of plant part and age, 107—109

storage of
root development, 111—113
root induction, 111
root initiation, 110—111
Energy acquisition, 90—94
environmental factors, 97—98
plant factors
leaf age, 96, 97
leaf chlorophyll concentration, 95
leaf geometry, 96
leaf starch and sugar concentration, 96—97
leaf stomata, 95—96
Energy allocation
recycling of carbon, 104
sink strength in relation to, 103—104
within plant, 103
Energy farming, 227
Energy transport, 98—99
modeling, 102
phloem loading, 99
phloem transport, 99—100
phloem unloading, 100—101
Environment
energy acquisition and, 97—98
energy allocation and, 103
factors affecting yield
length of growing season, 122—123
radiation, 122
rainfall, 123
temperature, 122
as production constraint in Asia and Pacific, 145—146
reproductive biology and, 81
requirements
light, 23
moisture, 23
temperature, 21—23
respiration and, 106
Erwinia chrysanthemi, 39
Ethanol, production of, 227—229
Ethylene, 126
Euscepes postfasciatus, 65

F

Feathery mottle virus (FMV), 49—51
Fermented products
alcohol production, 227—229
distilled spirit "Shochu", 226—227
Fertilizers
chemical, 14—19
excessive, 20
organic, 19—21
sweet potato tips and, 178
Fiber, dietary
nutritive value of, 154
proportions of, 157
in sweet potato tips, 181
Fibrous roots, 83, 85, 108
Fiji, disease and insect pests in, 145, 146

Firmness, factors affecting, 190—193
Flakes, sweet potato, 198—199
Flavor
development of variety with mild, 180
of processed products, 213—214
Flesh color, see also Color, 145
Flies, see Fruit flies; White flies
FMV, see Feathery mottle virus
Food composition tables, 149
Food crops, see also Crops
adaptation of, 5
annual production of, 4
calorie yields for, 6
Formed sweet potatoes, 211
Freezing, procedures for, 196—197
Fruit flies, 45
Fungicides, 3, 37
Fusarium
oxysporum, 44
solani, 44
Fusarium diseases, 37
Fusarium wilt, 44, 145
Fungal pathogens, 39
Ceratocystis fimbriata, 39, 40
Diplodia gossypina, 40—41
Fusarium oxysporum, 44
Fusarium solani, 44
leaf spot, 45
Macrophomina phaseolina, 41—42
Monilochaetes infuscans, 41, 42
Rhizopus nigricans, 44—45
Sclerotium rolfsii, 42
Sphaceloma batatas, 45

G

Garland chrysanthemum, nutritional value of, 152
Gelatinized urea-sweet potato meal (GUSP), 243
Glycolysis, 105
Grape, see also Food crops, 4
Green peach aphid, 72—73
Greens, sweet potato, see also Tips, 141, 145
desirable qualities of, 145
Growing period, effect on yield of, 122—123, 125
Growth analyses
area and volume measurements, 113—116
plant weight measurements, 116—118
rate measurements, 116, 118—122
Grubs, see *Plectris aliena*
GUSP, see Gelatinized urea-sweet potato meal

H

Harvest
nutritional components at, 154—159
rate of respiration after, 110
Harvest index, 116, 118
Harvesting
creeping growth habit and, 182

methods, 159—160
post-harvest handling
  curing, 160—162
  storage, 162—165
procedure, 160
Herbicides, 3, 27
Herse convolvuli, 73, 74
Homoptera
  Aphis gossypii, 71
  Bemisia tabaci, 71—72
  Myzus persicae, 72—73
Hormones, 126
Hydrolysis, partial starch, 207

I

Index system, nutrition, 149
India
  disease and insect pests in, 145, 146
  sweet potato production in, 140, 142
  utilization of sweet potato in, 140, 143
Indonesia
  consumption patterns in, 221
  disease and insect pests in, 145, 146
  sweet potato production in, 140, 142
  utilization of sweet potato in, 140, 143, 176
Industrial materials, varieties bred for, see also
    Processed products, 147
Industrial use, sweet potatoes for, 230—232
Infant food, sweet potato puree as, 212
Infection, prevention of, 36
Insect control
  green peach aphid, 73
  grubs, 69
  leaf beetle, 71
  moths, 73
  rootworms, 65
  scarabee, 68
  stem borers, 74
  striped sweet potato weevil, 60
  sweet potato weevil, 63—65
  tortoise beetle, 61
  vine borers, 74
  whitefly, 72
Insecticides, 3
  alternatives to, 60
  for sweet potato weevil protection, 64
Insect pests
  Coleoptera
    Alcidodes dentipes, 60, 61
    Aspidomorpha spp., 60—62
    Cylas spp., 61—65
    Diabrotica spp., 65, 67
    Euscepes postfasciatus, 65—68
    Plectris aliena, 68—69
    Typophorus spp., 70—71
  effect of yield of, 126
  Homoptera
    Aphis gossypii, 71
    Bemisia tabaci, 71—72
    Myzus persicae, 72—73
  Lepidoptera
    Herse convolvuli, 73, 74
    Omphisa anastomosalis, 73—75
"Insurance crop", 6
Internal cork disease, 51, 52
Internode length, 89
Ipomeamarone, 168
Ipomoea batatas, see Sweet potato
Ipomoein, 155
Iron, 5
  deficiency symptoms of, 13
  in nutrition value, 151
  in sweet potato tips, 181
Irrigation, see also Moisture
  effect on yield of, 124—125
  supplemental, 4
Ishku-byo, 38

J

Japan
  consumption patterns in, 221
  disease and insect pests in, 145, 146
  ethanol production in, 227—229
  pollution regulation in, 229
  starch production in, 221
  sweet potato consumption in, 257
  sweet potato production in, 140
  sweet potato yield in, 255
  utilization of sweet potato in, 140—142
Java black rot, 40, 41

K

Kale, nutritional value of, 152
Koji, 226
Korea
  consumption patterns in, 221
  disease and insect pests in, 145, 146
  sweet potato production in, 140—142
  sweet potato yield in, 255
  utilization of sweet potato in, 140—141, 143, 176
Krebs cycle, 105

L

LAD, see Leaf area duration
Lateral roots, 82, 86
Leaching losses, 20
Leaf area, measurement of, 113
Leaf area duration (LAD), 113—115
Leaf area index, 114, 121
Leaf beetle, sweet potato, 70—71
Leaf laminae, propagation from, 25
Leaf plastochron index, 97
Leaf spot pathogens, 45

Leaves
  age of, 96, 97
  characteristics of, 90
  chlorophyll concentration in, 95
  geometry of, 96
  respiratory rate for, 108, 109
  shapes of, 92
  starch and sugar concentration in, 96—97
  stomata, 95—96
Leek, Chinese, nutritional value of, 152
Lepidoptera
  Herse convolvuli, 73, 74
  Omphisa anastomosalis, 73—75
Liberia, utilization of sweet potato tips in, 176
Light, requirements for, 23
Lipids, 158
Little leaf disease, 38, 145
Livestock rations
  for aquaculture, 250
  for cattle, 247—248
  for microbiological feedstocks, 250
  for poultry, 248—250
  for sheep, 248
  for swine, 244—247
Lysine, 239, 242

M

Macrophomina phaseolina, 41—42
Magnesium
  deficiency, 12
  requirements, 18
Maize, see also Food crops, 4
Malaysia
  disease and insect pests in, 145, 146
  utilization of sweet potato tips in, 176
Manganese
  deficiency, 12
  requirements, 18
  toxicity, 21
Marketing oriented system, potential roles of sweet potato in, 256—257
Meloidogyne spp., 45—48
Microbiological feedstocks, sweet potatoes in, 250
Mineral nutrition, 158—159
  calcium, 18
  deficiency symptoms, 11—13
  magnesium, 18
  minor elements, 18
  nitrogen, 14—16
  organic fertilizers, 19—21
  phosphorus, 16—17
  potassium, 17
  response to adverse soil conditions and, 21
  sulfur studies, 18
Moisture
  requirements for, 23
  in sweet potato tips, 181
Monilochaetes infuscans, 41, 42
Mounds, 163

Münch hypothesis of phloem transport, 102
Münch pressure flow theory, 99
Myzus persicae, 72—73

N

Nematicides, 48
Nematode pathogens, 45—48
  Meloidogyne, 46—48
  Rotylenchulus reniformis, 48
Nematodes
  distribution in Asia and Pacific, 146
  effect on yield of, 126
  root-knot, 46, 48
New Guinea, utilization of sweet potato tips in, 176
New Zealand, disease and insect pests in, 145, 146
Niacin, 5, 158, 168
Nitrogen, 123
  balance, 158
  deficiency, 11, 12
  effect on yield of, 239
  requirements, 14—16
Njue, disease and insect pests in, 145, 146
Nonprotein nitrogen (NPN) fraction, 155
Noodles, 224
North Carolina Differential Host Test, 48
Nutrition, see Mineral nutrition
Nutritional quality
  digestible nutrients, 236, 240
  nutrient composition, 236, 237
Nutritive value, 3, 5
  calculation of, 150—151
  market price of, 151
  preparation methods and, 165—168
  of sweet potato tips, 178—180

O

Oceania, utilization of sweet potato in, see also Pacific Islands, 141
Omphisa anastomosalis, 73—75
Oxalic acid, in sweet potato tips, 181
Oxygen, in respiration process, 106

P

Pacific Islands, see also specific islands
  average yield in, 142
  desired varietal characteristics in, 144—145
  disease and insect pests in, 145, 146
  environmental constraints in, 145—146
  growing environment in, 141—144
  potential for sweet potato in, 147
  sweet potato production in, 140
  utilization of sweet potato in, 140—141, 143
  variety improvement programs in, 146—147
Palau Islands, disease and insect pests in, 145, 146
Papaya, nutritive value of, 151

Papua New Guinea (PNG)
  disease and insect pests in, 145, 146
  main staple crop of, 154
  sweet potato production in, 140, 142
  utilization of sweet potato in, 140, 143
Pasteur effect, 107
Pathogens, sweet potato
  bacterial, 38—39
  control of, 36—38
  Erwinia chrysanthemi, 38—39
  fungal, 39—45
  mycoplasma, 38—39
  nematode, 45—48
  Streptomyces ipomoea, 39
  viral, 48—53
  witches' broom agent, 38
Patties, sweet potato, production of, 211
Pectins, 165
  firmness and, 192
  nutritive value of, 154
Peeling, process of, 187—188
Pencil roots, 85—86
Penghu Islands, disease and insect pests in, 145, 146
PER, see Protein efficiency ratio
Petioles, 81
  consumption of, 176
  length of, 89
  respiratory rate for, 109
Phellogen activity, 84
Philippines
  disease and insect pests in, 145, 146
  ethanol production in, 227
  sweet potato production in, 140, 142
  utilization of sweet potato in, 140, 143, 176
Phloem
  loading, 99
  transport, 99
  unloading, 100—101
Phosphorus, 5
  deficiency, 12
  requirements, 16—17
Photorespiration, 104—105
Photosynthate transport, 94
Photosynthesis, see also Energy acquisition
  dark reactions of, 93
  influence of oxygen concentration on, 106
  light in, 92—93
  maximum rate of, 94
  process of, 90
Photosynthetic canopy, 81
Pies, sweet potato, 212
Pigs, sweet potato rations for, 224—247
Plant growth regulators, 126
Plectris aliena, 68—69
PNG, see Papua New Guinea
Potassium, 123
  deficiency, 12
  requirements, 17
Pox, 39
Potato, see also Food crops; Sweet potato, 4

Poultry, sweet potato rations for, 248—250
Preparation systems and nutritive value, 165—168
Processed products
  baked, 211—212
  chips, 210—211
  dehydrated sweet sweet potatoes, 208—210
  formed potatoes, 211
  needed research on, 258
  patties, 211
  problems with, 212—213
    color, 214
    flavor, 213—214
    texture, 214—215
  puree, 206—208
  specialty products, 212
Processing, see also Canning; Dehydration: Freezing
  cleaning in, 186
  for feed
    dry-heating, 238—239, 241, 242
    pelleting, 240, 243
    silage, 244
    starch urea (starea) for cattle, 243—244
  peeling and trimming, 187—188
  preheating, 186
  receiving and grading, 186
Production
  factors limiting, 255—256
  propagating materials, 24—25
  transplanting, 25—26
  weed control, 26—27
Protein, 5, 144
  content and quality, 154—155, 157—158
  effect of canning on, 194
  effect of cooking on, 166—167
  in nutrition value, 151
  in sweet potato tips, 181
Protein efficiency ratio (PER), 155
Pulp, starch, 229
Puree, sweet potato, 206—208

Q

Quality, see Nutritional quality
Quarantine regulations, 38

R

Radiation, effect on yield of, 122
Radish, nutritional value of, 152
Rainfall, effect on yield of, 123
Rate measurements
  crop growth, 119—121
  net assimilation rate, 121—122
  relative growth, 119—121
Raw materials, measurement of properties of
  dry matter content, 225—226
  starch content, 225
Recommended Daily Allowance, limitation of, 149
Relative growth rate, 119

Relative nutrient cost (RNC$_j$), calculation of, 149
Research
  needed, 257—258
  on storage, 164
Resistant sweet potato lines, see also Cultivars, 65, 66
Respiration
  dark, 105—107
  photo-, 104—105
  rate as function of plant part and age, 107—109
Respiratory quotient (RQ$_{10}$), 109
Retort schedules, 190, 191
Rhizoctonia bataticola, 41
Rhizopus nigricans, 44—45
Riboflavin, 5, 158, 168
Rice, see also Food crops, 4, 151
Rice culture, adaptability of sweet potatoes following, 21
RNC, see Relative nutrient cost
Rodents, see also Storage, 45
Rooted leaves, propagation from, 25
Roots, see also Animal feeds
  classification of
    adventitious, 82, 83
    fibrous, 83, 85
    lateral, 82, 86
    pencil, 85—86
    storage, 83—85
  distribution and architecture, 86—88
  enzymes in, 101
  insect control for, 64
  oxygen conditions for, 106—107
  pathology of, 36
  proximate analysis of, 238
  respiratory rate for, 109
  weights for, 12
Root system, 81
  carbon allocation within, 88
  primary components of, 85
Rootworms, 65, 67
Rotylenchulus, 45—46
Rotylenchulus reniformis, 48
Russet crack disease, 50—52

S

Salinity, soil, tolerance to, 21
Scab, sweet potato, 37, 46, 145
Scarabee, 65—68
Sclerotial blight, 42
Sclerotium rolfsii, 42—43
Scurf, 37, 41
Season, growing, effect on yield of, 122—123, 125
"Seed stock", disease tolerance of, 37
Seed supply, 6
Sexual reproduction, 81
Sheep, sweet potato rations for, 248
Shochu, 226
Sierra Leone, utilization of sweet potato tips in, 176
Skin, 145

Snacks, 212
Soft rot, 44—45
Soil
  adverse conditions of, 21
  disease of, 11
  fertility of, 123
  loss, 6
  mounding, 124
  oxygen concentration of, 124
  pH of, 4, 21, 124
  reaction (pH), 10
  ridging, 124
  root systems and, 83
  texture and drainage, 10
  type of, 124
Soil rot, 39
Sorghum, see also Food crops, 4
South American pasture beetle, 68—69
Southern stem rot, 42
Soybean, see also Food crops, 4
  nutritive value of, 151
  sweet potato vs., 143
Spacing, see Density, plant
Specialty products, 212
Sphaceloma batatas, 45, 46
SPMMV, 51, 53
Starch
  composition of, 154, 156
  manufacturing of, 149, 220
    in Japan, 221
    process of, 221—223
  partial hydrolysis of, 207
  properties of, 224—225, 256
  respiratory oxidation of, 105
  uses of, 223
Stele cells, 84
Stem borer, sweet potato, 73—75, 145
Stem rot, 44
Stems, respiratory rate for, 108, 109
Stomata, energy acquisition and, see also Leaves, 98
Storage
  clamp, 163
  long-term, 162
  research on, 164, 257—258
  short-term, 162
  sweet potato weevil and, 164
Storage roots, 83—85, 88
  development, 111—113
  induction, 111
  initiation of, 110—111
  measuring volume of, 115—116
  respiratory rate for, 108, 109
  susceptibility of, 107
  water content of, 109
Streptomyces ipomoea, 37
Stress metabolites, 168
Subsistance system, potential roles of sweet potato in, 255—256
Sugar
  kinds of, 154

respiratory oxidation of, 105
Sulfur
  deficiency, 12
  requirements, 18
Sweetness, 145
Sweet potato, see also Food crops, 4
  commercial potential of, 169
  as "cultigen", 81
  elemental concentrations of deficient and normal, 13—14
Sweet potato vein mosaic, 51
Swine, sweet potato rations for, 244—247
Syrup
  firmness and, 192
  starch, 223—224

T

Taiwan
  consumption patterns in, 221
  disease and insect pests in, 145, 146
  high-yield cultivars in, 237
  imports vs. sweet potato in, 143
  relative nutrient cost in, 150
  sweet potato production in, 140
  sweet potato yield in, 255
  utilization of sweet potato in, 140—141, 143
Tanzania, utilization of sweet potato tips in, 176
Technology input, 3
Temperature
  effect on yield of, 122
  energy acquisition and, 97
  requirements, 21—23
Textural properties
  of cooked sweet potato, 165—166
  of processed products, 214—215
Thailand
  ethanol production in, 227
  utilization of sweet potato tips in, 176
Thiamin, 5, 158, 168
Tips, sweet potato, see also Greens
  chemical constituents of, 180, 182
  consumer survey on, 176
  eating quality of, 182
  evaluation of, 177—180
  flavor of, 180
  morphological traits of, 177
  nutritional constituents of, 176, 181
  nutritional value of, 150, 178—180
  toughness of, 180
Tobacco whitefly, 71—72
Tonga, disease and insect pests in, 145, 146
Tortoise beetle, 60—62
Toughness, evaluation of, 180
Transplanting, 25—26
Tropics, adaptation to, 3
Trypsin inhibitors, 159, 241, 242, 256
  discovery of, 236
  effect of cooking on, 166
  protein and, 237—238
Typophorus spp., 70—71

U

United States, sweet potato consumption in, 257
Utilization, sweet potato
  by-product, 229
  expansion of, 206
  industrial, 230—232
  methods, of, 149
  needed research on, 258
  of roots, 140—141

V

Vanuatu, disease and insect pests in, 145, 146
Varietal characteristics, desired, see also Cultivars, 144—145
Vegetables, nutritional value of, 152
Vine borer, sweet potato, 73—75
Vine cuttings, see also Tips; Vines, 24—25
Vines, see also Animal feeds, 81
  branching of, 89, 91
  energy allocation and, 103
  length of, 89, 90
  proximate analysis of, 238
  relative nutrient cost of, 150
  weights for, 12
Viral pathogens, 48—53
  aphid-transmitted, 49—51
  distribution in Asia and Pacific, 146
  feathery mottle virus, 49—51
  mild mottle virus, 51, 53
  sweet potato vein mosaic, 51
  sweet potato virus N, 53
  white-fly-transmitted, 51—53
Virus N, sweet potato, 53
Vitamin A, 5, 144
  effect of canning on, 194
  effect of cooking on, 168
  nutritive value of, 151
  in sweet potato tips, 176, 181
Vitamin B
  effect of canning on, 194
  nutritive value of, 151
  in sweet potato tips, 176, 181
Vitamin C, 5, 158
  effect of canning on, 194
  effect of cooking on, 168
  nutritive value of, 151
  in sweet potato tips, 176, 181
Vitamins, 158

W

Waste water, disposal of, 229—230
Water
  energy acquisition and, 97
  loss, 6
Wavelength, photosynthesis and, 93

Weed control, 26—27
Weevil, striped sweet potato, 60, 61
Weevil, sweet potato, 61—65
   found in Asia and Pacific, 145
   storage and, 164
Weight measurements, 116—118
Wheat, see also Food crops, 4
Wheat flour, sweet potato flour as partial substitute for, 212
Whitefly, sweet potato, 71—72
White flies, viral pathogens transmitted by, 51—53
White grub, 68—69
Wind speed, energy acquisition and, 98
Witches' broom, 38, 145

Y

Yellow blight, 44
Yield
   effect of nitrogen application on, 239
   environmental factors affecting, 122—123
   per hectare, 255
   production factors affecting
      irrigation, 124—125
      length of growing period, 125
      plant population, 125
      soils, 123—124
Yield growth rate (YGR), 119

Z

Zaire, utilization of sweet potato tips in, 176